C000242524

Extreme Environment Astrophysics

Covering host systems of accreting, relativistic bodies, and the high-energy phenomena associated with them, this self-contained astrophysics textbook is ideal for advanced undergraduates.

The textbook introduces students to a unique blend of the astrophysical principles, including the evolutionary history of compact binary stars, the physics of accretion and accretion disc outbursts, the observed signatures of such discs in binary stars and active galactic nuclei, the X-ray emission of accreting compact bodies, and the physics of astrophysical jets and gamma-ray bursts. Worked examples, exercises with complete solutions, full-colour figures and informative chapter summaries guide students through their studies. Boxed equations and key facts highlight important points.

Produced by academics drawing on decades of experience delivering courses for The Open University and concentrating on supported learning, this textbook is an ideal guide for self study.

Ulrich Kolb is a Senior Lecturer in the Department of Physics and Astronomy, The Open University. He has authored Open University teaching materials in physics and astronomy, chairs The Open University residential school course in practical astronomy, and directs The Open University's robotic telescope project. His research interests focus on compact accreting systems, and he has published extensively in this area.

Extreme Environment Astrophysics

Author:

Ulrich Kolb

The Open University

CAMBRIDGE
UNIVERSITY PRESS

CAMBRIDGE UNIVERSITY PRESS

Cambridge, New York, Melbourne, Madrid, Cape Town, Singapore, São Paulo, Delhi, Dubai, Tokyo

Cambridge University Press
The Edinburgh Building, Cambridge CB2 8RU, UK

In association with THE OPEN UNIVERSITY

The Open University, Walton Hall, Milton Keynes MK7 6AA, UK

Published in the United States of America by Cambridge University Press, New York.

www.cambridge.org
Information on this title: www.cambridge.org/9780521187855

First published 2010.

Edited and designed by The Open University.

Typeset by The Open University.

Printed and bound in the United Kingdom by Latimer Trend and Company Ltd, Plymouth.

This book forms part of an Open University course S383 *The Relativistic Universe*. Details of this and other Open University courses can be obtained from the Student Registration and Enquiry Service, The Open University, PO Box 197, Milton Keynes MK7 6BJ, United Kingdom: tel. +44 (0)845 300 60 90, email general-enquiries@open.ac.uk

http://www.open.ac.uk

British Library Cataloguing in Publication Data available on request.

Library of Congress Cataloguing in Publication Data available on request.

ISBN   978-0-521-19344-3 Hardback
ISBN   978-0-521-18785-5 Paperback

Additional resources for this publication at www.cambridge.org/9780521187855

Cambridge University Press has no responsibility for the persistence or accuracy of URLs for external or third-party internet websites referred to in this publication, and does not guarantee that any content on such websites is, or will remain, accurate or appropriate.

1.1

EXTREME ENVIRONMENT ASTROPHYSICS

Contents

Introduction

Astronomy is a science that is not short of superlatives, and this textbook is in keeping with this tradition as it investigates examples of **extreme environments** in the cosmos, such as any that are extremely hot, extremely energetic, extremely fast, have extremely strong gravitational fields, or all of these combined. There is an abundance of phenomena in the sky that deserve the label *extreme environment* — too many, in fact, to be dealt with in a single textbook. The specific selection presented here is strongly driven by the research interests of the authors, and we hope that our enthusiasm is still apparent even when arguments and derivations become more involved.

The underlying theme of the book is *accretion power* — the sometimes staggering amount of energy released when matter approaches gravitating bodies. This is particularly potent in the vicinity of compact objects — white dwarfs, neutron stars and black holes big and small — where the gravitational potential well is steep and deep. Accreting matter heats up enormously and emits powerful high-energy radiation.

This is why this textbook studies accreting systems with such compact objects. The main focus of the text is indeed on stellar mass accretors, i.e. compact binaries, where the accretion flows and their emission can be studied in great detail, revealing a wealth of exciting insights into matter and radiation under extreme conditions. Furthermore, we choose mainly to discuss disc accretion — in fact, the theory of accretion discs, and observational signatures of discs, make up a third of the book. As a result, the low-mass X-ray binaries receive relatively more attention than the brighter high-mass X-ray binaries. Wherever appropriate we present links to the much larger analogues of accretion-powered binaries, active galactic nuclei (AGN) that host supermassive black holes. Yet we place the emphasis firmly on the stellar systems, and do not present AGN in detail.

This book focuses on a discussion of the physical concepts underpinning our current understanding of accreting systems, and supports this by a presentation of key observational facts and techniques.

Our journey into the wild and wonderful world of high-energy astrophysics begins with a short review of the basics of mass accretion, of black holes, compact binaries and AGN, and the thermal emission that we expect from them. In Chapter 2 we consider the host systems of stellar mass accretors and study their evolutionary history and current evolutionary state. We then move on to develop a simple theoretical description for a flat, Keplerian disc in a steady state (Chapter 3). Chapter 4 presents the disc instability model for the spectacular outbursts seen in soft X-ray transients and dwarf novae. Chapter 5 focuses on observational properties, mainly in the optical waveband, that constitute proof of the existence of accretion discs, and touches upon some of the more advanced indirect imaging techniques of accretion flows. This leads to Chapter 6 (main author Robin Barnard) on the X-ray properties of accreting objects. In the past decade the X-ray observatories XMM-Newton and Chandra have opened up a new window into the world of stellar mass accreting systems in galaxies other than the Milky Way (the cover image of this book is meant to encapsulate this), and some of the systems discovered in other worlds are stranger than fiction. Chapter 7 (main author Hara Papathanassiou) examines the physics of very fast (relativistic) outflows and jets that are present in X-ray binaries, AGN and gamma-ray bursts, paving the way for Chapter 8 (main author Hara Papathanassiou), the culmination of the book, on the most energetic and most extreme events known in the Universe to date: gamma-ray bursts.

This book is designed to be a self-study text, and can be used as a resource for distance teaching courses. Moreover, even though it is pitched at the advanced undergraduate level, by the nature of the selected topics and the depth of presentation, an attempt is made to build on familiar concepts and develop them further, with a minimum of higher-level mathematics, showing derivations in more detail than similarly advanced texts do.

Part of this book draws on teaching texts of an earlier Open University course (*Active Galaxies* by Carole Haswell and *Interacting Binary Stars* by Ulrich Kolb), which in turn referred heavily to other sources. In particular, the influence of the advanced textbook *Accretion Power in Astrophysics* by Juhan Frank, Andrew King and Derek Raine is still apparent in this new book. Another treasure trove of inspiration and facts has been unpublished lecture notes by Hans Ritter.

Thanks go to Carolin Crawford for critical comments on an earlier manuscript of this book and to Philip Davis for checking the exercises and numerical examples.

Ulrich Kolb
Hara Papathanassiou
Robin Barnard

Chapter 1   Accretion power

Introduction

In this chapter we shall consider the concept of mass accretion in astrophysics and its importance as a source of energy. We shall identify accretion-fed compact objects as powerful energy generators, and review the astrophysical context where sustained accretion can occur. In preparation for a closer look at compact binary stars and active galactic nuclei (AGN), we develop the Roche model for close binaries, and review the evidence for the existence of supermassive black holes in AGN. We conclude the chapter with a simple analysis of the continuum emission expected from a plasma that accretes with angular momentum to form an accretion disc.

1.1   Accretion as a source of energy

One of the most important astrophysical processes in the Universe is **mass accretion**, where a gravitating body grows in mass by accumulating matter from an external reservoir. The key importance of this process, which we shall henceforth simply refer to as accretion, is that it liberates gravitational potential energy, making accreting objects potentially very powerful sources of energy.

The concept of a test mass

Throughout this book we shall often study the physical characteristics of accreting systems by considering the fate of a test mass. We define this to be a gravitating body with a very small mass, so much so that the effect of the test mass on any existing gravitational field, such as from gravitating bodies in its neighbourhood, is negligible. The geometric size of the test mass is also considered to be negligible, and hence the test mass is treated as a point mass in most cases.

1.1.1   Accretion luminosity

The essence of accretion is most easily illustrated by considering a test mass m in the gravitational field of a spherically symmetric body with mass M and radius R ($m \ll M$). The **gravitational potential energy** of the test mass at a distance r from the central body is

$$E_{\mathrm{GR}}(r) = -\frac{GMm}{r}. \tag{1.1}$$

As the test mass moves from a very large distance $r \to \infty$ to the surface $r = R$ of the central body, the energy difference $\Delta E_{\mathrm{GR}} = E_{\mathrm{GR}}(r = \infty) - E_{\mathrm{GR}}(R)$ is released. With $r \to \infty$ we have $1/r \to 0$ and so $E(r = \infty) \to 0$, hence

$$\Delta E_{\mathrm{GR}} = \frac{GMm}{R}. \tag{1.2}$$

Suppose now that the central body, the **accretor**, is accreting mass continuously at a rate $\dot{M}$. In the time interval Δt it will therefore accrete the mass $\Delta M = \dot{M} \times \Delta t$, and according to Equation 1.2 it will liberate the energy $\Delta E_{GR} = GM\Delta M/R$. If *all* of this energy is radiated away at the same rate as it is liberated, the luminosity of the object due to the accretion process is $L_{acc} = \Delta E_{acc}/\Delta t$, which becomes

$$L_{acc} = \frac{GM\dot{M}}{R}. \tag{1.3}$$

The quantity L_{acc} in Equation 1.3 is called the **accretion luminosity**.

Notation for time derivatives

Often the time derivative of a quantity is denoted by a dot over the symbol representing this quantity. As an example, the mass accretion rate is the rate at which mass is added to an object, i.e. the rate by which the mass M of this object increases. Therefore it can be written as

$$\frac{dM}{dt} \equiv \dot{M}.$$

It is instructive to estimate the actual mass accretion rate needed to achieve a significant accretion luminosity for a normal star like our Sun, and confront this value with mass flow rates observed in the Universe. Mass accretion rates are often expressed in solar masses per year, $M_{\odot}\,\mathrm{yr}^{-1}$. Equation 1.4 below provides the conversion into SI units.

Worked Example 1.1

Calculate the mass accretion rate needed to power an accretion luminosity of $1.0\,L_{\odot}$ for the Sun.

Solution

We have

$$L_{acc} = \frac{GM\dot{M}}{R} = 1\,L_{\odot}$$

with $M = 1\,M_{\odot}$ and $R = 1\,R_{\odot}$. Solving this for the accretion rate, we obtain

$$\dot{M} = 1\,\frac{L_{\odot}R_{\odot}}{GM_{\odot}}$$

$$= \frac{3.83 \times 10^{26}\,\mathrm{J\,s}^{-1} \times 6.96 \times 10^{8}\,\mathrm{m}}{6.673 \times 10^{-11}\,\mathrm{N\,m^2\,kg}^{-2} \times 1.99 \times 10^{30}\,\mathrm{kg}}$$

$$= 2.01 \times 10^{15}\,\mathrm{kg\,s}^{-1}.$$

To convert this into $M_{\odot}\,\mathrm{yr}^{-1}$, we note that

$$1\,M_{\odot}\,\mathrm{yr}^{-1} = \frac{1.99 \times 10^{30}\,\mathrm{kg}}{60 \times 60 \times 24 \times 365.25\,\mathrm{s}}$$

$$= 6.31 \times 10^{22}\,\mathrm{kg\,s}^{-1}. \tag{1.4}$$

Hence the required mass accretion rate for the Sun is

$$\dot{M} = \frac{2.01 \times 10^{15}}{6.31 \times 10^{22}} \, M_{\odot} \, \mathrm{yr}^{-1} = 3.2 \times 10^{-8} \, M_{\odot} \, \mathrm{yr}^{-1}.$$

To put this value into perspective, we note that the Sun is continuously shedding mass at the much smaller rate of about a few times $10^{-14} \, M_{\odot} \, \mathrm{yr}^{-1}$ in the form of the **solar wind**, a stream of high-energy particles emanating from the Sun's atmosphere. Even though massive stars and giant stars have much stronger stellar winds than the Sun, it would be difficult for a star like our Sun to accrete mass from the stellar wind of another star, at a rate similar to the one calculated here — not least as in any reasonable setting only a small fraction of the wind would be captured by the accreting star.

For a fixed mass accretion rate, the accretion luminosity obviously increases with the compactness M/R of the accreting object. This reflects the fact that for a given mass M, the depth of its gravitational potential well increases with decreasing radius R.

Exercise 1.1   The Sun would have to accrete mass at a rate of $3.2 \times 10^{-8} \, M_{\odot} \, \mathrm{yr}^{-1}$ to generate an accretion luminosity that rivals its own energy output powered by core-hydrogen burning.

(a) Calculate the accretion luminosity that a white dwarf with mass $1 \, M_{\odot}$ and radius $10^{-2} \, R_{\odot}$ would have with the same accretion rate.

(b) Calculate the corresponding accretion luminosity for a neutron star with mass $1.4 \, M_{\odot}$ and radius $20 \, \mathrm{km}$. ∎

1.1.2   Accretion discs

Fluids and stellar plasma

Liquids and gases are collectively called **fluids**. In this book we are dealing with astrophysical fluids like stellar matter. Stellar matter is a gas, often referred to as stellar plasma or cosmic plasma. A **plasma** is a conducting fluid, e.g. an ionized gas, whose properties are determined by the existence of ions and electrons. Often stellar matter is fully ionized, but for low enough temperatures these ions and electrons may recombine to form neutral atoms and molecules.

Accreting matter hardly ever approaches the accretor in a straight-line trajectory. The conservation of angular momentum will in general lead to the formation of a flattened structure around the accretor in the plane perpendicular to the net angular momentum vector of the material. This plane could, for example, be the orbital plane of a binary system if the accreting material is donated from a companion star. Individual plasma blobs in such an **accretion disc** (Figure 1.1) orbit the accretor many times while maintaining a slow drift inwards, thereby losing angular momentum and gravitational potential energy. An accretion disc

Figure 1.1   An artist's impression of an accretion disc.

acts as an agent to allow the accreting plasma to settle gently on the mass accretor. In fact, accretion discs are like machines that extract gravitational potential energy and angular momentum from plasma.

In most cases, plasma at a distance r from a spherically symmetric accretor with mass M will orbit the accretor on a circular orbit with speed

$$v_{\mathrm{K}} = \left(\frac{GM}{r} \right)^{1/2}. \tag{1.5}$$

The plasma is said to execute a **Kepler orbit** (or Keplerian orbit), and the speed v_{K} is the Kepler speed.

- ● How does Equation 1.5 follow from the statement that the centripetal force on a blob of plasma with mass m arises from the gravitational force?

- ○ The gravitational force on a plasma blob with mass m is GMm/r^2, while the centripetal force is mv^2/r. Equating these, cancelling m, and rearranging for $v = v_{\mathrm{K}}$ reproduces Equation 1.5.

The orbit of a blob of plasma in an accretion disc is in fact a Kepler orbit with a slowly decreasing radius r. Hence the blob slowly drifts towards the accretor, losing gravitational potential energy while gaining kinetic energy.

The kinetic energy of a blob of plasma with mass m is

$$E_{\mathrm{K}} = \frac{1}{2}mv^2 = \frac{1}{2}\frac{GMm}{r},$$

while the gravitational potential energy is given by Equation 1.1. So we have

$$E_{\mathrm{K}} = -\tfrac{1}{2}E_{\mathrm{GR}}, \tag{1.6}$$

and the total energy of the system is

$$E_{\mathrm{tot}} = E_{\mathrm{K}} + E_{\mathrm{GR}} = \tfrac{1}{2}E_{\mathrm{GR}}. \tag{1.7}$$

The virial theorem

The **virial theorem** is a powerful diagnostic for a self-gravitating system in **hydrostatic equilibrium**, i.e. in a state where the system as a whole neither expands nor contracts as time goes by. The virial theorem relates the total gravitational potential energy E_{GR} of the system to its total kinetic energy E_{K} as

$$E_{\mathrm{K}} = -\tfrac{1}{2}E_{\mathrm{GR}}. \tag{Eqn 1.6}$$

Equations 1.6 and 1.7 show that half of the released gravitational potential energy is converted into kinetic energy of the blob, while the other half is available to heat the plasma and to power the emission of electromagnetic radiation such as visible or ultraviolet light, or X-rays.

When calculating the accretion luminosity of disc accretion, it is therefore appropriate to include the factor $\frac{1}{2}$ in Equation 1.3 to take account of the fact that only half of the energy expressed by Equation 1.2 is available to be radiated away by the disc itself:

$$L_{\text{disc}} = \frac{1}{2}\frac{GM\dot{M}}{R}. \tag{1.8}$$

1.1.3 Accretion efficiency

A useful measure that illustrates the power of accretion as an energy generator is the accretion efficiency η_{acc}, defined by the expression

$$L_{\text{acc}} = \eta_{\text{acc}}\dot{M}c^2, \tag{1.9}$$

where c is the speed of light. In general, the efficiency η expresses the amount of energy gained from matter with mass m, in units of its mass energy, $E = mc^2$.

Exercise 1.2   Estimate the efficiency of accretion onto a neutron star. Compare Equations 1.9 and 1.3, and use typical parameters of a neutron star, such as $M = 1\,\text{M}_\odot$ and $R = 10\,\text{km}$.

Exercise 1.3   (a) Compare the efficiency (= energy gain/mass energy of input nuclei) of nuclear fusion of hydrogen into helium with the result of Exercise 1.2. The mass defect of hydrogen burning is $\Delta m = 4.40 \times 10^{-29}$ kg.

(b) Explain why accretion can be regarded as the most efficient energy source in the Universe.   ■

The above exercises demonstrate that mass accretion has particularly significant consequences if it involves a compact accretor, i.e. an object with a higher density than that of normal stars or planets. Examples of stellar mass compact objects are stellar end-states such as white dwarfs or neutron stars, and the more exotic black holes. Both white dwarfs and neutron stars are subject to an upper mass limit. For white dwarfs this is the Chandrasekhar limit of $1.4\,\text{M}_\odot$, while for neutron stars the limit is less securely known but is thought to be $\lesssim 3\,\text{M}_\odot$. Black holes, on the other hand, apparently exist over a very wide range of masses. Stellar mass black holes are seen in binary systems, and supermassive black holes with masses up to $10^{11}\,\text{M}_\odot$ in the nuclei of active galaxies.

Black holes are the most exotic of the accreting compact objects that we discuss here, so in the next section we shall review some basic facts about black holes that are relevant for accretion physics.

1.2 Black hole accretors

A black hole forms when self-gravity causes material to collapse to such high densities that the escape velocity reaches the speed of light.

1.2.1 Schwarzschild black holes

Using Newtonian dynamics we can calculate the magnitude of the escape velocity v_{esc} from the surface of a spherically symmetric gravitating body with mass M and radius R by saying that the kinetic energy of a mass m travelling vertically upwards with speed v_{esc} must equal the change in gravitational potential energy, as given by Equation 1.2, required to completely escape from the body's gravitational field, i.e.

$$\tfrac{1}{2}mv_{esc}^2 = \frac{GMm}{R}.$$

Cancelling m and solving for v_{esc}, we have

$$v_{esc} = \left(\frac{2GM}{R}\right)^{1/2}. \tag{1.10}$$

To self-consistently calculate the magnitude of the escape velocity from an object with a density so high that the escape velocity reaches the speed of light, requires the use of general relativity. The relevant solution of Einstein's field equations is called the Schwarzschild solution, describing non-rotating black holes. These are therefore often referred to as **Schwarzschild black holes**. By a lucky coincidence, the correct general relativistic result for a non-rotating black hole is exactly what we obtain by setting $v_{esc} = c$ in Equation 1.10. A non-rotating black hole is formed when a mass M collapses to within a sphere of radius R_S, where

$$R_S = \frac{2GM}{c^2} \tag{1.11}$$

is the **Schwarzschild radius**, the radius of the sphere surrounding the collapsed mass at which the escape speed equals the speed of light. Within this sphere is a region of spacetime that is cut off from the rest of the Universe, since neither light nor any other form of information can escape from it. The sphere itself is known as the **event horizon**. Immediately outside the event horizon is a region of spacetime in which there is an extremely strong gravitational field.

A black hole forms at the end of the life of a massive star when there is no pressure source sufficient to oppose the self-gravitational contraction of the remnant stellar core. Similarly, if a much larger mass collapsed under self-gravity, a black hole would ultimately form, and indeed it is now thought that black holes of mass $M \gtrsim 10^6\,M_\odot$ are present at the cores of most (or possibly all) galaxies.

Accreting black holes offer the opportunity to study black hole properties since the energy generated through the accretion process makes these holes quite conspicuous. As we have discussed above, the accreting material will orbit the black hole before it crosses the Schwarzschild radius. In general relativity there is a **minimum stable circular orbit** close to the black hole, at about $3R_S$. Closer to the black hole, a stable orbit does not exist, and material will plunge towards the event horizon so swiftly that any energy is effectively trapped in the plasma and hence disappears with the matter down the hole. The framework of general relativity is needed to work out the accretion luminosity and accretion efficiency of an accreting black hole; this is beyond the scope of this book. Instead, we apply here the Newtonian expressions to obtain estimates for these quantities, and just note the general relativistic result.

Exercise 1.4   Estimate the accretion efficiency onto a non-rotating black hole by assuming that the accreting material executes Kepler orbits in an accretion disc and slowly drifts inwards. Assume that the inner edge of the accretion disc coincides with the last stable circular orbit at $3R_S$.   ∎

The correct general relativistic result for the accretion efficiency of a Schwarzschild black hole that accretes from a geometrically thin accretion disc is $\eta_{\mathrm{acc}} = 5.7\%$.

1.2.2 Rotating black holes

Most stars acquire angular momentum when they form, and, unless there is an effective braking mechanism at work that removes spin angular momentum, will keep it throughout their evolution. So the black hole remnant of a star is expected to rotate, too. The Schwarzschild solution of Einstein's field equations for non-rotating black holes can be generalized in the form of the more involved Kerr solution to describe rotating black holes, or **Kerr black holes** (Figure 1.2).

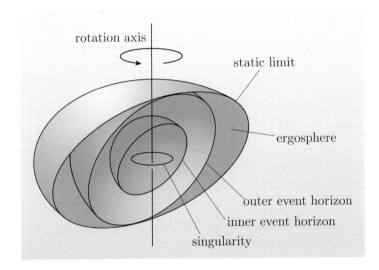

Figure 1.2 Critical surfaces of a Kerr black hole.

Both Schwarzschild and Kerr black holes represent gravitational singularities where the curvature of spacetime is infinite. In the case of Kerr black holes this singularity is a ring in the plane perpendicular to the rotational axis, while it is a single point for Schwarzschild black holes. Both types have a spherical event horizon where the escape speed is the speed of light. However, the (outer) event horizon of a Kerr black hole is surrounded by a second critical surface, the **static limit**, which has the shape of an oblate spheroid and touches the event horizon at its poles. The space between these two surfaces is called the **ergosphere**. Within the ergosphere the spacetime is dragged in the direction of the spinning black hole at a speed greater than c with respect to the outside Universe at rest, while at the static limit this speed equals c. As a consequence, matter inside the ergosphere cannot stay at rest. The material may even be ejected from the ergosphere by gaining energy from the black hole spin, thus spinning down the hole. If such a process could be sustained, the spinning black hole would eventually become a Schwarzschild black hole. There is also a maximum spin rate for a Kerr black hole.

In the context of accretion, a significant property of Kerr black holes is that the radius of the last stable orbit of matter orbiting the black hole outside of the event horizon decreases with increasing black hole spin (if the spin is in the same direction as the orbital motion of the accreting material). It is appropriate to use the last stable circular orbit in the expression for the accretion luminosity of a Kerr black hole.

● The last stable circular orbit for a Kerr black hole spinning at its maximum rate is $0.5R_S$. Using the same Newtonian method as in Exercise 1.4, estimate the accretion efficiency.

○ With

$$\eta_{\text{acc}} = \frac{1}{2}\frac{GM}{Rc^2}$$

and $R = 0.5R_S$, we obtain $\eta_{\text{acc}} = 0.5 = 50\%$. The correct general relativistic result is $\eta_{\text{acc}} = 0.32 = 32\%$.

1.3 Accreting systems

The accretion of mass is a very common phenomenon in the Universe. The Earth is constantly bombarded by meteorites and interplanetary dust particles (while also losing mass in the form of gas into space). Stars may sweep up interstellar matter as they cruise through clouds of hydrogen gas and dust. Such incidental, if not serendipitous, accretion rarely gives rise to appreciable accretion-powered emission. One notable exception is the high-energy emission from apparently isolated neutron stars that may accrete from the interstellar medium. For accretion to power sustained emission, a large enough mass reservoir must donate matter towards the accretor at a high enough rate.

This is the case, for example, in protoplanetary discs (proplyds). These circumstellar discs of dense gas surrounding newly formed stars (T Tauri stars) are the remnants of the star formation process and the birthplace of planetary systems.

Yet for accretion to power high-energy emission a compact accretor has to be present. There are principally two different groups of astrophysical systems with accreting compact objects, and throughout this book we shall look at each in more detail. The first group is compact binaries, systems with a compact star accreting matter from a companion, in most cases a normal star, either via the stellar wind of this star, or by a process called Roche-lobe overflow. The second group comprises supermassive black holes, with a mass in excess of $10^6\,M_\odot$, in the centres of galaxies, that swallow clouds of interstellar gas and dust, or even whole stars, from their vicinity.

1.3.1 Interacting binary stars

A binary star is a system consisting of two stars that orbit the common centre of mass. Binaries are very common as the star formation process involves the collapse and fragmentation of interstellar clouds, favouring the formation of protostars in close proximity to each other. Newly formed triple systems and higher multiples are gravitationally unstable and will eventually reduce to binary stars (or hierarchical binary stars) and single stars through the ejection of component stars.

In very wide binaries, the two stellar components will evolve just as they would do as single stars. If, on the other hand, two stars orbit each other in close proximity, neither of them can get arbitrarily large without feeling the restrictive presence of the second star. If one star becomes too large, the gravitational pull on its outer layers from the second star will become bigger than the pull towards its own centre of mass. Then mass is lost from one star and transferred to the other star. (See the box on 'Binary stars' below.) This is called **mass transfer**, and it will obviously give rise to accretion. It will also have an impact on the physical character of the two stars. If the mass accretor is an evolved, old star, it might rejuvenate when it acquires unspoilt, hydrogen-rich material. Conversely, once the mass donor has lost a significant amount of material, it may look much older than a star of its current mass usually would.

Binary stars

The binary component losing mass to the other component is called the mass **donor**, while the component on the receiving end is called the mass **accretor**. Here we refer to the accretor as the primary star, or just the **primary**, while the mass donor is the secondary star, or simply the **secondary**. This is because in many (but not all!) cases the accretor is more massive than the donor. Quantities carrying the index '1' usually refer to the primary, while those with index '2' refer to the secondary. The **mass ratio**

$$q = \frac{\text{donor mass}}{\text{accretor mass}} = \frac{M_2}{M_1} \tag{1.12}$$

is therefore the ratio of donor mass to accretor mass, and usually (but not always) less than unity. Unfortunately this is not a generally accepted convention, and other books or journal papers may define the mass ratio the other way round (M_1/M_2).

Pseudo-forces in a rotating frame

We shall now consider the physical foundation of the mass transfer process in greater detail. This is most conveniently discussed in a frame of reference that co-rotates with the binary. This frame rotates about the rotational axis of the orbital motion, i.e. an axis perpendicular to the orbital plane and intersecting this plane at the binary's centre of mass, with the same angular speed ω as the binary,

$$\omega = \frac{2\pi}{P_{\text{orb}}}, \tag{1.13}$$

where P_{orb} is the orbital period. The co-rotating frame does *not* constitute an **inertial frame**. **Pseudo-forces** (sometimes called fictitious forces) appear as a result of the rotational motion. To see why, we recall Newton's laws of motion. Any acceleration of a body in an inertial frame results from the action of a force on this body. In contrast, in a rotating frame a body may accelerate relative to the observer simply because the observer himself or herself is fixed to the rotating frame, while the body is not. The force causing this acceleration does not exist in the inertial frame, hence is called a pseudo-force. Nonetheless, for an observer in the rotational frame it can be very real.

In particular, there are two pseudo-forces in a rotating frame: the centrifugal force and the Coriolis force.

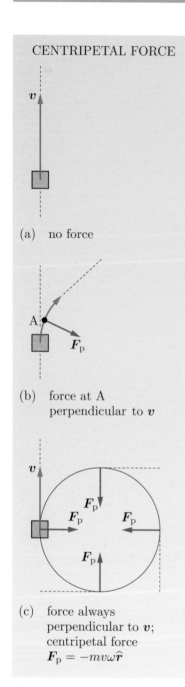

CENTRIPETAL FORCE

(a)   no force

(b)   force at A
      perpendicular to v

(c)   force always
      perpendicular to v;
      centripetal force
      $F_{\mathrm{p}} = -mv\omega\widehat{r}$

Figure I.3   The centripetal force. The observer is in a non-rotating frame of reference (inertial frame).

The **centrifugal force** is a familiar pseudo-force that, for example, the driver of a car experiences when following a sharp bend of the road at high speed. The driver is an observer fixed to the rotating frame, the car. The rotational axis is vertical and passes through the geometric centre of the bend. The driver is at rest in the driver's seat, but to stay there, muscle strength is needed to balance the centrifugal force pushing the driver radially outwards, away from the centre of the bend. For an observer in the inertial frame at rest with respect to the road — someone standing on the pavement — the driver is of course not at rest, but travelling on a circular path. The pedestrian concludes that there is a **centripetal force** acting on the driver, which is pulling the driver off the straight line (Figure 1.3). This force is mediated by the friction of the tyres on the road, the structural rigidity of the car, and the driver's muscle strength. In fact, the centripetal force has the same magnitude as the centrifugal force, but the opposite direction. The magnitude of the centrifugal force on a body with mass m and distance r from the rotational axis is

$$F_{\mathrm{c}} = m\frac{v^2}{r} = m\omega^2 r. \tag{1.14}$$

Here ω is the angular speed of the rotating frame, and $v = \omega \times r$ is the magnitude of the instantaneous velocity of a point fixed to the rotating frame, with respect to the non-rotating inertial frame.

The second pseudo-force in the rotating frame, the **Coriolis force**, acts *only* on bodies that are moving in this frame. The Coriolis force is always perpendicular to the direction of motion, and also perpendicular to the rotation axis. It is easy to see why such a force in addition to the centrifugal force must exist. A body at rest in an inertial frame would appear to move in a circle around the rotational axis in the rotating frame (Figure 1.4b). Hence the observer in the rotating frame concludes that there is a force at work that not only overcomes the outward centrifugal force, but also provides the inward centripetal force necessary to maintain the circular motion.

Worked Example I.2

Use the example of a body at rest in the inertial frame to show that the Coriolis force has the magnitude $2m\omega v$. Consult Figure 1.4.

Solution

The body of mass m is at rest in the non-rotating frame. Its distance from the rotational axis is r. In the frame rotating with angular speed ω, the same body appears to move on a circle with radius r and speed $v = r\omega$.

The observer in the rotating frame concludes that there is a net force, the centripetal force F_{p}, of magnitude $m\omega^2 r = m\omega v$ acting on the body. This force points towards the centre of the circle. The observer knows that there are two pseudo-forces acting on the body: the centrifugal force F_{c} of magnitude $m\omega^2 r = m\omega v$, pointing away from the centre, and the Coriolis force F_{Coriolis}. From the vector sum

$$F_{\mathrm{net}} = F_{\mathrm{p}} = F_{\mathrm{c}} + F_{\mathrm{Coriolis}}$$

(see Figure 1.4b), we have

$$F_{\mathrm{Coriolis}} = F_{\mathrm{p}} - F_{\mathrm{c}}.$$

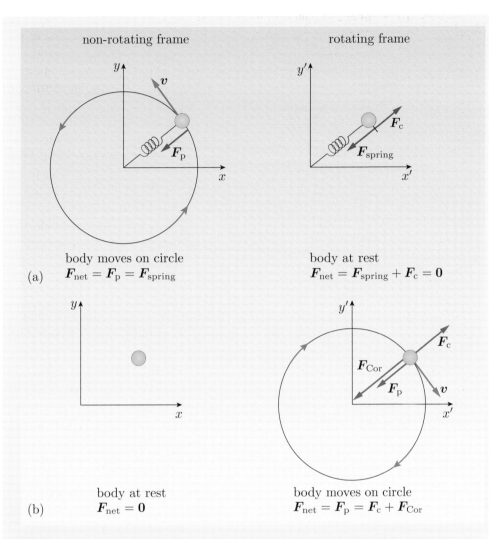

Figure 1.4 The Coriolis force. (a) A body is attached to a spring and moves on a circle with constant angular speed (as seen in the inertial frame). In the co-rotating frame the body is at rest; the net force on the body is zero — the spring force just balances the centrifugal force. (b) A body at rest in the inertial (non-rotating) frame is seen to move on a circle with constant angular speed in the rotating frame. The necessary centripetal force F_p for circular motion is given by the sum of the centrifugal force F_c and the Coriolis force F_{Coriolis}.

As F_p and F_c point in opposite directions, it is clear that the magnitude of the Coriolis force is just the sum of the magnitudes of F_p and F_c, i.e. $2m\omega v$, as required. (The expression for the Coriolis force is slightly more complicated if the velocity is not perpendicular to the rotation axis; it involves the vector product

$$F_{\text{Coriolis}} = -2m\boldsymbol{\omega} \times \boldsymbol{v}, \tag{1.15}$$

so only the velocity component perpendicular to the axis is involved.)

Figure 1.5   Earth as seen from space, with a cyclonic depression.

We do, in fact, live in a rotating frame of reference ourselves: on Earth. The Coriolis force is responsible for the motion of clouds around low-pressure weather systems, as seen in satellite images of the Earth (Figure 1.5). On the northern hemisphere, the Coriolis force deflects air moving towards a low-pressure region in a clockwise direction as seen from space. The air flow thus joins the anticlockwise, circular movement around the low-pressure area, a so-called cyclonic flow.

The magnitude and direction of the centrifugal force depends only on the position in the rotating frame. It can therefore be expressed as the gradient of a potential V, such that $\boldsymbol{F}_{\mathrm{c}} \propto \boldsymbol{\nabla} V$. In contrast, the Coriolis force depends on position *and* velocity, and cannot be derived from a potential. In the context of the Roche model below, the most important thing to remember about the Coriolis force is that it vanishes if $v = 0$ in the rotating frame!

The Roche model

To arrive at a useful and yet simple quantitative description of a close binary system with mass exchange, we now make three simplifying assumptions. These will allow us to express the force $\boldsymbol{F}$ on a test mass m in the system in terms of the **Roche potential** Φ_{R} in the co-rotating frame as

$$\boldsymbol{F} = -m\boldsymbol{\nabla}\Phi_{\mathrm{R}}.$$

● Express the physical meaning of this equation in words.

○ The direction of the force on the test mass, as seen in the co-rotating frame, is in the opposite direction to the gradient of the potential. In the x-direction the gradient of Φ_{R} is just the derivative $\mathrm{d}\Phi_{\mathrm{R}}/\mathrm{d}x$.

The first simplifying assumption is that the orbits of the binary components are circular. Close binaries with elliptical orbits do exist, but if one component has an extended envelope, then strong tidal forces within this envelope will act to reduce the eccentricity of the orbit on a short timescale. Most mass-transferring binaries do indeed have circular orbits.

The second assumption is that the two components are in effect point masses — which they clearly are not. However, the gas density inside stars increases markedly towards the centre. The bulk of the stellar mass is in fact concentrated in a small, massive core region that is practically unaffected by the presence of a companion, and hence the star can safely be approximated by a point mass.

The third assumption is that the outer layers of each of the stars rotate synchronously with the orbit. In close interacting binaries, tidal forces are indeed very effective in establishing the tidal locking of spin and orbit.

We consider now a test mass m fixed in the co-rotating frame (binary frame) at a position that we specify by the position vector $\boldsymbol{r}$ (see Figure 1.6). The primary with mass M_1 is at the position $\boldsymbol{r}_1$, the secondary with mass M_2 is at $\boldsymbol{r}_2$, and the centre of mass is at $\boldsymbol{r}_{\mathrm{c}}$. The test mass is subject to the gravitational force of the primary, the gravitational force of the secondary, and the centrifugal force. The Coriolis force vanishes as the test mass is at rest in the binary frame.

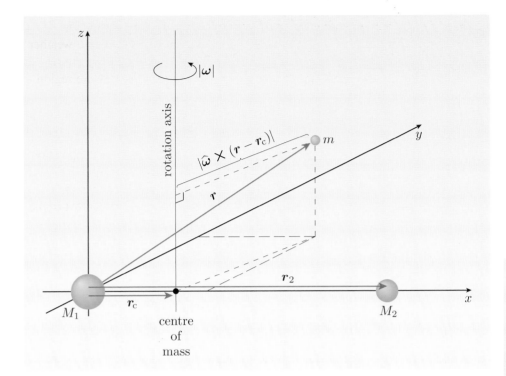

Figure 1.6 Definitions for the Roche model. The arrows representing r_c and r_2 are slightly displaced for clarity but they are coincident with the x-axis over the length of the vector in each case.

The Roche potential is then

$$\Phi_{\mathrm{R}}(\boldsymbol{r}) = -\frac{GM_1}{|\boldsymbol{r} - \boldsymbol{r}_1|} - \frac{GM_2}{|\boldsymbol{r} - \boldsymbol{r}_2|} - \tfrac{1}{2}\left(\boldsymbol{\omega} \times (\boldsymbol{r} - \boldsymbol{r}_{\mathrm{c}})\right)^2. \qquad (1.16)$$

Exercise 1.5 In the expression 1.16 for the Roche potential $\Phi_{\mathrm{R}}(\boldsymbol{r})$, explain the functional form of the three terms on the right-hand side.

Exercise 1.6 Assume that the x-axis goes through the centres of the two stellar components, and the origin is at the centre of the primary. Write down the Roche potential as a function of x, and determine the direction and magnitude of the force on a test mass m at the centre of mass of the system. ■

The significance of the Roche potential is that in equilibrium, for negligible fluid flow velocities, the surfaces of constant Roche potential, the Roche equipotentials, are also surfaces of constant pressure. In particular, the *surface* of a star, i.e. the layer with an optical depth of about 1 (see the box entitled 'Cross-section, mean free path and optical depth' in Section 6.2 for a definition of optical depth), coincides with a Roche equipotential. Hence the shape of the Roche equipotential determines the *shape* of the stellar components in binary systems.

Figure 1.7 illustrates the shape of the Roche equipotentials. Close to the centre of one of the stars, say the secondary, the equipotentials are nearly spherical, somewhat flattened along the rotational axis — in the z-direction in the figure (panel (a) in Figure 1.7). As long as the stellar radius is small compared to the orbital separation, the star adopts the characteristic shape of a single star rotating with the orbital period of the binary. With increasing distance from the stellar

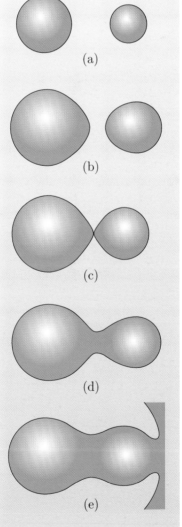

Figure 1.7 Roche equipotential surfaces for different values of Φ_{R}.

23

centre, the value of the Roche potential Φ_R increases (i.e. becomes less negative), and the corresponding equipotentials become more and more pointed towards the primary, while still excluding the primary's centre (panel (b) in Figure 1.7). The closed equipotential surface with the largest value of Φ_R (or smallest value of $|\Phi_R|$) that still excludes the primary's centre touches the corresponding equipotential surface that encloses the primary's centre in one critical point, the L_1 point, or **inner Lagrangian point** (panel (c) in Figure 1.7). The L_1 point is a saddle point of the Roche potential.

● Describe the characteristics of a saddle point (see Figure 1.8).

○ A saddle point of a potential is a point where the spatial gradient of the potential Φ vanishes, such that the potential is a maximum in one direction, e.g. along the x-axis, but a minimum in a direction perpendicular to the former, i.e. along the y-axis. Mathematically,

$$\frac{\partial \Phi}{\partial x} = \frac{\partial \Phi}{\partial y} = 0$$

and the second partial derivative $\partial^2\Phi/\partial x^2$ is negative, while $\partial^2\Phi/\partial y^2$ is positive. (See the box entitled 'Partial derivatives' in Subsection 3.2.2 for the meaning of the symbol ∂.)

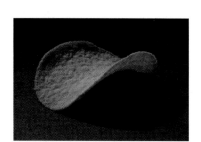

Figure 1.8   A familiar surface with a saddle point.

The two lobes of the critical Roche equipotential surface that contains the L_1 point are the **Roche lobe** of the secondary and the Roche lobe of the primary star, respectively. Mass exchange between these stars will proceed through the immediate vicinity of the L_1 point. A stellar component of the binary can expand only until its surface coincides with this critical Roche equipotential. If such a **Roche-lobe filling** star attempts to expand further, mass will flow into the direction of decreasing values of Φ_R, i.e. into the lobe of the second star. This is called **Roche-lobe overflow**.

For somewhat larger values of Φ_R (smaller values of $|\Phi_R|$), the equipotentials surround both stars, adopting a dumbbell-like shape (panel (d) in Figure 1.7), while at distances large compared with the orbital separation, the centrifugal component of the potential dominates, and near the orbital plane the equipotentials appear as nested cylinders aligned with the binary's orbital axis.

The values of the Roche potential along a line through the centres of the two binary star components provide an instructive illustration of Roche-lobe overflow (see Figure 1.9, where this line is the x-axis). The most notable features of the curve in this figure are the effect of the centrifugal repulsion at large distances from the binary's centre of mass (Φ_R falls off at large $|x|$), and the two deep valleys caused by the gravitational attraction of the corresponding star in the respective valley. A star can fill these valleys only up to the 'mountain pass' in between, the L_1 point. If the star attempts to grow further, mass flows over into the neighbouring valley. In a phase with a continuous flow of mass, the donor star will fill the maximum volume available to it, its Roche lobe. The mass flows to the less extended accretor, which resides well inside its own lobe. The binary is said to be **semi-detached**.

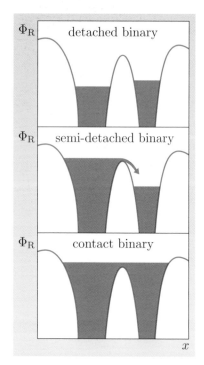

Figure 1.9   Schematic view of the potential wells of a detached, semi-detached and contact binary.

Roche-lobe overflow starts either when one of the stars attempts to grow beyond its lobe, or because the lobe closes in on the star. The former can occur simply as a result of the star's nuclear evolution, e.g. when the star expands to become a giant. The latter can occur if the orbit shrinks by losing orbital angular momentum. We shall come back to both possibilities in Chapter 2.

There is yet another way to establish mass transfer, as indicated in the lower panel of Figure 1.10. Massive stars and giant stars display rather strong stellar winds (see also Chapter 7). The accretor can capture some fraction of the matter lost by the other star in its wind. Hence mass is transferred even though the mass-losing star is well inside its Roche lobe. This mode of mass transfer is called **wind accretion**. Most of the mass in the wind is lost from the binary, however.

Figure 1.11 depicts an artist's impression of a semi-detached binary with a white dwarf accretor, while Figure 1.12 presents a sketch of the black hole binaries known at the time of writing (2008), drawn to scale. The images show that matter leaving the donor star through the L_1 point settles into an accretion disc around the compact star. We now turn to accreting compact objects on much larger scales.

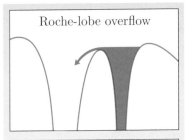
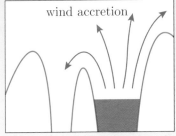

Figure 1.10 Schematic view of Roche-lobe overflow and wind accretion.

Figure 1.11 An artist's impression of a **cataclysmic variable star** — a compact binary where a white dwarf accretes from a Roche-lobe filling normal star.

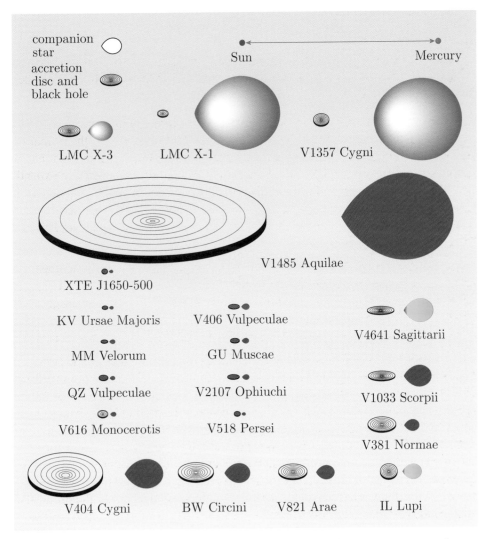

Figure 1.12 Known compact binaries with a black hole accretor, on a scale based on the distance between the Sun and Mercury, indicated at the top of the figure. The colour of the companion (donor) star indicates its surface temperature: dark red is cool, bright yellow is hot. (Courtesy of Jerry Orosz.)

1.3.2   Active galactic nuclei

An **active galaxy** contains a bright, compact nucleus that dominates its host
galaxy's radiation output in most wavelength ranges. These **active galactic
nuclei** (or **AGN**) are thought to be powered by a supermassive black hole (the
engine) that accretes from a large hot accretion disc. The disc is the source of the
continuum emission in the ultraviolet and X-ray bands, while an obscuring dust
torus surrounding the disc emits in the infrared (Figure 1.13).

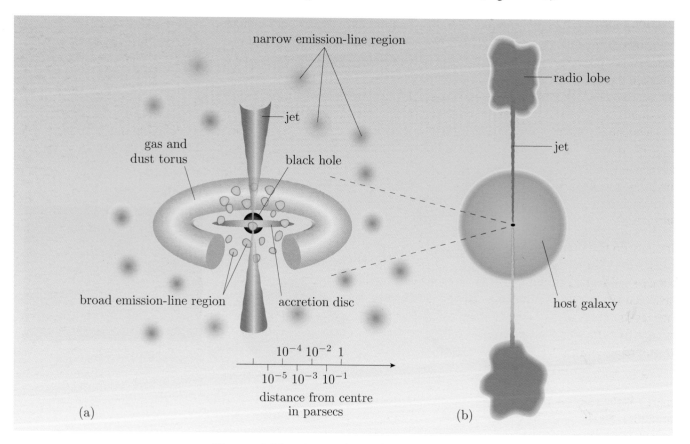

Figure 1.13   A generic model for an active galaxy. (a) A supermassive black
hole is surrounded by an accretion disc; jets emerge perpendicular to it. An
obscuring torus of gas and dust encloses the broad-line region (a few light-days
across) with the narrow-line region (a few hundred parsecs across) lying further
out. (b) The entire AGN appears as a bright nucleus in an otherwise normal
galaxy, while jets (hundreds of kiloparsecs in length) terminate in radio lobes.

Active galaxies come in many disguises, and consequently they can be grouped in
numerous classes with quite diverse observed properties. Unified models attempt
to explain this range of AGN on the assumption that they differ only in luminosity
and the angle at which they are viewed.

One broad classification criterion is based on the observed activity in the
radio band; there are weak and strong radio emitters. The most important
representatives of AGN that display no or only a very weak radio emission are
Seyfert galaxies and **quasars** (although $\lesssim 10\%$ of quasars are strong radio
sources; it is these that created the name quasars — *quasi-stellar radio sources*).

Seyfert galaxies look like normal galaxies, but with an unusual luminous nucleus. The host galaxies of quasars are so distant and so much fainter than the point-like (quasi-stellar) nucleus that they are seen only on deep images taken with the most powerful telescopes. The point-like nucleus, on the other hand, can be easily detected.

Among the strong radio-emitting active galaxies are **radio galaxies** and **blazars** which, like quasars, appear star-like. Many radio-bright active galaxies display prominent, narrowly focused jets that emanate from the AGN in opposite directions and often extend to distances exceeding the size of the host galaxy.

The active nucleus of Seyfert galaxies easily outshines its entire host galaxy, demonstrating that the intrinsic AGN luminosity must be very large indeed. Values in the range 10^{11}–10^{15} $L_\odot$ are commonly derived from the observed flux and the large distances implied by the observed cosmological redshift of AGN emission lines. The evidence for the existence of an accreting supermassive black hole that could generate the enormous emitting power of AGN is circumstantial but very compelling. The AGN central engine must fit in a very small volume of space, and this small volume must contain a very large amount of gravitating mass. We now look at both of these facts in detail.

Compactness

Not only do the active nuclei of even the closest Seyfert galaxies appear as unresolved point sources of light, but the luminosities of some AGN are also seen to vary significantly over a few days. This means that the time Δt for light to travel across the entire source must be only a few days, because otherwise the changes in luminosity would be smoothed out by the delayed arrival times of the photons from the more distant regions of the source. This can be expressed by the general requirement that

$$l \lesssim \Delta t \times c,$$

where l is the size of the emitting source and Δt is the timescale for observed variability. Using this to work out the size limit corresponding to a **light travel time** of a few days, we have

$$l \lesssim 10 \times 24 \times 60 \times 60 \times 3 \times 10^8 \, \text{m},$$

where we have adopted a typical value of $\Delta t = 10$ days, converted this into seconds, and used an approximate value for the speed of light: $c \approx 3 \times 10^8 \, \text{m s}^{-1}$. Evaluating, and retaining only 1 significant figure, we have

$$l \lesssim 3 \times 10^{14} \, \text{m},$$

which can be converted into length units more convenient for astronomical objects:

$$l \lesssim \frac{3 \times 10^{14} \, \text{m}}{1.5 \times 10^{11} \, \text{m AU}^{-1}}, \quad \text{i.e.} \quad l \lesssim 2 \times 10^3 \, \text{AU}. \qquad (1.17)$$

Thus the observations require that a luminosity of perhaps a 100 times that of the entire Milky Way galaxy be generated within a region with diameter only about 1000 times that of the Earth's orbit!

Exercise 1.7   Convert the length scale in Equation 1.17 into parsecs (pc). ■

Mass

The second piece of evidence for the existence of a supermassive black hole as the engine of AGN is based on the **virial theorem** (see the box on page 14). In the AGN context this can be recast in terms of the mean (or typical) velocity of a large number of individual bodies that are all part of the gravitating system. If the total mass of the gravitating system is M and its radial extent is r, then Equation 1.6 can be written as

$$\frac{1}{2}m\langle v^2\rangle \simeq \frac{1}{2}\frac{GMm}{r}, \tag{1.18}$$

where m is a typical mass of these bodies, and $\langle v^2\rangle$ is the mean value of the squares of their speeds with respect to the centre of mass. Hence

$$\langle v^2\rangle \simeq \frac{GM}{r}. \tag{1.19}$$

The motion of these bodies — stars and clouds or blobs of gas — will give rise to **Doppler broadening** of the spectral lines that they emit. As the AGN emitting region is so small, the observer will only see the superposition of the emission lines from individual emitters, all Doppler-shifted by their corresponding radial velocity. The combined emission line will therefore have a line profile width that reflects the average velocity $\langle v\rangle$ of the individual emitters. This is also called the **velocity dispersion**.

The velocity dispersion measured for the so-called broad-line region of AGN, which is contained within the torus of infrared emitting dust (see Figure 1.13), is typically observed to be 10^3–$10^4\ \mathrm{km\ s^{-1}}$. The observed variability of the line-emitting region implies that it is a few tens of light-days across. (We shall discuss the broad-line region in more detail in Section 5.4 below.) Hence the virial theorem gives the mass of the central black hole as approximately 10^7–$10^{11}\ \mathrm{M_\odot}$. The label *supermassive* seems to be well justified.

Exercise 1.8   Confirm the above statements by calculating the mass of the central black hole if the emitting region is ≈ 30 light-days across, and displays Doppler broadening of $6000\ \mathrm{km\ s^{-1}}$. ■

Supermassive black holes in galactic nuclei

It is now thought that most galaxies harbour a supermassive black hole (with mass $\gtrsim 10^6\ \mathrm{M_\odot}$) in the centre. In active galaxies this black hole is accreting and a strong power source, while in normal galaxies such as the Milky Way the black hole lies dormant. The crucial evidence comes from the observation of the motion of stars near the centre of the galaxy.

Perhaps the most dramatic example is the case of the supermassive black hole at the centre of our own Milky Way galaxy. This region is impossible to study in optical light because there is a lot of gas and dust in the plane of the galaxy, which obscures our view of the central regions. At other wavelengths, however, the optical depth is less, and it has long been known that the centre of our galaxy harbours a compact radio source, which is called Sgr A* and is shown in Figure 1.14. Apart from Sgr A*, the radio emission apparent from Figure 1.14 is diffuse and filamentary. The stars near the centre of the Galaxy are not visible because they are not strong radio sources. The infrared view shown in the left

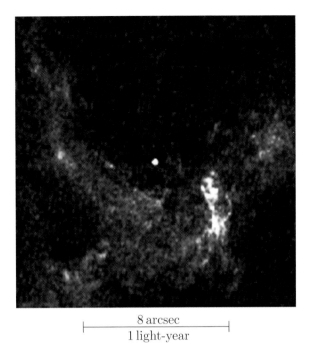

8 arcsec
1 light-year

Figure 1.14 A radio image of the centre of the Milky Way. White areas indicate intense radio emission, and the red and black areas are progressively less intense. This image was taken with the Very Large Array (VLA) by Jun-Hui Zhao and W. M. Goss. The white dot at the centre of the image is the Sgr A* compact radio source.

panel of Figure 1.15 is very different. The image is **diffraction-limited**, and gives a resolution of 0.060 arcseconds. The blobs are individual stars within 60 light-days of the Sgr A* radio source, whose position is marked with the small cross at the centre of Figure 1.15.

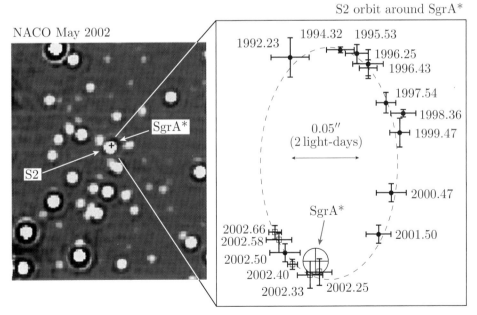

Figure 1.15 (a) An infrared image from May 2002 at 2.1 μm wavelength of the region near Sgr A* (marked by the cross). The image is about 1.3 arcseconds wide, corresponding to about 60 light-days. (b) The orbit of S2 as observed between 1992 and 2002, relative to Sgr A* (marked with a circle). The positions of S2 at the different epochs are indicated by crosses, with the dates (expressed in fractions of the year) shown at each point. The solid curve is the best-fitting elliptical orbit — one of the foci is at the position of Sgr A*.

● How do the scales of the images in Figure 1.14 and Figure 1.15 (left panel) compare?

○ The bar in Figure 1.14 represents 8 arcseconds, while the image in Figure 1.15 is less than 2 arcseconds across.

The left panel of Figure 1.15 is only one frame of a series of high spatial resolution infrared images of the centre of our galaxy, which were taken starting in the early 1990s. The motions of individual stars are clearly apparent when subsequent frames are compared. The right panel of Figure1.15 shows the example of the star S2, which can be clearly seen to orbit Sgr A* with a period of about 15 years! A number of such stellar orbits have now been measured, and from Kepler's law the gravitating mass inside the orbit can be determined. This is analogous to the determination of the Sun's gravitational field (and hence the Sun's mass) by studying the orbits of the planets in the solar system. The stars at the centre of the Galaxy are not neatly aligned in a plane analogous to the ecliptic in the solar system. Instead, the stars follow randomly oriented orbits. The observed motions require the presence of a dark body with mass $4 \times 10^6 \, \mathrm{M_\odot}$ at the centre of our galaxy. This dark central body is almost certainly a black hole.

In galaxies where the orbits of stars or clouds cannot be mapped in this detail, the virial theorem is used to deduce the gravitational field from the observed dispersion of the velocity of the detected individual moving sources.

1.4   Radiation from accretion flows

As accretion flows often take the form of a disc-like structure, we now investigate the basic appearance of such accretion discs, as indicated by the temperature of the disc. We shall return to the physics of accretion discs in greater detail in Chapters 3 and 4.

1.4.1   Temperature of an accreting plasma

We wish to arrive at a quantitative estimate for the temperature of the accretion disc plasma as it approaches the accretor. To this end we make two assumptions. First, all of the locally liberated gravitational potential energy is instantly converted into thermal energy. Second, photons undergo many interactions with the local stellar plasma and are thermalized before they emerge from the surface of the disc. In other words, the plasma is **optically thick** and radiates locally like a black body.

As we shall see in Subsection 3.4.4, the accretion disc surface temperature T_{eff} of a geometrically thin, optically thick steady-state accretion disc varies with distance r from the accretor as

$$T_{\mathrm{eff}}^4(r) \simeq \frac{3GM\dot{M}}{8\pi\sigma r^3}. \tag{1.20}$$

This relation holds if r is large compared to the inner disc radius. The exact form of the profile peaks very close to the inner disc radius R, at a temperature

$$T_{\text{peak}} \simeq 0.5 \times \left(\frac{3GM\dot{M}}{8\pi\sigma R^3} \right)^{1/4}. \tag{1.21}$$

In a steady-state disc, the local mass accretion rate at each radius r is the same in each disc annulus, and in fact equals the mass accretion rate $\dot{M}$ onto the central object.

It is easy to see that a relation of the form $T_{\text{eff}}^4 \propto GM\dot{M}/r^3$ must apply. Consider a disc annulus at radius r and width Δr (hence area $\propto r\Delta r$). The gravitational potential energy (Equation 1.1) changes across the annulus per unit time by

$$\frac{\mathrm{d}E_{\text{GR}}}{\mathrm{d}r} \propto \frac{\mathrm{d}(GM\dot{M}/r)}{\mathrm{d}r} \propto \frac{GM\dot{M}}{r^2}.$$

The energy dissipated per unit area in the annulus is therefore $\propto GM\dot{M}/r^3$. By assumption, the liberated energy heats up the disc annulus to a temperature T_{eff}, which in turn radiates as a black body through thermal emission. The flux F emitted by a black body source, i.e. the energy emitted per unit time per unit area, is given by the Stefan–Boltzmann law

$$F = \sigma T_{\text{eff}}^4, \tag{1.22}$$

where σ is the Stefan–Boltzmann constant. Now F must equal the rate of energy generation, so $F \propto T_{\text{eff}}^4 \propto GM\dot{M}/r^3$. (A more detailed derivation is presented in Chapter 3.)

Equation 1.20 states that accreting plasma heats up with decreasing distance from the centre as $T_{\text{eff}} \propto r^{-3/4}$, so material will become very hot as it approaches a compact object. Temperatures in excess of 10^5 K in the case of white dwarfs and 10^7 K for neutron stars are the norm.

Exercise 1.9   Calculate the peak temperature of an accretion disc:
(a) around a white dwarf with mass $0.6\,\mathrm{M_\odot}$ and radius $R = 8.7 \times 10^6$ m for an accretion rate of $10^{-9}\,\mathrm{M_\odot\ yr^{-1}}$ (which is typically observed in cataclysmic variables with a few hours orbital period);
(b) around a neutron star with mass $1.4\,\mathrm{M_\odot}$ and radius $R = 10$ km for an accretion rate of $10^{-8}\,\mathrm{M_\odot\ yr^{-1}}$ (observed in some bright neutron star X-ray binaries).

Exercise 1.10   (a) Express the peak temperature of an accretion disc around a Schwarzschild black hole in terms of the mass accretion rate in units of $\mathrm{M_\odot\ yr^{-1}}$ and the black hole mass in units of $\mathrm{M_\odot}$. Assume that the inner edge of the accretion disc is at a radius $3R_{\text{S}}$.

(b) Calculate the peak disc temperature for a black hole with mass $10\,\mathrm{M_\odot}$ and accretion rate $10^{-7}\,\mathrm{M_\odot\ yr^{-1}}$, as in bright low-mass X-ray binaries.

(c) Calculate the peak disc temperature for a black hole with mass $10^7\,\mathrm{M_\odot}$ and accretion rate $1\,\mathrm{M_\odot\ yr^{-1}}$, as in AGN.   ∎

1.4.2   Continuum emission

Stellar matter at such high temperatures emits electromagnetic waves of very high frequencies, so compact binaries are powerful sources of high-energy radiation.

The brightest sources in the X-ray sky are in fact accreting neutron star and black hole binaries (see Figure 1.16). Several hundreds of these **X-ray binaries** reside in the Milky Way, and many more are known in distant, external galaxies.

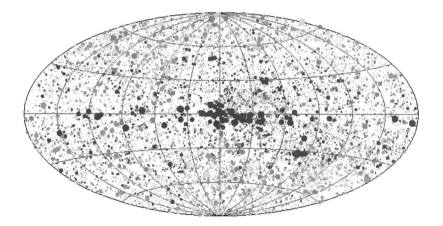

Figure 1.16   An all-sky map of the X-ray sky as seen by the X-ray satellite ROSAT. The colour of the dots indicates the 'X-ray colour', i.e. the spectral characteristics of the X-ray source. The size of the dots indicates the intensity of the emitted X-rays. The celestial sphere has been mapped into the plane of the page such that the Galactic Equator (the band of the Milky Way on the sky) appears as the horizontal line in the middle of the diagram. Along the Galactic Equator the Galactic longitude changes from $-180°$ at the left to $+180°$ at the right. The Galactic Centre is in the middle. The vertical line in the middle joins the Galactic North Pole (top) with the Galactic South Pole (bottom).

The mass M_2 of the compact object's companion star can be used to separate X-ray binaries into two main groups with distinct properties, which we shall discuss in Chapters 2 and 6. In **low-mass X-ray binaries**, or **LMXBs** (Figure 1.17), the companion star is a low-mass star ($M_2 \lesssim 2\,\mathrm{M_\odot}$), while in **high-mass X-ray binaries**, or **HMXBs** (Figure 1.18), this is a massive star ($M_2 \gtrsim 10\,\mathrm{M_\odot}$). Companion stars with masses in between these limits are much less commonly observed but theoretically implied, and are sometimes referred to as **intermediate-mass X-ray binaries**.

The discs in AGN are much larger than in X-ray binaries, but they are not quite as hot (see Exercise 1.10). The observed AGN X-ray emission is likely to be reprocessed thermal emission from the underlying accretion discs. This will be discussed further in Chapter 6 (Subsection 6.3.2 and Section 6.7). AGN still emit about 10% of their total energy budget in the X-ray band, and the fact that the emission is highly variable demonstrates that it is generated in the innermost regions near the compact object. Therefore AGN are much more powerful X-ray emitters than X-ray binaries, but the nearest AGN is so far away that the X-ray flux we receive from it is smaller than the received flux from bright X-ray binaries in the Galaxy. Yet AGN are ubiquitous across the X-ray sky and are the dominant X-ray source group for faint X-ray fluxes. In nearby galaxies it is often difficult to tell if an X-ray source is an X-ray binary residing in this galaxy, or an unrelated, distant AGN that happens to be in the same line of sight as the nearby galaxy.

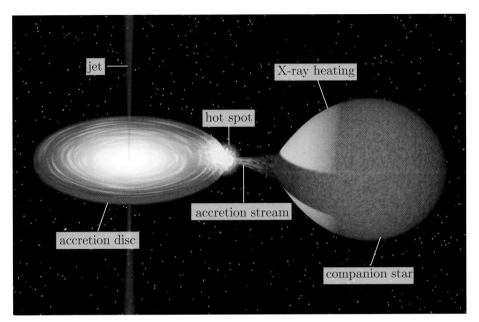

Figure 1.17 Artist's impressions of the low-mass X-ray binary V1033 Scorpii (also known as GRO J1655-40), which is a superluminal jet source (see also Chapter 7). The black hole accretes matter from a Roche-lobe filling low-mass or intermediate-mass companion star. The orbital period is 2.6 days. (Courtesy of Rob Hynes.)

Figure 1.18 Artist's impression of the high-mass X-ray binary Cygnus X-1. The black hole accretes from the wind of the massive companion star. The orbital period is 5.6 days. (Courtesy of Rob Hynes.)

To better understand the continuum emission of accretion discs with compact accretors, we now recall the properties of black body radiation. In the idealized case considered above, each disc annulus radiates as a black body with the surface temperature T_{eff} of the annulus.

Black body radiation

Black body radiation is in thermal equilibrium with matter at a fixed temperature. Often the emission from astronomical objects is a close approximation to this **thermal radiation**. Many thermal sources of radiation, for instance stars, have spectra which resemble the black body spectrum, which is described by the Planck function

$$B_\nu(T) = \left(\frac{2h\nu^3}{c^2}\right)\frac{1}{\exp(h\nu/kT)-1}. \qquad (1.23)$$

The quantity B_ν is the power emitted by per unit area per unit frequency per unit solid angle (and has the units $\text{W m}^{-2}\,\text{Hz}^{-1}\,\text{sr}^{-1}$); k is the Boltzmann constant, and h is Planck's constant.

Figure 1.19 illustrates the way that black body spectra peak at wavelengths that depend on temperature. This is quantified by the **Wien displacement law**, which states that the maximum value of B_ν shown in Figure 1.19 occurs at a wavelength λ_max determined by

$$\lambda_\text{max} T = 5.1 \times 10^{-3}\,\text{m K}. \tag{1.24}$$

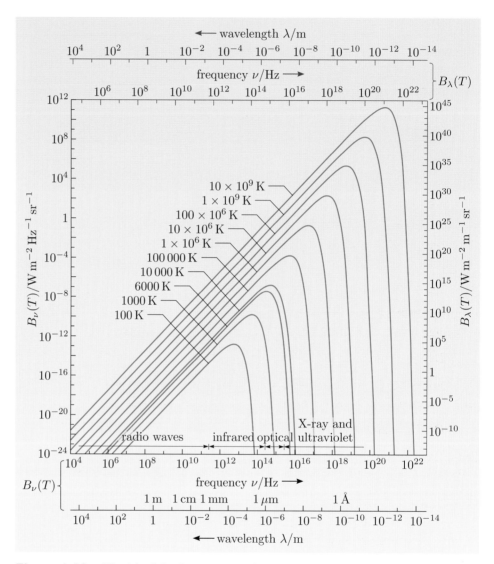

Figure 1.19   The black body spectrum for various temperatures. The peak emission occurs at a wavelength described by the Wien displacement law. The shape at substantially longer wavelengths is known as the Rayleigh–Jeans tail; at substantially shorter wavelengths it is the Wien tail.

The figure also shows that the emitted power of a black body increases at all wavelengths as the temperature increases. So a hotter black body will be brighter than a cooler black body even at the peak wavelength of the latter.

A useful way of characterizing the black body emission is in terms of its mean photon energy

$$\langle E_\text{ph} \rangle = 2.70 k T. \tag{1.25}$$

The typical thermal energy of particles in a gas with temperature T is the same as the mean photon energy $\langle E_{\mathrm{ph}} \rangle$.

Photon temperature

Photon energies are often expressed in electronvolts rather than joules, where $1\,\mathrm{eV} = 1.602 \times 10^{-19}\,\mathrm{J}$. In the context of the high-energy emission of astrophysical bodies, it is common practice to quote the photon energy as a temperature, $T \simeq E_{\mathrm{ph}}/k$, obtained by inverting Equation 1.25. This concept of a radiation temperature is enormously useful when attempting to estimate the radiation expected from a gas or plasma with a certain temperature. The rule of thumb

$$1\,\mathrm{eV} \text{ corresponds to } 10^4\,\mathrm{K} \tag{1.26}$$

is worth remembering.

Exercise 1.11   Verify Equation 1.26.

Exercise 1.12   (a) Calculate the typical photon energies for the accretion discs considered in Exercises 1.9 and 1.10. Express the results in eV.

(b) Calculate the corresponding wavelengths, and compare them to the wavelength range of visible light.   ■

The shape of the Planck function at substantially shorter wavelengths than the peak (high energies) is known as the **Wien tail**, which is described by

$$B_\nu(T) = (2h\nu^3/c^2)\exp(-h\nu/kT). \tag{1.27}$$

The shape of the Planck function at substantially longer wavelengths than the peak (low energies) is known as the **Rayleigh–Jeans tail**, which is described by

$$B_\nu(T) = 2kT\nu^2/c^2. \tag{1.28}$$

These 'tails' at both extremes of wavelength are sometimes referred to as the long-wavelength (or low-energy) cut-off and the short-wavelength (or high-energy) cut-off.

Exercise 1.13   Show that the Planck function (Equation 1.23) depicted in Figure 1.19 adopts the functional form expressed in Equation 1.27 for large frequencies, and the functional form expressed in Equation 1.28 for small frequencies.   ■

Multi-colour black body spectrum

Accretion discs can be thought of as composed of a series of annuli with different radii r, all emitting locally as a black body (Equation 1.23) with temperature $T(r) = T_{\mathrm{eff}}(r)$ as calculated in Equation 1.20. The resulting continuum emission spectrum is a sum of black body spectra at different T, but the dominant contribution will come from the region where the accreting plasma is hottest, i.e. from the vicinity of the inner edge of the disc. We consider this now in detail.

The total output from the disc is obtained by summing the contributions of all disc annuli, i.e. by the integral

$$F_\nu \propto \frac{1}{D^2} \int_{r_{\rm in}}^{r_{\rm out}} \left(\frac{2h\nu^3}{c^2}\right) \frac{1}{\exp(h\nu/kT_{\rm eff}(r)) - 1} \, 2\pi r \, {\rm d}r \qquad (1.29)$$

from the inner disc radius $r_{\rm in}$ to the outer disc radius $r_{\rm out}$. The flux F_ν per unit frequency received by the observer scales as $1/D^2$, where D is the distance between observer and emitter. The azimuthal part of the integral in Equation 1.29 has already been carried out and gave the factor 2π.

We have $T_{\rm eff}^4 \propto r^{-3}$, so the hottest black body that contributes has approximately the temperature $T_{\rm eff}(r_{\rm in}) \equiv T_{\rm in}$, while the coolest black body that contributes has the temperature $T_{\rm eff}(r_{\rm out}) \equiv T_{\rm out}$.

We consider now the shape of the disc spectrum F_ν in three different regimes.

For $h\nu \ll kT_{\rm out}$, i.e. for low-energy photons, cooler than the coolest part of the disc, the Planck function adopts the form of the Rayleigh–Jeans tail (Equation 1.28), and the integral can be written as

$$F_\nu \propto \int \nu^2 T_{\rm eff}(r) r \, {\rm d}r \propto \nu^2 \int T_{\rm eff}(r) r \, {\rm d}r, \qquad (1.30)$$

i.e. the disc spectrum at low frequencies also has the characteristic Rayleigh–Jeans tail shape, $F_\nu \propto \nu^2$, as the integral in Equation 1.30 is independent of ν.

For $h\nu \gg kT_{\rm in}$, i.e. for high-energy photons, hotter than the hottest part of the disc, the Planck functions adopts the form of the Wien tail (Equation 1.27), and the integral can be written as

$$F_\nu \propto \nu^3 \int \exp(-h\nu/kT_{\rm eff}(r)) r \, {\rm d}r. \qquad (1.31)$$

The integral is proportional to the difference in the values of $\exp(-h\nu/kT_{\rm eff}(r))$ at the inner and outer disc radii. As $h\nu/kT_{\rm out}$ is much larger than $h\nu/kT_{\rm in}$, the term with $T_{\rm out}$ is negligible. So the integral scales as $\exp(-h\nu/kT_{\rm in})$, and we have

$$F_\nu \propto \nu^3 \exp(-h\nu/kT_{\rm in}), \qquad (1.32)$$

i.e. the disc spectrum has a Wien tail that corresponds to the temperature of the innermost disc.

For the intermediate range of photon energies, much larger than the thermal energies at the outer disc but much smaller than those at the inner disc, i.e. for $kT_{\rm out} \ll h\nu \ll kT_{\rm in}$, we define

$$x = \frac{h\nu}{kT_{\rm eff}(r)} = \varepsilon\nu r^{3/4}, \qquad (1.33)$$

$\varepsilon = (h/k)(8\pi\sigma/3GM\dot{M})^{1/4}$ can be obtained from Equation 1.20 when $T_{\rm eff}$ is expressed in terms of r.

where ε is a constant for a system with a given mass and mass accretion rate. We therefore have $r = (x/\varepsilon\nu)^{4/3}$ and

$$\frac{{\rm d}r}{{\rm d}x} = \frac{4}{3} \frac{x^{1/3}}{(\varepsilon\nu)^{4/3}}. \qquad (1.34)$$

Expressing Equation 1.29 in terms of x, we obtain

$$F_\nu \propto \int_{x_{\rm in}}^{x_{\rm out}} \nu^3 \frac{1}{e^x - 1} \times \left(\frac{x}{\nu}\right)^{4/3} \times \frac{x^{1/3}}{\nu^{4/3}} \, {\rm d}x$$

$$\propto \nu^{1/3} \int_{x_{\rm in}}^{x_{\rm out}} \frac{x^{5/3}}{e^x - 1} \, {\rm d}x.$$

As $x_{in} \ll 1$ and $x_{out} \gg 1$, the integral is approximately equal to

$$\int_0^\infty \frac{x^{5/3}}{e^x - 1}\, \mathrm{d}x,$$

and is therefore independent of ν. So we have

$$F_\nu \propto \nu^{1/3}, \tag{1.35}$$

which is a spectral shape that is often quoted as characteristic for an accretion disc. The width of the frequency range over which the disc spectrum does indeed follow the $\nu^{1/3}$ relation depends on the difference between the inner and outer disc temperatures; see Figure 1.20.

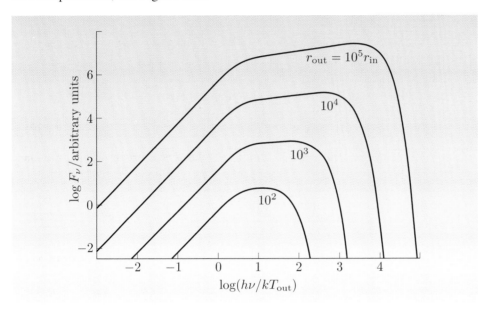

Figure 1.20 The spectrum of an accretion disc that emits locally like a black body, for different ratios of outer to inner disc radius.

Summary of Chapter 1

1. The process where a gravitating body grows in mass by accumulating matter from an external reservoir is called mass accretion. The accretion luminosity of a body with mass M and radius R is

$$L_{acc} = \frac{GM\dot{M}}{R}, \tag{Eqn 1.3}$$

where $\dot{M}$ is the accretion rate. A useful unit for the accretion rate is

$$1\,\mathrm{M_\odot\,yr^{-1}} = 6.31 \times 10^{22}\,\mathrm{kg\,s^{-1}}. \tag{Eqn 1.4}$$

2. The conservation of angular momentum implies that the accreting material will in general settle into an accretion disc around the accretor. The orbital motion of disc material can be approximated well by Kepler orbits, with a slow, superimposed radial drift towards the accretor. As the accreting plasma slowly drifts inwards, gravitational potential energy is lost. Half of this energy is converted into kinetic energy, while the other half is available to heat the plasma and to power the emission of electromagnetic radiation.

3. The accretion efficiency η_{acc}, defined by

$$L_{acc} = \eta_{acc}\dot{M}c^2, \qquad \text{(Eqn 1.9)}$$

expresses the rate of energy gained by the accretion of matter, in units of the mass energy of that matter. Accretion onto compact objects (white dwarfs, neutron stars and black holes) returns a very large accretion efficiency, much larger than the efficiency for hydrogen burning, the energy source powering main-sequence stars.

4. A non-rotating (Schwarzschild) black hole is formed when a mass M collapses to within a sphere of radius R_S, where

$$R_S = \frac{2GM}{c^2}. \qquad \text{(Eqn 1.11)}$$

The Schwarzschild radius R_S is the radius of the sphere surrounding the collapsed mass at which the escape speed equals the speed of light. The general relativistic result for the accretion efficiency of a disc-accreting Schwarzschild black hole is $\eta_{acc} = 5.7\%$. For a Kerr black hole with maximum pro-grade spin this efficiency is $\eta_{acc} = 32\%$.

5. Both Schwarzschild and Kerr black holes represent gravitational singularities where the curvature of spacetime is infinite. In the case of Kerr black holes this singularity is a ring in the plane perpendicular to the rotational axis, while it is a single point for Schwarzschild black holes. Both types have a spherical event horizon where the escape speed is the speed of light. A Kerr black hole is surrounded by a second, larger critical surface, the static limit, which has the shape of an oblate spheroid and touches the event horizon at its poles. The space between these two surfaces is called the ergosphere.

6. A binary star is a system consisting of two stars that orbit the common centre of mass. A compact binary is a close binary system where one component is a compact star. If the orbital separation is of order the stellar radii, the binary components can interact by exchanging mass.

7. Mass transfer in binaries is best described in a frame of reference that co-rotates with the binary orbital motion. In such a frame there are two types of pseudo-forces: the centrifugal force of magnitude

$$F_c = m\frac{v^2}{r} = m\omega^2 r, \qquad \text{(Eqn 1.14)}$$

and the Coriolis force

$$\boldsymbol{F}_{Coriolis} = -2m\boldsymbol{\omega} \times \boldsymbol{v}, \qquad \text{(Eqn 1.15)}$$

which vanishes for a test mass at rest in the co-rotating frame.

8. The force on a test mass m in the binary frame can be obtained as $\boldsymbol{F} = -m\boldsymbol{\nabla}\Phi_R$, where the Roche potential Φ_R is given by

$$\Phi_R(\boldsymbol{r}) = -\frac{GM_1}{|\boldsymbol{r} - \boldsymbol{r}_1|} - \frac{GM_2}{|\boldsymbol{r} - \boldsymbol{r}_2|} - \tfrac{1}{2}\left(\boldsymbol{\omega} \times (\boldsymbol{r} - \boldsymbol{r}_c)\right)^2. \quad \text{(Eqn 1.16)}$$

The surface of a binary stellar component coincides with a Roche equipotential surface.

9. The inner Lagrangian point (L_1 point) is the point between the two stars where the force on a test mass vanishes. The L_1 point is a saddle point of the Roche potential. The Roche equipotential surface that contains L_1 consists of two closed surfaces that meet at L_1. The enclosed volume is the Roche lobe of the respective star. A stellar component of the binary can expand only until its surface coincides with this critical Roche equipotential. If such a Roche-lobe filling star attempts to expand further, mass will flow into the direction of smaller values of Φ_R, i.e. into the lobe of the second star. This is called Roche-lobe overflow. The binary is said to be semi-detached.

10. An active galaxy contains a bright, compact nucleus that dominates its host galaxy's radiation output in most wavelength ranges. These active galactic nuclei (or AGN) are thought to be powered by accretion onto supermassive black holes. AGN are very compact. From the timescale of AGN variability, the light-crossing time can be deduced to be only a few days.

11. The velocity dispersion in the broad-line region of AGN is typically observed to be several 10^3 km s^{-1}, while the emitting region is seen to be smaller than a few tens of pc. Hence the virial theorem suggests that the mass of the central black hole is approximately 10^8–10^{11} M$_\odot$.

12. Most galaxies harbour a supermassive black hole (with mass $\gtrsim 10^6$ M$_\odot$) in the centre. In active galaxies, this black hole is a strong power source, while in normal galaxies such as the Milky Way it lies dormant.

13. The radial temperature profile of an optically thick, steady-state accretion disc is approximately

$$T_{\text{eff}}^4(r) \simeq \frac{3GM\dot{M}}{8\pi\sigma r^3}. \qquad \text{(Eqn 1.20)}$$

14. Accretion onto compact objects leads to very high plasma temperatures, 10^5–10^7 K. The corresponding black body emission peaks in the ultraviolet and X-ray regimes. The brightest sources in the X-ray sky are accreting neutron star and black hole binaries. In low-mass X-ray binaries, the companion star is a low-mass star ($M_2 \lesssim 2$ M$_\odot$), while in high-mass X-ray binaries, the companion star is a massive star ($M_2 \gtrsim 10$ M$_\odot$).

15. The thermal emission from an optically thick accretion disc is like a stretched-out black body. At high frequencies, the flux distribution F_ν has a Wien tail that corresponds to the temperature of the innermost disc. At low frequencies, the disc spectrum has the familiar Rayleigh–Jeans tail shape. The disc spectrum has a characteristic flat part $F_\nu \propto \nu^{1/3}$ at intermediate frequencies.

Chapter 2   Formation and evolution of accretion-powered binaries

Introduction

Among the multitude of binary star classes, we focus here on compact binaries, i.e. systems where at least one component is a compact star — a white dwarf, neutron star or black hole. These end-states of stellar evolution represent the burnt-out, exposed core regions of formerly much more massive and extended stars.

As we have seen in Chapter 1, compact binaries are powered by the very potent energy source of mass accretion. They dominate the X-ray sky, and are in fact the most accessible laboratories for studying the physics of accretion flows. In order to meaningfully interpret the detected high-energy emission, we first have to develop an understanding of the physical state of this laboratory, i.e. we have to study the evolutionary state and hence formation history of the host systems.

A main subject of this chapter is the quantitative description of mass transfer, the feeding process of the accretion flow, and its relation to the orbital characteristics of the binary. We shall discuss the driving mechanisms for mass transfer, slow and rapid orbital angular momentum losses, and the obstacles that stand in the way of forming compact stars in short-period binaries.

2.1   Binary stars

Stars seem to prefer life as a member of a binary system over solitude. Statistical counts suggest that more stars reside in binaries than there are single stars. Estimates for the incidence of binarity range from 20% to more than 90% and suggest that this **binary fraction** depends on the masses of the stellar components (higher mass stars showing a higher binary fraction). The true binary fraction is difficult to measure because some of the characteristics of binaries work against their discovery, for example, wide orbits with correspondingly rather slow orbital velocities. Except for the few visual binaries, where both component stars can be clearly seen and their orbital motion monitored, a binary in general appears as a point-like light source even when observed with the largest telescopes. The binary nature of a star is usually recognized through a periodic Doppler shift of the spectral lines of at least one of the components due to the orbital motion. If this is the case, the system is a **spectroscopic binary** (Figure 2.1). The velocity of a star with mass M_1 in a circular orbit with radius a around another star with mass M_2, as measured in an inertial frame where the centre of mass of the binary is at rest, is given by

$$v_1 = \left(\frac{GM_2}{a} \frac{M_2}{M} \right)^{1/2} , \tag{2.1}$$

where $M = M_1 + M_2$ is the total binary mass. If the separation is large, the Doppler shift (which is proportional to v_1) is small and difficult to measure.

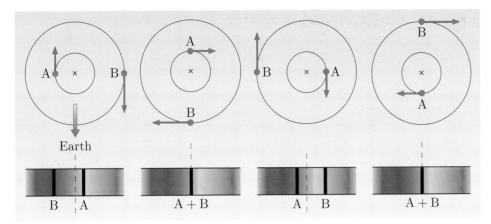

Figure 2.1 A spectroscopic binary cannot be resolved in telescopes, but reveals its identity through the periodic Doppler shift of spectral lines. Here the composite spectrum, at the bottom of the figure, of the two component stars, A and B, shows the absorption line of a particular atomic transition for both stars, indicated by the black bars in the continuum spectrum. The line position is blueshifted if the emitter is moving towards Earth, and redshifted when it is moving away from Earth. The spectral lines of stars A and B fall on top of each other in the second and fourth panels as both binary components then have the same (vanishing) radial velocity. The difference of the line positions is maximal in the first and third panels where the stars move straight towards or away from the observer.

- Equation 2.1 is similar to Equation 1.5, but includes the additional factor M_2/M to take into account the fact that the star orbits the centre of mass, not the centre of the other star. Show how Equation 2.1 arises.

○ The star with mass M_1 (or M_2) orbits the centre of mass with an orbital radius a_1 (or a_2), where $a_1 + a_2 = a$. From the definition of the centre of mass we have $M_1 a_1 = M_2 a_2$ and so $M_1 a_1 = M_2 \times (a - a_1)$, or

$$a_1 = \frac{M_2}{M} a. \qquad (2.2)$$

The corresponding expression for a_2 is $a_2 = (M_1/M)a$. The orbital speed can then be determined by noting, as in Section 1.1, that the gravitational force $GM_1 M_2/a^2$ provides the centripetal force for the circular motion, $M_1 v_1^2/a_1$. Equating these expressions, solving for v_1^2, and substituting for a_1 gives

$$v_1^2 = \frac{a_1}{M_1} \times \frac{GM_1 M_2}{a^2} = \frac{GM_2 a_1}{a^2} = \frac{GM_2^2 a}{Ma^2}.$$

This becomes Equation 2.1.

The very existence of binary stars is a stroke of luck for astrophysicists. The stellar companion effectively constitutes a *measuring device*. Stellar parameters that otherwise are accessible only in a rather indirect way, or inaccessible altogether, suddenly become measurable. One example is the mass of a star. We can *weigh* the stars in a binary by observing the orbital motion (Equation 2.1 contains the mass of the system). Another example refers to **eclipsing binaries**, where the line of sight to the observer on Earth lies in the orbital plane. As a result, one component passes in front of the other once or twice per orbit. From the eclipse **light curve** it is possible to deduce the radii of the two stars. This is analogous to transiting **exoplanets**, where a tiny fraction of the host star is eclipsed by its planet when it moves in front of the star as seen from Earth.

There is a bewildering variety of binary stars — all possible combinations of mass, spectral type end evolutionary state seem to exist. Although here we are mainly interested in interacting compact binaries, we begin with general considerations about the nature of mass transfer in the context of the Roche model. Our focus will be on Roche-lobe overflow as it is a mode of mass transfer that any type of donor star can be subject to, regardless of the stellar wind strength. If one component is Roche-lobe filling, the mass transfer rate by Roche-lobe overflow is always higher than the wind accretion rate.

Sustained mass transfer by Roche-lobe overflow does not just occur by itself; a driving mechanism is required that continuously expands the mass donor relative to its Roche lobe. This can be powered either by an intrinsic expansion of the star itself, or by a contraction of the Roche lobe through a systemic mechanism unrelated to mass transfer. Alternatively, there are situations where mass transfer is self-accelerating, and hence unstable, but a binary system is unable to remain in such a phase for long without a significant, sometimes catastrophic change of the system's properties.

2.2   The Roche-lobe radius

For a quantitative determination of the mass transfer rate via Roche-lobe overflow, it is necessary to calculate the size of the Roche lobe, as this sets the maximum volume available to the donor star. A convenient measure is the **Roche-lobe radius**, which is defined as the radius of a sphere with the same volume as the Roche lobe. Calculating the volume enclosed by a Roche equipotential surface requires the numerical integration of the Roche potential (Equation 1.16). Although the result of this integration is not a simple analytical function, it can be approximated by such a function to an accuracy that is quite sufficient for most practical applications.

Kepler's law

Kepler's third law,

$$\frac{a^3}{P_{\text{orb}}^2} = \frac{G}{4\pi^2}\,M,$$ (2.3)

relates the binary's orbital separation a and its orbital period P_{orb}, involving the total mass $M = M_1 + M_2$ of the system. Kepler's law was conceived to describe the orbits of planets in the Solar System, but it is a general consequence of Newton's law of gravity for any bound system of two point masses. Kepler's law can also be written in terms of the orbital angular speed $\omega = 2\pi/P_{\text{orb}}$ as

$$a^3\omega^2 = GM.$$ (2.4)

In the context of close binary systems it is convenient to rewrite Equation 2.3 with the separation expressed in solar radii, the orbital period in days and the mass in solar units:

$$\left(\frac{a}{R_\odot}\right)^3 = 74.382 \times \left(\frac{P_{orb}}{d}\right)^2 \left(\frac{M}{M_\odot}\right). \tag{2.5}$$

To gain a better understanding of the functional form of these approximations, we first consider what variables will determine the size of the Roche lobe. The gravitational interaction and dynamical behaviour of a binary system consisting of two point-like stars in a circular orbit are fully determined by the stellar masses M_1 and M_2 and the orbital separation a.

Worked Example 2.1

Rewrite the Roche potential

$$\Phi_R(r) = -\frac{GM_1}{|r - r_1|} - \frac{GM_2}{|r - r_2|} - \tfrac{1}{2}\left(\omega \times (r - r_c)\right)^2 \tag{Eqn 1.16}$$

in such a way that all length scales are normalized to the orbital separation a, and all masses to the total binary mass $M = M_1 + M_2$.

Solution

We multiply each length (such as r) by a factor a/a, and each mass (such as M_2) by a factor M/M, and obtain

$$\Phi_R(r) = -\frac{GM(M_1/M)}{a\,|r/a - r_1/a|} - \frac{GM(M_2/M)}{a\,|r/a - r_2/a|} - \tfrac{1}{2}\omega^2 a^2 \left(\frac{\omega}{\omega} \times \left(\frac{r}{a} - \frac{r_c}{a}\right)\right)^2,$$

where ω is the magnitude of the vector ω, so ω/ω is a unit vector in the direction of the orbital axis. Using Kepler's law in the form of Equation 2.4 to substitute $\omega^2 a^2 = GM/a$, and factoring out GM/a, we obtain

$$\Phi_R(r) =$$
$$-\frac{GM}{a}\left\{-\frac{M_1/M}{|r/a - r_1/a|} - \frac{M_2/M}{|r/a - r_2/a|} - \tfrac{1}{2}\left(\frac{\omega}{\omega} \times \left(\frac{r}{a} - \frac{r_c}{a}\right)\right)^2\right\}.$$

All terms enclosed by the large curly brackets are dimensionless, and depend only on the mass ratio. The factor GM/a in front of the brackets sets the magnitude scale of the Roche potential.

As the worked example demonstrates, a sets the length scale of the system, while the relative shape of the Roche equipotentials is determined by the mass ratio $q = M_2/M_1$. The pattern of Roche equipotentials for a system with, say, $M_1 = 4\,M_\odot$ and $M_2 = 2\,M_\odot$ is the same as for a system with $M_1 = 2\,M_\odot$ and $M_2 = 1\,M_\odot$.

Therefore the Roche-lobe radius must scale as $R_L \propto a$, and the constant of proportionality f depends only on the mass ratio $q = M_2/M_1$. We write

$$R_{L,2} = f(q)\,a \tag{2.6}$$

for the Roche-lobe radius of the secondary, where $f(q)$ is a dimensionless function of the mass ratio q.

● Show that the Roche-lobe radius of the primary is given by $R_{L,1} = f(1/q)\,a$.

○ The Roche-lobe radius for either binary component is $R_L \propto a$. The constant of proportionality f is a unique function of the ratio of the mass of the star whose Roche-lobe radius is to be determined and the mass of the other star. So $R_{L,2}/a = f(M_2/M_1) = f(q)$ while $R_{L,1}/a = f(M_1/M_2) = f(1/q)$.

Eggleton (1983) published the following useful analytic approximation for the function f:

$$f(q) = \frac{R_{L,2}}{a} \simeq \frac{0.49q^{2/3}}{0.6q^{2/3} + \log_e(1 + q^{1/3})}, \qquad (2.7)$$

which is accurate to better than 1% for all values of q. The simpler approximation given by Paczyński (1971),

$$f(q) = \frac{R_{L,2}}{a} \simeq 0.462 \left(\frac{M_2}{M}\right)^{1/3} = 0.462 \left(\frac{q}{1+q}\right)^{1/3}, \qquad (2.8)$$

is often preferred for mass ratios $q \lesssim 0.8$, where it is accurate to within 2%.

Exercise 2.1 Calculate the quantity f for mass ratios 0.5, 1.0 and 2, using both Eggleton's and Paczyński's approximations. Determine the relative difference between the two approximations. ■

A rather remarkable property of Roche-lobe filling stars is that their mean density $\overline{\rho}$ is effectively fixed by the orbital period. As the next exercise demonstrates, this is a consequence of Kepler's third law.

Exercise 2.2 (a) Use Kepler's law and Paczyński's approximation for the Roche-lobe radius to derive

$$\frac{P_{orb}}{h} \simeq 10.5 \left(\frac{\overline{\rho}}{10^3\,\text{kg m}^{-3}}\right)^{-1/2}. \qquad (2.9)$$

Hint: Remember that $\overline{\rho} = M_2/(4/3)\pi R_2^3$.

(b) For what range of mass ratios is Equation 2.9 valid? ■

Another useful way of expressing Equation 2.9 is

$$\log_{10}\left(\frac{P_{orb}}{h}\right) \simeq 0.9472 + \tfrac{3}{2}\log_{10}\left(\frac{R_2}{R_\odot}\right) - \tfrac{1}{2}\log_{10}\left(\frac{M_2}{M_\odot}\right). \qquad (2.10)$$

Exercise 2.3 Derive Equation 2.10. ■

The radius of low-mass main-sequence stars (with mass $\lesssim 1\,M_\odot$) scales roughly as the mass, $R_2 \propto M_2$. The mean density therefore scales as $\overline{\rho} \propto M_2/R_2^3 \propto M_2^{-2}$. In this case the period–mean density relation (Equation 2.9) simplifies to the rule of thumb

$$P_{orb} \simeq 8.8\,\text{h} \left(\frac{M_2}{M_\odot}\right). \qquad (2.11)$$

This expression can be used only for low-mass main-sequence stars and mass ratios less than unity.

2.3 Steady-state mass transfer

As mass transfer commences, mass and angular momentum are redistributed in the binary system, or may even be lost altogether into space, depending on the physical mechanisms at work. Therefore both the orbital separation and the mass ratio may change, so the Roche-lobe radius will in general change as well. In addition, the radius of the donor star changes when mass is transferred. The loss of mass perturbs the equilibrium structure of the star, and triggers adjustment processes that could either expand or contract the star.

In addition, both the Roche-lobe radius and the stellar radius may also continuously change due to systemic mechanisms unrelated to mass transfer. Examples are gravitational wave emission and magnetic stellar wind braking, both of which decrease the orbital angular momentum, or the normal nuclear evolution of the star that usually expands the star.

The mass transfer rate

The secondary star is the Roche-lobe filling component of the semi-detached binary. The mass of the secondary is M_2. As a result of Roche-lobe overflow this mass decreases, so the derivative dM_2/dt is negative. The **mass transfer rate**, the rate at which matter flows from the secondary into the primary's Roche lobe, is usually taken to be positive. Therefore the mass transfer rate is just $-dM_2/dt > 0$. Using dot notation, we have

$$-\frac{dM_2}{dt} \equiv -\dot{M}_2.$$

The key to understanding how the mass transfer rate adjusts to these changes is the insight that the donor's stellar radius R_2 and Roche-lobe radius $R_{L,2}$ must *move in step* if mass transfer is to be maintained at a steady rate.

In the language of Figure 1.10 in Chapter 1, this is rather obvious. The star is always just as big as its Roche lobe — if it were bigger, the surplus matter would quickly flow off to the primary star. A quantitative consideration of the nature of the plasma flow in the Roche potential near the inner Lagrangian point delivers the following expression for the instantaneous mass transfer rate:

$$-\dot{M}_2(\text{inst}) \simeq \dot{M}_0 \exp\left(\frac{R_2 - R_{L,2}}{H_2}\right), \tag{2.12}$$

where $\dot{M}_0 \approx 10^{-8}\,M_\odot\,\text{yr}^{-1}$ is only a weak function of stellar and binary parameters. This shows that the instantaneous transfer rate $-\dot{M}_2(\text{inst})$ is a sensitive function of the difference between Roche-lobe radius and stellar radius (Figure 2.2 overleaf).

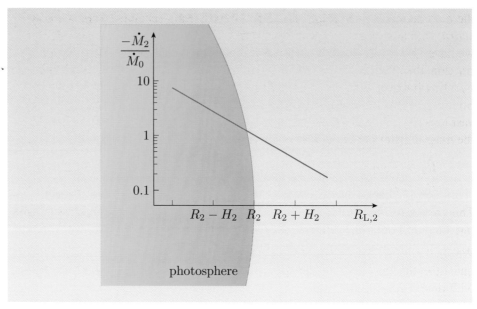

Figure 2.2   Mass transfer rate, in units of the rate $\dot{M}_0$, as a function of Roche-lobe radius $R_{L,2}$. Note that the $\dot{M}$-axis is in logarithmic units. The large circle indicates the position of the secondary star's photosphere ('stellar surface').

The quantity H_2 that appears in Equation 2.12 is the photospheric **scale height** and is a measure of how *sharp* the stellar rim is. It is defined as the length scale over which the stellar pressure P in the outermost layers of the star, close to the stellar photosphere (where $P = P_{ph}$), drops by a factor $e \approx 2.7$ (Figure 2.3).

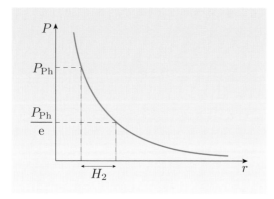

Figure 2.3   Photospheric pressure scale height.

If the difference between R_2 and $R_{L,2}$ changes only by about 2–3 H_2, the transfer rate will change by an order of magnitude.

● Why is this the case?

○ Suppose that initially the difference $R_2 - R_{L,2}$ is 0. Then the transfer rate is $-\dot{M}_2(\text{inst}) = \dot{M}_0 \exp(0) = \dot{M}_0$. If the new difference is $2.3 \times H_2$, then the new transfer rate is $-\dot{M}_2(\text{inst}) = \dot{M}_0 \exp(2.3) \approx 10 \times \dot{M}_0$.

Most stars do have a fairly sharp rim, in the sense that their photospheric scale height is much smaller than their radius, i.e. $H_2/R_2 \ll 1$. There are exceptions, for example very extended asymptotic giant branch stars where $H_2/R_2 \simeq 0.1$, but the most abundant species of stars, the low-mass main-sequence stars, are *very sharp*, with a photospheric scale height that is only 0.01% of the stellar radius. If we have $H_2/R_2 \ll 1$ and the donor star underfills its Roche lobe by only a tiny fraction, then the transfer rate is negligibly small, while if the star overfills its lobe by a tiny fraction, then the transfer rate would become too large to be sustainable. So in a phase of a sustained mass transfer we can assume that $R_2 \approx R_{\mathrm{L},2}$ in the vast majority of cases. The radii R_2 and $R_{\mathrm{L},2}$ are equal and remain so throughout the mass transfer phase, and we can write

$$\frac{\dot{R}_2}{R_2} = \frac{\dot{R}_{\mathrm{L},2}}{R_{\mathrm{L},2}}. \tag{2.13}$$

This expression is the foundation for any estimate of the equilibrium mass transfer rate in a semi-detached binary system.

We shall now use Equation 2.13 to obtain such an estimate by introducing known quantities that describe how the donor star's radius and Roche-lobe radius change with time.

Donor star radius

The rate of change $\dot{R}_2/R_2$ of the secondary star radius will in general be the sum of two terms:

$$\frac{\dot{R}_2}{R_2} = \left(\frac{\dot{R}_2}{R_2}\right)_{\mathrm{nuc}} + \left(\frac{\dot{R}_2}{R_2}\right)_{\dot{M}}. \tag{2.14}$$

The first term, $(\dot{R}_2/R_2)_{\mathrm{nuc}}$, describes the radius expansion or contraction that exists regardless of mass transfer. The subscript 'nuc' signifies that this is usually set by the nuclear evolution of the star.

The second term, $(\dot{R}_2/R_2)_{\dot{M}}$, quantifies the direct consequence of mass transfer. We express it as the product of a coefficient ζ with the logarithmic transfer rate $\dot{M}_2/M_2$, to highlight its explicit dependence on the mass loss rate.

If upon mass loss the donor radius evolves with donor mass as a power law of the form

$$R_2 \propto M_2^{\zeta}, \tag{2.15}$$

where ζ is the **mass–radius index**, then $(\dot{R}_2/R_2)_{\dot{M}}$ can be expressed as

$$\left(\frac{\dot{R}_2}{R_2}\right)_{\dot{M}} = \zeta \left(\frac{\dot{M}_2}{M_2}\right)$$

by taking the logarithmic derivative of the mass–radius relation (Equation 2.15).

● Explain why ζ is also called the (logarithmic) *slope* of the mass–radius relation.

○ From Equation 2.15 we find $\log_{10} R_2 = \zeta \log_{10} M_2 + \text{constant}$, so there is a linear relation between $\log_{10} R_2$ and $\log_{10} M_2$, with slope ζ.

Logarithmic derivatives

A rather convenient way to work out the derivative of a product or quotient is by the use of logarithmic derivatives. For obvious reasons, the derivative of the form

$$\frac{\dot{a}}{a} = \frac{\mathrm{d}}{\mathrm{d}t}(\log_e a)$$

is called the **logarithmic derivative** of a. If a quantity a is given as $a = b^n/c^m$, where n and m are positive numbers, then the logarithmic derivative of a is

$$\frac{\dot{a}}{a} = n\frac{\dot{b}}{b} - m\frac{\dot{c}}{c}. \tag{2.16}$$

This is a simple consequence of the chain and product rule:

$$\dot{a} = \frac{\mathrm{d}}{\mathrm{d}t}(b^n c^{-m}) = \left[nb^{n-1}\dot{b} \times c^{-m} \right] + \left[b^n \times (-m)c^{-m-1}\dot{c} \right]$$

$$= n\frac{b^n}{bc^m}\dot{b} - m\frac{b^n}{cc^m}\dot{c} = n\frac{a}{b}\dot{b} - m\frac{a}{c}\dot{c},$$

which reproduces Equation 2.16 when we divide both sides by a. This can be generalized to expressions with more factors in the numerator and denominator. For each factor in the numerator, add the logarithmic derivative of this factor; for each factor in the denominator, subtract the logarithmic derivative of this factor.

Roche-lobe radius

Similarly, the expression for the rate of change of the Roche-lobe radius, $\dot{R}_{L,2}/R_{L,2}$, will in general contain two terms: one that is independent of mass transfer, and one that is proportional to the logarithmic mass transfer rate.

This can be obtained by considering the expression for the total orbital angular momentum of the binary system:

$$J = M_1 M_2 \left(\frac{Ga}{M} \right)^{1/2}. \tag{2.17}$$

Exercise 2.4   Derive Equation 2.17 from first principles.   ■

Using Equation 2.6 we can replace a in Equation 2.17 with the term $R_{L,2}/f(M_2/M_1)$, and then solve the resulting equation for $R_{L,2}$. This gives the Roche-lobe radius as a function of J and the component masses in the system. Taking the logarithmic derivative of this delivers an expression of the form

$$\frac{\dot{R}_{L,2}}{R_{L,2}} = 2\frac{\dot{J}_{\mathrm{sys}}}{J} + \zeta_L\frac{\dot{M}_2}{M_2}, \tag{2.18}$$

where the quantity ζ_L, the **Roche-lobe index** of the donor, is a function of the mass ratio only.

● Why is there a factor of 2 in front of the orbital angular momentum term?

○ This is because Equation 2.17 has to be squared for it to be solved for a, and hence $R_{L,2}$.

To work out the quantity ζ_L explicitly requires one to express the derivatives of M_1 and M in terms of $\dot{M}_2$. To this end we introduce two dimensionless parameters that make the expressions simple and usable while maintaining the capability to describe a complex physical picture. The first parameter is defined by

$$\dot{M}_1 = -\eta \dot{M}_2, \tag{2.19}$$

expressing that the mass accretor captures a fraction η of the transferred mass. The remaining fraction $(1 - \eta)$ is lost from the binary system into space. The total mass of the system $M = M_1 + M_2$ changes as

$$\dot{M} = \dot{M}_1 + \dot{M}_2 = -\eta \dot{M}_2 + \dot{M}_2 = (1 - \eta)\dot{M}_2.$$

If $\eta < 1$, then mass is lost from the system ($\dot{M} < 0$). This mass will carry some angular momentum, so orbital angular momentum is lost also. We quantify this by defining the parameter ν_J as the specific angular momentum (angular momentum per unit mass) of the matter that is lost into space, in units of the mean specific orbital angular momentum, J/M. Then the total orbital angular momentum loss due to mass loss is

$$\dot{J} = \nu_J \frac{J}{M} \dot{M}. \tag{2.20}$$

With these definitions, the Roche-lobe index becomes, after some lengthy but straightforward algebra,

$$\zeta_L = (1 - \eta)(2\nu_J + 1)\left(\frac{q}{1+q}\right) + 2(q\eta - 1) + \tfrac{1}{3}\left(\frac{1+\eta q}{1+q}\right). \tag{2.21}$$

(To obtain this last equation we used Paczyński's approximation for $f(q)$, which implies that $(q/f)(\mathrm{d}f/\mathrm{d}q) = -(1/3)M_1/M$.)

Mass transfer rate

Equipped with these definitions we now finally obtain an expression for the steady-state mass transfer rate from equating Equations 2.14 and 2.18, and solving for $\dot{M}_2/M_2$:

$$\frac{\dot{M}_2}{M_2} = \frac{2\dot{J}_{\mathrm{sys}/J} - (\dot{R}_2/R_2)_{\mathrm{nuc}}}{\zeta - \zeta_L}, \tag{2.22}$$

where we note that both $\dot{M}_2$ and $\dot{J}$ are negative.

By far the most commonly considered case is **conservative mass transfer**, where both the orbital angular momentum and the total binary mass are constant throughout the evolution. The primary accretes all of the mass that the secondary loses. In other words, we have $\eta = 1$, and ν_J is irrelevant as there is no mass lost from the system. Then the Roche-lobe index becomes (still using Paczyński's approximation for $f(q)$)

$$\zeta_L = 2q - \tfrac{5}{3}. \tag{2.23}$$

● Verify that Equation 2.23 is consistent with the more general Equation 2.21.

○ The right-hand side of Equation 2.21 consists of a sum with three terms. For $\eta = 1$ the first term vanishes, and the third term becomes simply $+1/3$. Hence $\zeta_L = 2(q - 1) + 1/3 = 2q - 5/3$.

Exercise 2.5   Derive Equation 2.23 explicitly by taking the logarithmic time derivative of Equation 2.17. Replace a with the Roche-lobe radius, and use Paczyński's approximation for $f(q)$. Express time derivatives of M_1 and M in terms of $\dot{M}_2$.   ■

2.4   Nuclear-driven mass transfer

The most common driving mechanism for mass transfer is indeed the expansion of the donor star, which in turn mirrors the sequence of nuclear burning phases in the stellar core region. We refer to this simply as the nuclear evolution of the star.

As an example, Figure 2.4 shows the radius evolution of a single star with mass $5\,\mathrm{M_\odot}$. This star undergoes three different epochs with a significant radius increase. The first epoch of gentle growth (between the points labelled A and B in the figure) is the core-hydrogen burning phase, or main-sequence phase of the star. The second epoch, with a much more rapid growth (C–E), is the phase of expansion towards the giant branch and along the first giant branch, but before ignition of core-helium burning (at E). The third phase (F–I) is the expansion along the asymptotic giant branch after termination of core-helium burning. Mass transfer driven by the expansion of the donor star in one of these three phases is usually called **case A**, **case B** or **case C mass transfer**, respectively. In between the second and third phases of radius expansion, the star burns helium in its core. In this phase the star settles at a radius that is significantly smaller than the radius it had at the tip of the first giant branch, immediately before the ignition of helium burning. So a binary system that is wide enough to escape Roche-lobe overflow when the primary is on the first giant branch will then continue to evolve as a detached binary through the primary's helium burning. Only on the asymptotic giant branch, after the termination of core-helium burning, does the primary then expand further and possibly reach its Roche lobe.

The physical difference between the three cases of mass transfer is immediately obvious from Figure 2.4. First, the radii of potential donor stars in a case C mass transfer are larger than in a case B, and these in turn are larger than in a case A. A larger radius implies a larger orbital separation, to accommodate the donor star, and, from Kepler's law, a longer orbital period. So case A mass transfer binaries have short orbital periods, case B mass transfer binaries have intermediate orbital periods, and case C binaries have long orbital periods when mass transfer starts.

Exercise 2.6   Use Figure 2.4 to obtain typical orbital periods for a binary that begins mass transfer in a case A, case B and case C situation. Assume a mass ratio just larger than 1, and explain why $q < 1$ is unphysical if this is the first phase of mass transfer in the life of the binary.   ■

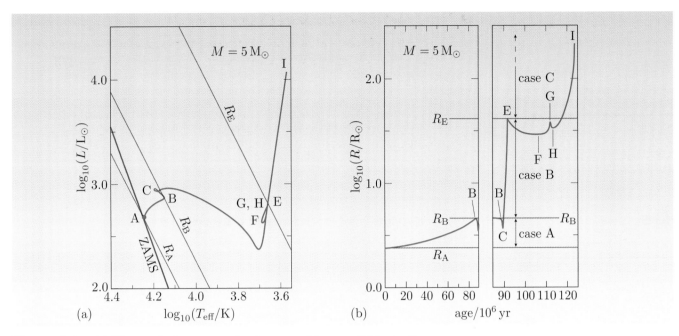

Figure 2.4 (a) The evolutionary track of a star with mass $5\,M_\odot$ in the Hertzsprung–Russell diagram. (b) The corresponding radius evolution of this star as a function of time.

A second difference between the cases of mass transfer is the rate of radius increase. For case A, the radius grows on the nuclear time t_{nuc} of the donor star (the main-sequence lifetime), while in case B the radius increase is on the donor star's thermal time t_{th}. A good approximation for the thermal time is the so-called Kelvin–Helmholtz time

$$t_{\mathrm{KH}} = \frac{GM^2}{RL} \tag{2.24}$$

(where M, R and L are the star's mass, radius and luminosity, respectively). The Sun's Kelvin–Helmholtz time is about 3×10^7 yr. The thermal time t_{th} is typically a factor 1000 shorter than t_{nuc}, hence the case B mass transfer rate must be larger than the case A rate by about this factor. For case C evolution the radius increase and hence the transfer rate are set by the growth time of the core mass of the mass donor.

Exercise 2.7   Calculate the Kelvin–Helmholtz time for the Sun, and for $0.5\,M_\odot$ and $5\,M_\odot$ main-sequence stars, assuming that these follow a mass–radius relation $R \propto M$ and mass–luminosity relation $L \propto M^4$. ∎

Worked Example 2.2

Use Equation 2.22 describing steady-state mass transfer to estimate the mass transfer rate $-\dot{M}_2$ for a donor star in a case A mass transfer phase, assuming that the timescale for the radius change is the nuclear time of the donor star. Consider $0.5\,M_\odot$, $1\,M_\odot$ and $5\,M_\odot$ donor stars and a $10\,M_\odot$ black hole accretor. (Note: This is not the first phase of mass transfer in this system.)

Solution

To determine the denominator in Equation 2.22 we note that main-sequence stars subjected to gentle mass loss follow an approximate mass–radius relation $R_2 \propto M_2$, so $\zeta \simeq 1$. Assuming conservative mass transfer we also have $\zeta_L = 2q - 5/3$. With $q = 0.05, 0.1$ and 0.5 this becomes $\zeta_L = -1.57$, -1.47 and -0.67, respectively, so for the order of magnitude estimate that we are attempting here, it is sufficient to say that $\zeta - \zeta_L \simeq 2$ in all three cases.

To work out the numerator in Equation 2.22 we assume that the angular momentum loss rate is negligible, so that we only need to estimate the nuclear radius expansion $(\dot{R}_2/R_2)_{\text{nuc}}$. The radius of a star roughly doubles during the main-sequence phase. So we make the estimate

$$\left(\frac{\dot{R}_2}{R_2}\right)_{\text{nuc}} \simeq \frac{1}{R_2} \frac{2R_2 - R_2}{t_{\text{MS}}} = \frac{1}{t_{\text{MS}}}.$$

In other words, the radius expansion term is the inverse of the main-sequence lifetime; this is often also called the nuclear timescale of the star. The main-sequence lifetime is proportional to the fuel reservoir ($\propto M_2$) and inversely proportional to the rate of burning, i.e. the luminosity. Hence,

$$t_{\text{MS}} \propto \frac{M_2}{L} \propto \frac{M_2}{M_2^4} \propto M_2^{-3}.$$

For the Sun, $M = 1\,M_\odot$ and $t_{\text{MS}} = 10^{10}$ yr, so we can write

$$t_{\text{MS}} \approx 10^{10}\,\text{yr} \left(\frac{M_2}{M_\odot}\right)^{-3}.$$

(This form follows when the luminosity of main-sequence stars is assumed to scale with mass as M^4. In reality the power index of the mass–luminosity relation varies slightly with mass.) Taken all together, we obtain from Equation 2.22

$$-\dot{M}_2(\text{case A}) \simeq \frac{M_2}{2t_{\text{MS}}} \simeq \frac{M_2}{2 \times 10^{10}\,\text{yr}\,(M_2/M_\odot)^{-3}}$$

or

$$-\dot{M}_2(\text{case A}) \simeq 5 \times 10^{-11} M_\odot\,\text{yr}^{-1} \times \left(\frac{M_2}{M_\odot}\right)^4.$$

So the approximate case A transfer rate for $M_2 = 0.5\,M_\odot$, $1\,M_\odot$ and $5\,M_\odot$ is $0.5^4 \times 5 \times 10^{-11} M_\odot\,\text{yr}^{-1} \approx 3 \times 10^{-12}\,M_\odot\,\text{yr}^{-1}$, $5 \times 10^{-11}\,M_\odot\,\text{yr}^{-1}$ and $5^4 \times 5 \times 10^{-11}\,M_\odot\,\text{yr}^{-1} \approx 3 \times 10^{-8}\,M_\odot\,\text{yr}^{-1}$, respectively.

Exercise 2.8   Repeat the reasoning of the worked example above to estimate the mass transfer rate $-\dot{M}_2$ for a donor star in a case B mass transfer phase, assuming that $\zeta = 0$ and that the timescale for radius change is the thermal time of the donor star, i.e. $(\dot{R}_2/R_2)_{\text{nuc}} = 1/t_{\text{th}}$. ∎

2.5 Mass transfer driven by angular momentum losses

Any case A mass transfer involving a low-mass donor star would proceed at a very low rate because the star's nuclear evolution is very slow. In Worked Example 2.2 we have calculated that a $0.5\,M_\odot$ donor star would only support a transfer rate of order $10^{-12}\,M_\odot\,\mathrm{yr}^{-1}$. Yet there are numerous and in fact rather conspicuous semi-detached systems with a low-mass main-sequence donor that are seen to be subject to a mass transfer rate that is up to 3–4 orders of magnitude larger than this value. These systems form the class of **cataclysmic variables**, or **CVs**, where a white dwarf accretes mass from a low-mass star, in most cases via an accretion disc, as shown in the artist's impression in Figure 1.11 in Chapter 1.

The bulk of the luminosity of CVs is due to accretion, and typically implies mass accretion rates, and hence mass transfer rates, of order $10^{-9}\,M_\odot\,\mathrm{yr}^{-1}$ to $10^{-8}\,M_\odot\,\mathrm{yr}^{-1}$ for systems with orbital periods longer than about 3 h, and of order a few times $10^{-11}\,M_\odot\,\mathrm{yr}^{-1}$ for systems with shorter periods.

Here it is not the donor star expansion that maintains mass overflow, but a continuous contraction of the donor's Roche lobe relative to the star, caused by mechanisms such as the emission of **gravitational waves** and **magnetic stellar wind braking**. In Equation 2.22 the orbital angular momentum loss term dominates the term describing the nuclear expansion of the donor star.

Gravitational wave radiation

One of the most dramatic predictions of Einstein's theory of general relativity is the existence of gravitational waves. In general relativity, space and time make up the unified four-dimensional spacetime. Mass curves spacetime. A variation of the mass distribution in a system causes a corresponding variation in the spacetime curvature. These changes can propagate outwards in the form of gravitational waves, carrying away energy and angular momentum. The 'ripples' in spacetime travel with the speed of light (Figure 2.5).

For a binary star with component masses M_1 and M_2 in a circular orbit with orbital period P_{orb}, the loss rate $\dot{J}_{\mathrm{GR}}$ of orbital angular momentum J is given by

$$\frac{\dot{J}_{\mathrm{GR}}}{J} = -1.27 \times 10^{-8}\,\mathrm{yr}^{-1}\frac{M_1 M_2}{(M_1 + M_2)^{1/3}\,M_\odot^{5/3}}\left(\frac{P_{\mathrm{orb}}}{\mathrm{h}}\right)^{-8/3}. \qquad (2.25)$$

The emission of gravitational waves leads to a slow spiral-in of the binary. The predictions made by the more general form of this *quadrupole formula* for elliptical orbits have been beautifully confirmed to very high accuracy by observations of the orbital decay of the **Hulse–Taylor binary pulsar** and other systems.

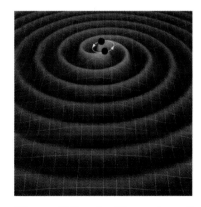

Figure 2.5 An illustration of gravitational waves emitted by a binary system.

Exercise 2.9 Use Equations 2.22 and 2.25 to estimate the mass transfer rate in a CV with a two-hour orbital period and a white dwarf with mass $1\,M_\odot$. (*Hint*: You will need to estimate the mass of the secondary star.) ∎

Magnetic braking

Observations of CVs with orbital periods longer than about 3 h suggest that
there is an orbital angular momentum loss mechanism at work with 10 to
100 times higher efficiency than gravitational waves. Many researchers
believe that this is magnetic stellar wind braking (Figure 2.6), the same
effect that causes the observed gradual spin-down of single main-sequence
stars. Such stars have a stellar wind, a constant flow of stellar plasma away
from the star, *and* a large-scale magnetic field that essentially co-rotates with
the star — as the magnetic field of a rotating bar magnet would. Through
interaction with the magnetic field, the material in the wind acquires a
large specific angular momentum. Thus the braking is very efficient,
although there is very little mass lost in the wind. The secondary is tidally
locked to the orbit, i.e. **tidal forces** ensure that the secondary in CVs is
rotating synchronously with the orbit. Hence any loss of rotational angular
momentum leads to a net loss of orbital angular momentum.

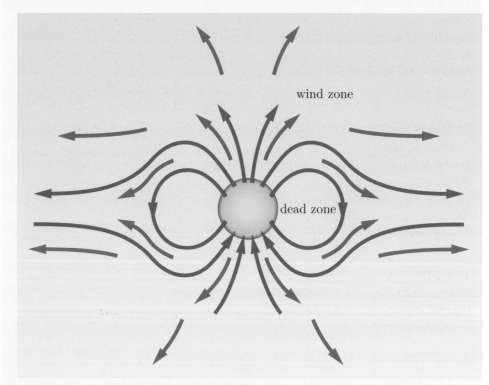

Figure 2.6   An illustration of magnetic stellar wind braking. Magnetic
field lines are shown in blue, plasma flow in red.

The quantitative description of magnetic braking is marred by the complexity of
the process, so predicting the braking strength from first principles is currently not
feasible. Instead, the observed slow-down of stars with age that is attributed to the
action of magnetic braking is used to obtain a quantitative description, often
mixed with analytical dependencies that emerge from simplified models of the
process.

A popular dependence is the **Skumanich law**, which states that the rotational angular speed ω of solar-type single stars decreases with stellar age t as $\omega \propto t^{-1/2}$. Therefore the rate of change of ω is $\dot{\omega} \propto -(1/2) \times t^{-3/2} \propto -\omega^3$. The spin angular momentum of the star is $J \propto \omega$, so also

$$\dot{J} \propto \omega^3.$$

This form of angular momentum loss is often applied to magnetic stellar wind braking in binary systems, with J interpreted as the orbital angular momentum. However, there is evidence that the rotational braking rate increases less steeply with ω for fast rotators, such as the binary components in short-period binaries. It is conceivable that the braking rate scales with a lower exponent of ω, perhaps $\dot{J} \propto \omega$, if the rotational speed exceeds a saturation value of order ten times the Sun's rotational speed.

2.6 Mass transfer stability

The very existence of close compact binaries is puzzling — their current orbital separation is much smaller than the radius that the compact star's progenitor must have had at some point in the past, so the binary's orbital separation must have decreased markedly at some point. It is commonly assumed that this is achieved through an episode of unstable mass transfer, leading to very high mass transfer rates and causing dramatic changes in the system's configuration. We shall discuss the ensuing common envelope phase in more detail below, but first we study the underlying cause, unstable, runaway mass transfer, in a more general context.

To this end we have to consider the role of the denominator $\zeta - \zeta_L$ in the expression for steady-state mass transfer. Clearly, for Equation 2.22 to work, the denominator has to be positive, as both the numerator and the left-hand side are in general negative. (In cases where the term $(\dot{R}_2/R_2)_{\mathrm{nuc}}$ is negative and the dominant term in the numerator, there will be no mass transfer.) Otherwise $\dot{M}_2$ would become positive, which is clearly inconsistent with mass loss from the star. Physically, $\zeta - \zeta_L < 0$ signals mass transfer instability.

To see this, we consider a Roche-lobe filling star, and assume that there are no systemic angular momentum losses, and that there is no (or only a negligible) internal stellar evolution. The star will lose mass at the instantaneous rate $-\dot{M}_2$ set by Equation 2.12 to its companion via the L_1 point.

Then from Equations 2.14 and 2.18 we have

$$\frac{\dot{R}_2}{R_2} - \frac{\dot{R}_{L,2}}{R_{L,2}} = (\zeta - \zeta_L) \times \frac{\dot{M}_2}{M_2} \tag{2.26}$$

or, introducing the difference $\Delta R = R_2 - R_{L,2}$, and its rate of change $\Delta \dot{R} = \dot{R}_2 - \dot{R}_{L,2}$, and noting that $R_2 \simeq R_{L,2}$ for a Roche-lobe filling star,

$$\frac{\Delta \dot{R}}{R_2} = (\zeta - \zeta_L) \times \frac{\dot{M}_2}{M_2}. \tag{2.27}$$

If $\zeta - \zeta_L < 0$ we have also $\Delta \dot{R} > 0$ (as $\dot{M}_2 < 0$), i.e. the star expands with respect to its Roche lobe. This in turn leads to an increase in the instantaneous mass transfer rate (Equation 2.12) — despite the absence of any mechanism driving the mass transfer. The increased transfer rate would lead to a further increase of $\Delta\dot{R}$, increasing the transfer rate even further, and so on. In such a runaway situation the system does not settle into a steady state.

Conversely, consider the same Roche-lobe filling star, but with $\zeta - \zeta_L > 0$. This time, purely as a result of mass transfer, the Roche lobe expands with respect to the star, i.e. the difference $R_2 - R_{L,2}$ decreases. The mass transfer rate will continue to decrease until it effectively ceases. Mass transfer is stable and can be maintained only if there is an additional driving mechanism that continually increases $R_2 - R_{L,2}$.

A stable system is also stable against a perturbation of its steady-state configuration. If the binary is settled at the steady-state mass transfer rate given by Equation 2.22, then a sudden, small increase of the transfer rate will lead to an instantaneous decrease of ΔR, and a consequent decrease of the instantaneous mass transfer rate given by Equation 2.12, re-establishing the original steady-state rate.

We summarize this finding once more as follows:

$$\text{if } \zeta - \zeta_L > 0, \text{ then mass transfer is stable.} \tag{2.28}$$

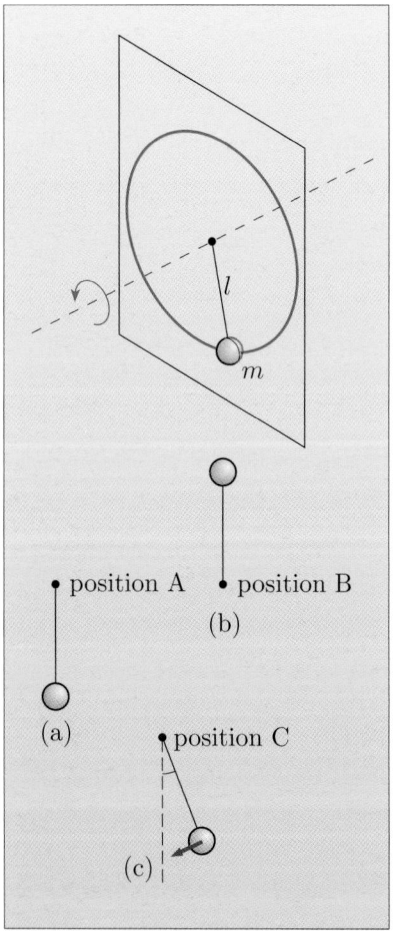

Figure 2.7   Equilibrium positions of a pendulum.

Stability, instability and stability analysis

The meaning of a statement like 'a physical system is stable' can be illustrated with a simple example: the pendulum. Consider an idealized pendulum on Earth, consisting of a weight with mass m that is attached to one end of an essentially massless rod with length l (Figure 2.7). The other end of the rod is fixed to a horizontal axis. The pendulum is allowed to rotate freely around this axis in a fixed vertical plane. Assume that the only forces acting on the pendulum are the gravity $m\mathbf{g}$ on the weight (g is the gravitational acceleration on the surface of the Earth), and a certain amount of friction, so that the swinging pendulum would eventually come to rest.

There are exactly two equilibrium positions of the pendulum, i.e. positions where the weight is at rest and stays at rest (at least in principle). The first such position (A) is when the weight is at the lowest point (Figure 2.7a). The second position (B) is when the weight is at the highest point (Figure 2.7b). In practice the pendulum would stay at rest only in equilibrium A. Even if the pendulum could be made to rest in the upside-down position B, the slightest vibration or air breeze would perturb the equilibrium. The pendulum would inevitably swing round and eventually end up in the equilibrium position A. This is the basis for saying that equilibrium A is *stable*, while equilibrium B is *unstable*.

More formally, a physical system is said to be stable, or in a stable state, if a small perturbation, after it has been applied to this state, will die away by itself, thus re-establishing the original state. This is the case for the pendulum in position A. If we displace the weight by a small angle and then release it (Figure 2.7c), there is a net force on the pendulum that causes an acceleration towards position A. The pendulum will oscillate around the equilibrium position A for a short while, but due to friction it will eventually come to rest at position A.

Conversely, a physical system is said to be unstable, or in an unstable state, if a small perturbation, after it has been applied to this state, will grow by itself, and lead the system further away from the original state. This is the case for the pendulum in position B. If we displace the weight by a small angle and then release it, there is a net force on the pendulum that causes an acceleration *away* from position B.

Although the stability behaviour is intuitively clear for the pendulum, there are plenty of physical systems where this is not the case. A formal stability analysis can show if a physical system is in a stable state or not. In such a stability analysis, one considers small perturbations to the state — like the small displacements applied to the pendulum — and *calculates* how the system would react to it. *The system is stable if it opposes the initial perturbation; the system is unstable if it amplifies the initial perturbation.*

For conservative mass transfer and the standard case $\zeta = 1$, the condition for stable mass transfer is a simple upper limit on the mass ratio.

- Determine this limit.
- With $\zeta_L = 2q - 5/3$ from Equation 2.23 and $\zeta = 1$, the condition for mass transfer stability, $\zeta - \zeta_L > 0$, becomes $1 - 2q + 5/3 > 0$ or $q < 4/3$.

In general, the denominator in Equation 2.22 is positive as long as the mass ratio is less than some critical value of order unity.

Exercise 2.10 What is the longest period for stable mass transfer in a low-mass X-ray binary with a main-sequence star donor and a $1.4\,M_\odot$ neutron star accretor? ■

If mass transfer is unstable, then depending on the nature of the donor star, the binary may either completely disrupt or enter into a transient, short-lived phase with a very high transfer rate. In the latter case the system can stabilize and revert to the normal rate given by, for example, Equation 2.22 once the mass ratio is smaller than the critical limit.

2.7 Common envelope evolution

Unstable mass transfer is the key for understanding the conundrum of short-period binary stars with compact components that must have been the core region of giant stars in the past. In the standard theory for the formation of these compact binaries, the progenitor binary forms with a massive star and a second, less massive star as its companion (which will become the donor star in the later compact binary phase). The orbital separation is large enough to accommodate the massive primary star, even when this becomes a giant star. The primary fills its Roche lobe for the first time (Figure 2.8(a) overleaf) either on the giant branch (case B), or on the asymptotic giant branch (case C), and as the mass ratio is much larger than unity, the ensuing mass transfer is unstable. On a very short timescale the mass transfer rate becomes so large that the companion star, which is still on the main sequence, is unable to accrete and accommodate the incoming mass into its envelope structure. The companion star begins to expand and fills its own Roche lobe in turn while mass transfer is continuing at an ever higher rate. The envelope material of the giant eventually spills over the Roche lobes of both stars and thus engulfs the whole binary. A **common envelope** is formed (Figure 2.8(b)).

The binary now consists of a proto-compact star, the high-density core of the former giant, and the largely unaltered normal main-sequence star, which orbit their common centre of mass within a tenuous common envelope of hydrogen-rich plasma. The envelope material exerts a drag force on the orbiting components, thus removing orbital angular momentum from the embedded binary. The orbit therefore tightens, i.e. the orbital separation decreases while the orbital speed increases. It is thought that this in-spiralling process is very effective and leads to a dramatic decrease of the orbital separation by a factor of perhaps 10^3 within an (astronomically speaking) extremely short time of perhaps less than 1000 years. In this process orbital energy and angular momentum is transferred to the common envelope until the envelope has acquired enough energy to be ejected from the system, thus terminating the common envelope phase (Figure 2.8(c)). For a certain time the remnant of this evolution will look like a planetary nebula (the ejected envelope) with a close, compact binary in its centre. There are indeed quite a few double cores of planetary nebulae known (see, for example, Figure 2.9). The short-lived nature of the common envelope phase itself implies that it is quite unlikely that we would catch a system right in the middle of it.

The detailed physics of the common envelope phase is still not well understood. Large-scale numerical computations of the corresponding plasma flows are required to model the phase. A major challenge is the need to resolve effects on drastically different length scales and timescales while keeping the run time reasonable. The gas density in a giant star's envelope increases from perhaps $10^{-5}\,\mathrm{kg\,m^{-3}}$ near the stellar surface to $\sim 10^8\,\mathrm{kg\,m^{-3}}$ near the core/envelope interface in the star's interior. Although the initial phases and early evolution can be traced by numerical schemes, a detailed computation of the final phase is still beyond the capabilities of current computer models.

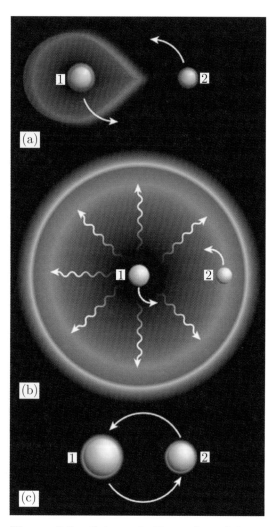

Figure 2.8   Schematic illustration of the common envelope phase (not to scale).

For estimates, and in statistical population studies, a simple energy budget argument is often used to link the pre- and post-common envelope binary parameters quantitatively. This assumes that orbital energy released by the embedded binary is immediately deposited into the envelope, with an efficiency α_{CE}, and that the common envelope terminates when the deposited energy equals the original envelope's binding energy E_{env}, i.e. the energy it needs to acquire to escape from the system. So we have

$$E_{\mathrm{env}} = \alpha_{\mathrm{CE}}\left(E_{\mathrm{orb,\,before}} - E_{\mathrm{orb,\,after}}\right),$$

which becomes an expression of the form

$$\frac{GM_1 M_{\mathrm{env}}}{R} = \alpha_{\mathrm{CE}} \left(\frac{GM_{\mathrm{core}} M_2}{2a_{\mathrm{f}}} - \frac{GM_1 M_2}{2a_{\mathrm{i}}} \right), \qquad (2.29)$$

where M_1 is the mass of the Roche-lobe filling giant star at the onset of the common envelope phase, R is its radius at this point, M_2 is the companion star mass, and M_{env} and M_{core} are the envelope and core mass of the giant star. (We have $M_1 = M_{\mathrm{env}} + M_{\mathrm{core}}$, and M_{core} is also the mass of the compact object remnant of the giant.) The orbital separation at the beginning of the common envelope phase is a_{i}, while a_{f} is the separation at the end of it.

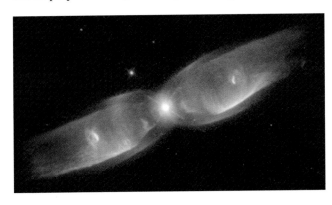

Figure 2.9   Hubble Space Telescope image of the Butterfly Nebula M2-9, a planetary nebula with a binary star at its core.

Worked Example 2.3

A giant star with mass $2.4\,\mathrm{M}_\odot$ and core mass $0.6\,\mathrm{M}_\odot$ fills its Roche lobe at an orbital period of 550 days. The companion star is a $0.5\,\mathrm{M}_\odot$ main-sequence star. Calculate the post-common envelope orbital period if $\alpha_{\mathrm{CE}} = 0.2$.

Solution

We begin by rearranging Equation 2.29 to obtain an expression for the orbital shrinkage factor, $a_{\mathrm{i}}/a_{\mathrm{f}}$. To this end we note that at the beginning of the common envelope phase, the giant star just fills its Roche lobe, so that its radius can be expressed as $R = f(M_1/M_2) \equiv f_1 a_{\mathrm{i}}$ (Equation 2.6). We therefore have

$$\frac{GM_1 M_{\mathrm{env}}}{\alpha_{\mathrm{CE}} f_1 a_{\mathrm{i}}} = \left(\frac{GM_{\mathrm{core}} M_2}{2a_{\mathrm{f}}} - \frac{GM_1 M_2}{2a_{\mathrm{i}}} \right).$$

Cancelling G and rearranging gives

$$\frac{M_{\mathrm{core}} M_2}{2a_{\mathrm{f}}} = \frac{M_1 M_{\mathrm{env}}}{\alpha_{\mathrm{CE}} f_1 a_{\mathrm{i}}} + \frac{M_1 M_2}{2a_{\mathrm{i}}}.$$

Multiplying both sides by $2a_{\mathrm{i}}/M_{\mathrm{core}} M_2$ gives

$$\frac{a_{\mathrm{i}}}{a_{\mathrm{f}}} = \frac{2M_1 M_{\mathrm{env}}}{\alpha_{\mathrm{CE}} f_1 M_{\mathrm{core}} M_2} + \frac{M_1 M_2}{M_{\mathrm{core}} M_2},$$

so finally

$$\frac{a_{\mathrm{i}}}{a_{\mathrm{f}}} = \frac{M_1}{M_{\mathrm{core}}} \left(\frac{2M_{\mathrm{env}}}{\alpha_{\mathrm{CE}} f_1 M_2} + 1 \right). \qquad (2.30)$$

Now with $M_{env} = M_1 - M_{core} = 1.8\,M_\odot$, $\alpha_{CE} = 0.2$, and an initial mass ratio of $q = 2.4/0.5 = 4.8$, we have $f_1 \equiv f(q = 4.8) = 0.517$ (using Equation 2.7), so

$$\frac{a_i}{a_f} = \frac{2.4}{0.6}\left(\frac{2 \times 1.8}{0.2 \times 0.517 \times 0.5} + 1\right) \approx 283.$$

The initial orbital period is 550 days, so Kepler's law (Equation 2.5) gives the initial separation

$$\frac{a}{R_\odot} = \left(74.382 \times 550^2 \times 2.9\right)^{1/3} \approx 403,$$

and therefore $a_f = (403/283)\,R_\odot \approx 1.42\,R_\odot$. Using Kepler's law in reverse, we obtain the final orbital period

$$P_{orb,\,f} = \left(\frac{2.82^3}{74.382 \times (0.6 + 0.5)}\right)^{1/2}$$
$$= 0.187\,d = 4.5\,h.$$

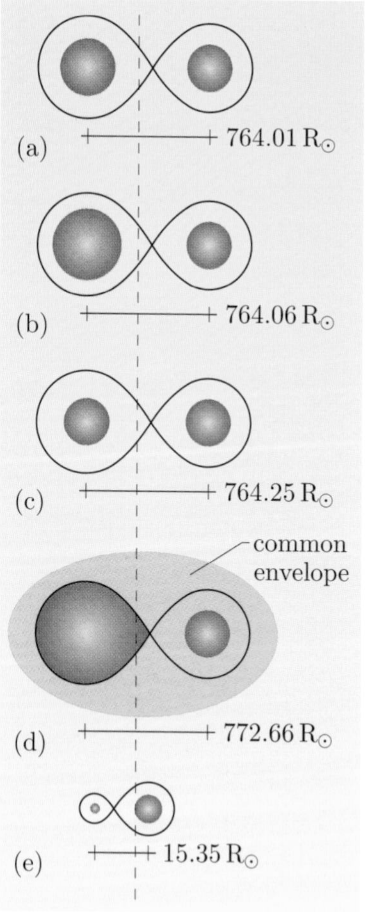

The evolutionary sequence shown in Figure 2.10 illustrates the common envelope paradigm for the formation of compact binaries with a white dwarf primary. As we shall discuss below, progenitors of neutron star and black hole X-ray binaries are also thought to undergo a common envelope phase.

2.8   Neutron star and black hole formation

2.8.1   Neutron star binaries

The formation of a close neutron star binary has yet another obstacle to overcome, apart from that faced by white dwarf binaries. In addition to a common envelope event needed to tighten the orbit, the neutron star forms in a supernova explosion, which is accompanied by a sudden and significant mass loss. The ejected supernova envelope has easily more mass than the remaining binary. However, a spherically symmetric explosion would leave the remnant binary unbound if more than half of the original binary mass is ejected.

(a) 764.01 $R_\odot$
(b) 764.06 $R_\odot$
(c) 764.25 $R_\odot$
common envelope
(d) 772.66 $R_\odot$
(e) 15.35 $R_\odot$

Figure 2.10   Characteristic stages in the evolution of a white dwarf binary. (a) Detached binary with both stars on the zero-age main sequence (primary mass $M_1 = 3.50\,M_\odot$, secondary mass $M_2 = 2.50\,M_\odot$). (b) Primary has evolved to a giant star ($M_1 = 3.50\,M_\odot$); binary is still detached. (c) Primary starts core-helium burning phase ($M_1 = 3.43\,M_\odot$); binary is still detached. (d) Onset of the common envelope phase ($M_1 = 3.43\,M_\odot$). (e) Binary emerges from the common envelope; only the core of the primary, a white dwarf, remains $M_{core} = 0.78\,M_\odot$, $M_2 = 2.50\,M_\odot$.

Exercise 2.11   Show that this statement is correct. Calculate the orbital speed of the companion before the explosion, and compare this with its escape speed after the explosion. Assume that the exploding star's mass is much larger than the companion star's mass, i.e. $M_1 \gg M_2$.

The mass range of main-sequence stars that will evolve to explode in a supernova and leave a neutron star remnant is thought to be $10 \lesssim M/M_\odot \lesssim 20$. A typical neutron star, on the other hand, has the canonical mass $1.4\,M_\odot$, while the maximum physically possible mass of a neutron star is just above $3\,M_\odot$. Therefore the mass loss in the supernova event is in general in excess of $\simeq 10\,M_\odot$. If the ejection is spherically symmetric, this will not leave a neutron star binary with a low-mass companion ($M_2 \lesssim 2\,M_\odot$) intact. Yet there are hundreds of low-mass X-ray binaries with neutron star primaries known in our Galaxy, and many in other nearby galaxies. It is of limited help that massive stars are likely to experience significant wind losses on the main sequence and in later phases of their evolution, reducing the pre-explosion mass by a few solar masses. Instead, it is commonly accepted that a common envelope evolution must precede the supernova event, so that the bulk of the primary's envelope has already been lost at the time of the explosion. Specifically, it is thought that the common envelope phase occurs after the core-hydrogen burning phase but before the helium-burning phase of the primary star. The common envelope remnant is then a helium star with mass $3 \lesssim M/M_\odot \lesssim 5$ and a low-mass companion. The helium star continues to evolve until it collapses to a neutron star in a type Ib or Ic supernova (see also Section 8.2.2, specifically the discussion headed *Association with supernovae*), indicating hydrogen deficiency. Even in this case the associated mass loss would unbind the large majority of systems, so that a new mechanism is required to guarantee the survival of a large enough number of systems that is consistent with the observed population of neutron star low-mass X-ray binaries (LMXBs).

The detailed working of this mechanism is not yet fully understood, but it appears to impart a **kick velocity** to the neutron star at birth. The existence of kick velocities can be deduced from the observed large proper motions of radio pulsars. Pulsars are isolated, rapidly rotating neutron stars that emit beamed radiation. This is registered as a pulse each time the beam sweeps past the Earth. As radio sources with an extremely accurate clock, set by the pulse, the astrometric positions of these sources on the celestial sphere can be measured very accurately, so the change of this position over time, the proper motion, can be measured. Figure 2.11 (overleaf) shows the space velocities of the observed sample of pulsars.

There is no clear relation between the direction of the kick (proper motion) and other physical parameters of the star and supernova remnant, so for now it is best to assume that the kick velocity imparted to the neutron star at birth has a magnitude distribution similar to the one shown in the figure, but the direction is isotropically distributed in space.

If the nascent neutron star in a binary receives the same sort of kick as isolated neutron stars, then a subset of systems will receive a suitably directed kick of the right magnitude to compensate for the loss of mass and change in escape speed, and the binary remains bound. The post-supernova binary will in general have an elliptical orbit, but tidal interactions with the normal star are likely to circularize the orbit on a short timescale.

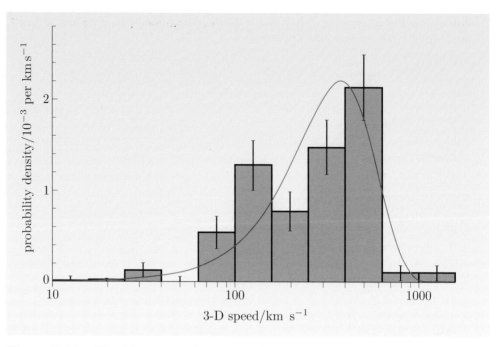

Figure 2.11   Blue histogram: the space velocity distribution of pulsars, deduced from their observed proper motions. Red curve: best-fitting Maxwellian distribution.

2.8.2   Black hole binaries

The formation of black holes in binaries is even less well understood. Standard wisdom has it that black holes form from initially yet more massive stars than the neutron star progenitors discussed above, say from stars with mass in excess of 20–30 $M_\odot$. Such massive stars display very strong stellar winds, so their pre-collapse mass is much reduced, perhaps halved. The collapse to the black hole may or may not be accompanied by the ejection of mass — and it is unclear if there are any kick velocities imparted to the black hole or not.

The spatial distribution of the known black hole binary systems in the Galaxy suggests that the kick is much weaker or absent for black holes. As with all young stellar populations in our Galaxy, the spatial distribution of low-mass X-ray binaries (LMXBs) follows the Galactic disc, with a concentration towards the Galactic midplane. In the vertical direction away from the midplane, the number density of systems drops off quickly, with a scale height of a few 100 pc. The black hole systems show a smaller scale height than the neutron star systems. This could be a result of kicks that the neutron star systems receive which make them move faster and hence cruise to larger distances from the midplane on their orbits around the Galactic centre.

2.9   Double degenerates and mergers

There are yet more extreme representatives of the class of compact binaries which we wish to mention here in passing. They will be the target of highly sensitive gravitational wave detectors: exciting, exotic astronomical measuring devices that are currently being refined to become more sensitive.

Detecting gravitational waves

Verifiable evidence for the existence of gravitational waves remains elusive. So far only upper limits have been determined. There are chiefly two kinds of detectors: the pioneering resonant devices, and the more sensitive laser interferometric devices. The former employ massive metal bodies such as cylinders or spheres that will vibrate when exposed to gravitational waves with frequencies near the devices' resonance frequencies. The latter aim to detect the relative movement of individual, small masses when a gravitational wave passes through. A monochromatic laser beam bounces off mirrors attached to these masses, and is brought to interference. Tiny changes in the distance between the mirrors will manifest themselves as changes in the interference pattern. There are several such detectors in operation worldwide. Currently the largest interferometric observatory is LIGO (Laser Interferometer Gravitational-wave Observatory), which operates two sites in the USA, while the proposed, ambitious space-based device LISA (Laser Interferometer Space Antenna) would significantly increase the sensitivity and probe an as yet inaccessible frequency range that should be populated by compact binary sources.

The local strength of a gravity wave is commonly expressed as the wave amplitude h. This measures the relative deformation experienced by an arrangement of independent test masses, such as a ring of test masses, in a vacuum when the gravity wave passes by. A binary system with component masses M_1 and M_2 with orbital angular frequency ω at a distance D from Earth gives rise to gravitational waves with an amplitude $h \propto M_1 M_2 / D$ and angular frequency 2ω. Figure 2.12 (overleaf) shows the expected wave amplitude as a function of frequency for various astrophysical gravitational wave sources, together with the sensitivity limits of LIGO and LISA.

Merging compact stars are expected to stand out. The signal from the final minutes of the merger sweeps through the 10–1000 Hz frequency range, growing in amplitude. The amplitude growth and frequency modulation are specific to the gravitational source. Hence it is *pattern recognition* techniques that underpin the hopes of detecting such an event.

As we have discussed in Section 2.5, the orbital evolution of a short-period compact binary with a low-mass stellar companion is driven by the emission of gravitational waves. The angular momentum loss rate scales inversely with a large power of the orbital period P_{orb} (Equation 2.25). The rate becomes large for very short-period systems, say for those with periods of a few minutes or shorter. Such a short period implies a very small orbital separation, of order the distance between the Earth and the Moon, but involving components with masses similar to the Sun. In order to fit into such a small orbit, both the stellar mass components have to be compact objects. This could be two white dwarfs (a so-called **double degenerate**) or two neutron stars, or two black holes, or in fact any pairing of white dwarf, neutron star and black hole.

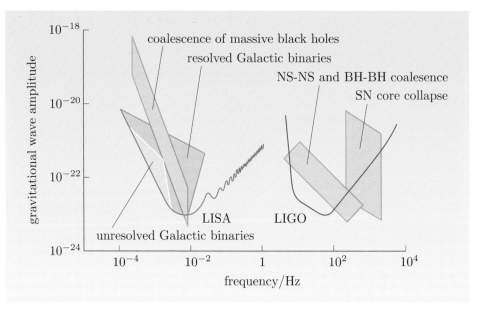

Figure 2.12 Expanded parameter space for various astrophysical gravitational wave sources in wave amplitude versus frequency space. The sensitivity limits for LIGO (blue curve) and the proposed LISA mission (red curve) are also shown.

The best studied group is double degenerates, some of which are in fact undergoing mass transfer as one white dwarf is filling its Roche lobe. The closest of these 'AM CVn' systems (after the prototype AM Canes Venaticorum) emit a gravitational wave signal strong enough to be detectable by a future space-based gravitational wave detector, such as the proposed LISA project. A double degenerate in a tight orbit is thought to have evolved through two common envelope phases of some kind, thus multiplying the uncertainties that we have in the description of their formation.

Even less well understood is the formation of a double neutron star or a double black hole, as their progenitors must survive two supernova explosions. These systems are intrinsically very rare, not least because stars form preferentially with a low mass and do not become a neutron star or black hole.

Initial mass function

The initial mass function (commonly abbreviated as IMF) quantifies the relative number of stars forming per unit mass interval, as a function of stellar mass M. The observationally deduced form of the initial mass function is

$$\text{IMF}(M) \propto M^{-2.7} \tag{2.31}$$

for stars with mass $\gtrsim 1\,\text{M}_\odot$, and is somewhat less steep (i.e. with a power index that is less negative) for smaller masses.

But if a binary with two massive stars does form, and succeeds in evolving to a close neutron star/black hole pair, its final evolution will be most dramatic. Once the orbital period is short enough, the emission of gravitational waves leads to a rapid in-spiralling of the components. The binary will merge, and the related catastrophic event releases a vast amount of energy. Compact binary mergers are thought to be the central engines of some gamma-ray bursts, a phenomenon that we shall discuss in detail in Chapter 8.

Summary of Chapter 2

1. The binary fraction ranges from 20% to more than 90% and may depend on the masses of the stellar components.

2. We can *weigh* components of binary systems by observing the orbital motion; see Equation 2.1. In eclipsing binaries it is sometimes even possible to measure directly the radii of the two stars.

3. There are two chief driving mechanisms for sustained mass transfer by Roche-lobe overflow: nuclear expansion of the donor due to its own stellar evolution, and orbital angular momentum losses due to processes such as the emission of gravitational waves or magnetic stellar wind braking.

4. Kepler's third law relates the orbital separation, orbital period and total mass of a binary system:

$$\frac{a^3}{P_{\text{orb}}^2} = \frac{G}{4\pi^2} M. \qquad \text{(Eqn 2.3)}$$

5. The Roche-lobe radius

$$R_{\text{L}} = f(q)\, a \qquad \text{(Eqn 2.6)}$$

is the radius of a sphere with the same radius as the Roche lobe. The dimensionless factor f is a function of the mass ratio q (i.e. the mass of the star for which the Roche lobe is to be determined, divided by the mass of the other star). A good approximation for $f(q)$ in the range $q \lesssim 0.8$ is

$$f(q) = \frac{R_{\text{L},2}}{a} \simeq 0.462 \left(\frac{M_2}{M}\right)^{1/3}. \qquad \text{(Eqn 2.8)}$$

6. In a semi-detached system, the orbital period is given by the mean density of the Roche-lobe filling star:

$$\log_{10}\left(\frac{P_{\text{orb}}}{\text{h}}\right) \simeq 0.9472 + \tfrac{3}{2}\log_{10}\left(\frac{R}{R_\odot}\right) - \tfrac{1}{2}\log_{10}\left(\frac{M}{M_\odot}\right). \qquad \text{(Eqn 2.10)}$$

7. The instantaneous mass transfer rate in a semi-detached binary is

$$-\dot{M}_2(\text{inst}) \simeq \dot{M}_0 \exp\left(\frac{R_2 - R_{\text{L},2}}{H_2}\right), \qquad \text{(Eqn 2.12)}$$

where the coefficient $\dot{M}_0 \approx 10^{-8}\,M_\odot\,\text{yr}^{-1}$ is only a weak function of stellar and binary parameters.

8. The total orbital angular momentum of the binary system is

$$J = M_1 M_2 \left(\frac{Ga}{M}\right)^{1/2}. \qquad \text{(Eqn 2.17)}$$

9. For steady-state mass transfer, the radii R_2 and $R_{L,2}$ move in step:

$$\frac{\dot{R}_2}{R_2} = \frac{\dot{R}_{L,2}}{R_{L,2}}. \qquad \text{(Eqn 2.13)}$$

The steady-state mass transfer rate can be expressed as

$$\frac{\dot{M}_2}{M_2} = \frac{2\dot{J}_{sys}/J - (\dot{R}_2/R_2)_{nuc}}{\zeta - \zeta_L}. \qquad \text{(Eqn 2.22)}$$

The secondary's Roche-lobe index ζ_L for conservative mass transfer, where both the orbital angular momentum and the total binary mass are constant throughout the evolution, is

$$\zeta_L = 2q - \tfrac{5}{3}. \qquad \text{(Eqn 2.23)}$$

The stellar index ζ is the exponent in the donor star's mass–radius evolution under mass loss:

$$R_2 \propto M_2^{\zeta}. \qquad \text{(Eqn 2.15)}$$

10. The nuclear evolution of a star has three main phases of radius expansion: the main-sequence phase, the first giant phase, and the asymptotic giant branch phase. Mass transfer driven by the nuclear expansion of a donor star in these phases is called case A, case B or case C, respectively.

11. In binary systems with low-mass donors, mass transfer is driven by systemic orbital angular momentum losses.

12. Mass transfer is unstable if the Roche lobe closes in on the star as a result of the transfer of a small amount of mass. Mass transfer is stable if $\zeta - \zeta_L > 0$ (Equation 2.28). This translates into an upper limit of order unity on the mass ratio of binaries with stable mass transfer.

13. Compact binaries are thought to have experienced a common envelope phase as a result of unstable mass transfer from the giant star progenitor of the compact object to its companion star. The giant's core and the companion orbit in a common envelope that is spun up and ejected from the system, thus tightening the remnant binary. An estimate of the effect on the binary's orbital parameters is obtained from the energy budget argument

$$\frac{GM_1 M_{env}}{R} = \alpha_{CE} \left(\frac{GM_{core} M_2}{2a_f} - \frac{GM_1 M_2}{2a_i} \right), \qquad \text{(Eqn 2.29)}$$

where the efficiency is $\alpha_{CE} \lesssim 1$.

14. Neutron stars form in a supernova explosion that ejects a large amount of mass into space. If the explosion is spherically symmetric, the binary will be unbound if more than half of the original binary mass is ejected. Neutron star binaries with low-mass companions can survive as bound systems in large enough numbers if the neutron star receives a suitably directed kick at birth. This is consistent with the large observed proper motion of radio pulsars. Black hole binaries may experience less significant mass loss at the formation of the black hole.

15. Double degenerates (compact binaries with two white dwarf stars), double neutron stars or black hole/neutron star binaries can evolve to very short orbital periods and become strong sources of gravitational waves, particularly in their final evolutionary stages before they merge.

Chapter 3 Steady-state accretion discs

Introduction

We have discussed in some detail the characteristics of compact binary evolutionary phases that set up an environment where a compact object accretes hydrogen-rich material from a mass reservoir, the companion star. We have also seen that the most common form of accretion flow is an accretion disc. In a semi-detached binary the mass flow is effectively confined to the orbital plane and leads quite naturally to the formation of an accretion disc around the accretor. The inner edge of the disc is near the surface of the accreting star or at the last stable circular orbit, while the outer edge is near the Roche-lobe radius of the accreting object.

In this chapter we shall take a closer look at the structure of accretion discs and develop a quantitative model for them. To simplify our task we consider here only steady-state discs. 'Steady state' means that the accretion flow pattern is the same at all times — or, in technical terms, the quantities in the disc equations that we shall derive have no explicit dependence on time. (We shall relax the steady-state assumption in Chapter 4.) We furthermore assume that an accretion disc really is what the name suggests — flat — and that the vertical and radial structure of the disc are decoupled and can be treated separately. We finally assume that there are no external magnetic fields that could interfere with the accretion flow.

Physical models

The purpose of a physical model is to check our understanding of the physics governing the system that we are trying to describe. We shall use the model to predict certain properties of the system and test these predictions against observations or experiment. Usually such a reality check will quickly reveal the general usefulness of the model. The feedback from experiment usually allows one to refine and extend the model, and thus to improve on the understanding of the underlying physics. This way we may uncover a piece of physics that was missing from the original model, but now proves vital to make it work. Sometimes such a discovery has consequences far beyond the immediate context of the model and thus contributes another piece to the puzzle of the physical world that we live in. This is research at work!

As almost always in physics, most insight is gained when the model is set up in as simple a way as possible. That is, a good model should capture the essential physical ingredients of the process, but not dwell on unnecessary details. Once the essentials are understood, more details can be added. Einstein said: physics should be made as simple as possible, but not simpler.

First we establish the frame of reference in which we shall cast the mathematical description of accretion discs.

3.1 A coordinate system for accretion discs

We describe the accretion disc in a frame of reference centred on the accretor, using cylindrical coordinates (r, ϕ, z), as shown in Figure 3.1. The disc mid-plane coincides with the $z = 0$ plane. The radial coordinate is the distance of a point *from the rotation axis* (the z-axis) of the disc. The distance d of a point *from the central object* is $d = (r^2 + z^2)^{1/2}$, i.e. we have $d = r$ only in the disc mid-plane ($z = 0$). The coordinate ϕ is the azimuth angle, and the coordinate z is the height above the disc mid-plane. We use ω to denote the angular speed of disc plasma orbiting the accreting object.

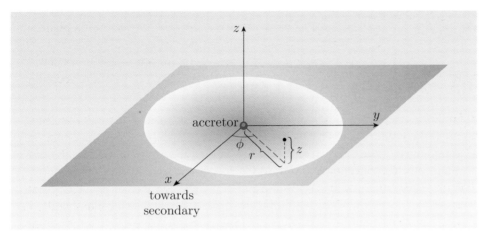

Figure 3.1 Accretion disc and cylindrical coordinates.

We shall make use of the concept of vertically integrated quantities, which proves rather useful in the study of the radial structure of accretion discs. This reduces the 3-dimensional problem to a 2-dimensional one, and exploits the fact that discs are *geometrically thin*. The disc is effectively confined to the orbital plane, and the plasma density falls off rapidly away from the plane. Another way of saying this is to state that the characteristic disc scale height H perpendicular to the disc mid-plane at a radial distance r from the accretor is small, $H \ll r$. We shall confirm this quantitatively later.

The most prominent of the integrated quantities is the **surface density** Σ (of mass).

Formally, Σ is defined by

$$\Sigma(r) = \int_{-\infty}^{+\infty} \rho(r, z)\, \mathrm{d}z. \tag{3.1}$$

> Σ is the Greek capital letter sigma. Note that here Σ is *not* used to denote summation.

This integrates the gas density ρ in a direction perpendicular to the disc mid-plane from very far below ($z \to -\infty$) to very far above ($z \to \infty$) the disc mid-plane. The third coordinate, the azimuth ϕ, does not appear explicitly because the disc is axisymmetric, so that neither ρ nor Σ depends on ϕ. The surface density tells us how much mass is contained in a disc ring with unit surface area at radius r.

● What are the SI units of Σ?

○ As Σ is mass per surface area, the unit must be $\mathrm{kg\,m^{-2}}$. Another way of seeing this is by recognizing that integration over z contributes a length (m), and ρ obviously has the unit $\mathrm{kg\,m^{-3}}$. Then $\mathrm{m} \times \mathrm{kg\,m^{-3}} = \mathrm{kg\,m^{-2}}$.

It is useful to relate the surface density Σ to the familiar volume mass density ρ in a simple way. If ρ is constant in the z-direction, and the full disc thickness is just H, then Equation 3.1 reduces to

$$\Sigma = H\rho. \tag{3.2}$$

Even if the density varies with height z, it is still possible to write down an equation like Equation 3.2. In that case H is a characteristic vertical scale height of the disc, and ρ a typical density in the disc at radius r. This is the so-called *one-zone model*, a model that we shall apply frequently below.

Now, to develop the accretion disc model description we first must discuss an effect of key importance for accretion physics, namely the transport phenomenon of viscosity.

3.2 Viscosity and its causes

Accretion in general, and in a disc in particular, works only if the matter to be accreted sheds excess angular momentum so that it can spiral in towards the accreting mass at the disc's centre. The key behind the physical mechanism that mediates this angular momentum transport is that particles in the gas stream interact. They undergo close or distant collisions, thereby exchanging energy and momentum. Interactions on scales that are small compared to the radial and vertical extent of the disc, or even on microscopic scales, can nevertheless lead to an efficient transport of energy and linear or angular momentum over macroscopic scales. One such **transport phenomenon** is **viscosity**.

Viscosity is a familiar effect in everyday life. For example, a cup of tea can be easily stirred with a spoon — there is very little resistance to the motion of the spoon. A lot more effort is needed in a cup full of honey. The honey is sticky, it has a much higher viscosity than the tea.

3.2.1 Stress, strain and viscosity

To quantify viscosity — the degree of 'stickiness' — we need to consider small elements of the fluid (tea, honey or, here, stellar plasma) and how linear or angular momentum changes across the flow. This will involve *derivatives* of quantities that describe the plasma flow. Depending on the direction of the momentum transport relative to the flow direction, we distinguish various types of viscosity. The **bulk viscosity** refers to the case where the flow velocity varies *along* the direction of the flow (Figure 3.2a). When the flow velocity varies *orthogonal* to the direction of the flow — in the presence of a shear flow (see, for example, Figure 3.2b) — there is an analogous shear viscous effect.

An accretion disc, where matter orbits the central object with approximately Keplerian speed (Equation 1.5), represents a *rotating* fluid. The angular speed $\omega(r)$ of disc plasma at a distance r from the central accretor with mass M is then given by the Keplerian value

$$\omega_{\mathrm{K}} = \left(\frac{GM}{r^3}\right)^{1/2}. \tag{3.3}$$

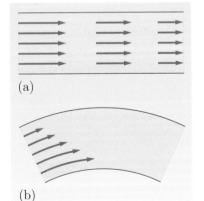

(a)

(b)

Figure 3.2   Velocity varying (a) along the flow and (b) orthogonal to the flow.

The flow in such a *Keplerian disc* is in fact a shear flow, and hence we have to consider the **shear viscosity** in some detail.

- Why is the flow in an accretion disc a shear flow?
- Consider two adjacent gas rings in a Keplerian disc. As ω drops with increasing r (Equation 3.3), the inner ring slowly slides past (overtakes) the outer ring.

More generally, if the angular velocity ω varies with distance r from the rotational axis, we speak of **differential rotation**, as opposed to the more familiar solid-body rotation where the angular velocity is the same everywhere. In the case of a differentially rotating body, those parts of the body that are far away from the rotational axis may take longer (or less time) to complete one revolution than the parts that are closer to the axis. The body becomes distorted as a result.

Exercise 3.1   Give examples of differential rotation in everyday life.   ■

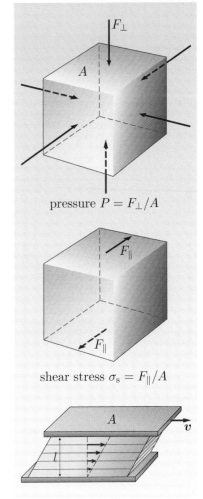

pressure $P = F_\perp / A$

shear stress $\sigma_s = F_\parallel / A$

Figure 3.3   The definition of pressure and shear stress, and the velocity profile of a viscous fluid between two moving plates.

'Stress causes strain'

Stress measures the *force* applied over the surface of a body, while **strain** is a measure of the *deformation* that the stress is causing. In the case of a fluid flow — such as accreting plasma — the 'deformation' manifests itself as a change in the velocity field in the flow.

In the simplest case, the applied stress is proportional to the resulting strain, and the viscosity is just the constant of proportionality:

stress = viscosity × strain.

The physical quantity *stress* denotes the force exerted per unit area on a surface. As the force F can have different directions with respect to the orientation of the area A, there are different types of stress F/A. (Sometimes these different types are called stress components.) For our purposes, the most important types of stress are **pressure**, usually denoted by P, and **shear stress**. These are illustrated in Figure 3.3 for a small, cubic volume of the fluid flow. In the case of pressure the force is always perpendicular to the surface, while in the case of shear stress the force is applied parallel to the surface. Imagine a fluid between two parallel plates, one of them stationary, the other moving with velocity v parallel to the other plate. A viscous fluid would develop a velocity profile like the one shown in Figure 3.3. The shear stress σ_s is the force $F_\parallel$, divided by the plate area A, needed to pull the plate with constant velocity v against the resistance of the fluid. A measure of the corresponding shear strain — the 'deformation' — is the gradient $\partial v / \partial z \approx v/l$ of the velocity in the shear flow perpendicular to the direction of motion of the plate.

We have

$$\sigma_s = \frac{F_\parallel}{A} = -\nu_{\mathrm{vis}}\, \rho\, \frac{\partial v}{\partial z}, \tag{3.4}$$

where ν_{vis} is the **kinematic viscosity** and ρ is the density of the fluid. The product $\nu_{\mathrm{vis}} \times \rho$ is also called the dynamical viscosity, and is the constant of

See also the box entitled *Partial derivatives* in Subsection 3.2.2 below.

proportionality introduced in the box entitled 'Stress causes strain'. Below we consistently use the kinematic viscosity ν_{vis} to highlight that the viscous effects scale with the density of the fluid.

3.2.2   Viscous torque and dissipation

The kinematic viscosity due to the turbulent motion of particles or gas blobs with a characteristic speed v_{c} over a characteristic length λ_{c} is just

$$\nu_{\text{vis}} \simeq \lambda_{\text{c}} \times v_{\text{c}}. \tag{3.5}$$

This can be understood by considering a simple shearing motion where fluid planes slide parallel to one another, as in Figure 3.3. Assume that the fluid flows in the x-direction, with a velocity $u(z)$ that increases linearly with distance z from the $z = 0$ plane (see Figure 3.4).

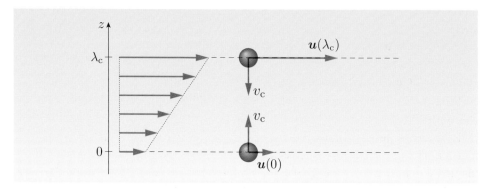

Figure 3.4   Shearing flow with exchange of fluid elements.

Consider now the turbulent motion of fluid elements (blobs of plasma), superimposed onto the general flow in the x-direction. In particular, assume that these blobs will move up (in the z-direction) or down with a characteristic speed v_{c}, in a straight line over the length λ_{c}, before they interact or collide with other blobs and exchange linear momentum. The time between collisions is $\Delta t = \lambda_{\text{c}}/v_{\text{c}}$.

An upward-moving blob that starts at $z = 0$ will deposit the linear momentum corresponding to the fluid flow at $z = 0$ at the height $z = \lambda_{\text{c}}$. Conversely, a fluid element starting at $z = \lambda_{\text{c}}$ will deposit its linear momentum at $z = 0$.

In a volume $V = A\lambda_{\text{c}}$ with base area A, parallel to the $z = 0$ plane, and height λ_{c}, the total mass of blobs is $\rho A\lambda_{\text{c}}$, where ρ is the fluid density. The linear momentum flowing up then is $A\lambda_{\text{c}}\rho\, u(0)$. To conserve mass there must be the same flow of blobs downwards as upwards, so the linear momentum flowing down is $A\lambda_{\text{c}}\rho\, u(\lambda_{\text{c}})$.

In the time Δt there is therefore a net flow $\Delta p = A\lambda_{\text{c}}\rho(u(0) - u(\lambda_{\text{c}}))$ of linear momentum in the z-direction, which according to Newton's second law implies a force $F = \Delta p/\Delta t$ and a corresponding stress

$$\sigma_{\text{s}} = \frac{F}{A} = \frac{\Delta p}{\Delta t\, A} = \frac{\Delta p\, v_{\text{c}}}{\lambda_{\text{c}} A}.$$

This becomes

$$\sigma_s = \rho v_c (u(0) - u(\lambda_c)) = \rho v_c \lambda_c \frac{u(0) - u(\lambda_c)}{\lambda_c}.$$

Comparing this with Equation 3.4 confirms Equation 3.5 as we have $-\partial u/\partial z \simeq (u(0) - u(\lambda_c))/\lambda_c$.

An important application of Equation 3.5 is the so-called **α-viscosity**, which describes the turbulent viscosity in accretion discs. The maximum possible value for the length scale that turbulent eddies (blobs) would be able to travel is the vertical disc scale height H, and their maximum possible speed is the local sound speed c_s. Hence the turbulent viscosity satisfies $\nu_{\text{vis}} < Hc_s$. The α-viscosity is defined as

$$\nu_{\text{vis}} = \alpha H c_s, \tag{3.6}$$

with a dimensionless factor $\alpha \leq 1$. The parameter α is unknown and encapsulates all of the complex microphysics that influences the small-scale turbulence that is thought to operate in accretion discs. The accretion disc models that we shall study below make the assumption that α is constant throughout the disc. This is necessarily a crude assumption, but it allows one to establish simple analytical accretion disc models. Observational data should then be able to constrain the value of α required for describing real astrophysical discs, and this in turn will shed light on the nature of the physical mechanism giving rise to this viscosity.

Currently, the most promising viscosity mechanism is magneto-hydrodynamic (MHD) turbulence. This involves weak magnetic fields in the accretion disc plasma. The interaction of the shearing plasma flow and the initially weak magnetic field leads to an amplification of the field, and to turbulent motion in the disc. The researchers Steven Balbus and John Hawley rediscovered the importance of this instability for accretion discs in the early 1990s. MHD turbulence continues to be an area of active research.

Viscous torque

The presence of a viscous shearing flow in the accretion disc implies an associated flow, or exchange, of angular momentum in the radial direction. We shall now derive an expression for the rate of change of angular momentum, i.e. the viscous torque G_{vis}, in the accretion disc.

Partial derivatives

In the cylindrical coordinate system, a quantity like ω could in principle depend on all three coordinates r, z and ϕ. This dependence can be characterized by the three partial derivatives $\partial \omega/\partial r$, $\partial \omega/\partial z$ and $\partial \omega/\partial \phi$ (see Figure 3.5, which illustrates the meaning of partial derivatives using the example of a function $V(x,y)$ of variables x and y). The partial derivative $\partial \omega/\partial r$ expresses the local change of ω in the direction of r, i.e. along a line with $z =$ constant and $\phi =$ constant. In the framework of the one-zone model, ω does not depend on z. As the disc is also axisymmetric, ω does not depend on ϕ either. Therefore in our case ω depends *only* on r. Hence in this case there is no difference between the partial derivative $\partial \omega/\partial r$ and the standard notation $\mathrm{d}\omega/\mathrm{d}r$.

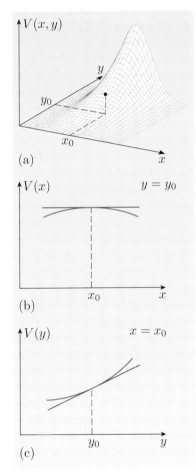

Figure 3.5 Diagram illustrating the meaning of partial derivatives. The quantity V is a function of both x and y.

We follow the standard notation in the literature and denote the viscous torque with the capital letter G, but to distinguish it from the gravitational constant we add the index vis to it.

Consider two adjacent gas rings in an accretion disc at radius r. The magnitude of the torque exerted by the outer ring on the inner ring is given as force times lever arm. As the force is a viscous force, it can be written as

$$F = \text{contact surface area} \times \text{viscous stress} = A \times \sigma_{\text{s}}.$$

The contact surface area of the two disc rings is $A = 2\pi r H$ (see Figure 3.6). To calculate the viscous stress we need to apply Equation 3.4 in the context of a rotating fluid. As the orbital speed of gas in the ring is just $v = r\omega$, we simply replace the gradient of the velocity v with the product $r\,\partial\omega/\partial r$, which involves the gradient of the angular speed. Then the shear stress becomes

$$\sigma_{\text{s}} = -\nu_{\text{vis}}\,\rho r\,\frac{\partial \omega}{\partial r}. \tag{3.7}$$

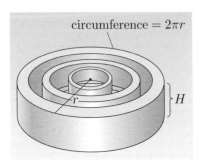

Figure 3.6   Accretion disc rings.

Exercise 3.2   Show that it is not appropriate to use the expression $\partial(r\omega)/\partial r$ in Equation 3.7 instead of $r\,\partial\omega/\partial r$. Use the product rule to relate the two expressions, and discuss what happens in the absence of shear. ∎

Now using the expressions for A and σ_{s}, and observing that $\Sigma = H\rho$, we find that the magnitude of the torque is

$$G_{\text{vis}} = -Fr = -A\sigma_{\text{s}} \times r = 2\pi r H \times \nu_{\text{vis}}\,\rho r\,\frac{\partial\omega}{\partial r} \times r$$

and hence

$$G_{\text{vis}} = 2\pi r\,\nu_{\text{vis}}\,\Sigma r^2\,\frac{\partial\omega}{\partial r}. \tag{3.8}$$

Note the sign of G_{vis}: we introduced a minus sign in the above calculation, which cancelled with the minus sign of Equation 3.7. Implicit in our derivation was the assumption that the torque vector is parallel or antiparallel to the rotational axis of the accretion disc. If it is parallel and in the same direction as the vector $\boldsymbol{\omega}$, then G_{vis} is a positive quantity, and the torque acts in the same direction as the fluid rotates. In other words, the rotating fluid speeds up. Conversely, if the torque is antiparallel to the vector $\boldsymbol{\omega}$, then the quantity G_{vis} is negative, and the torque brakes the rotation. In the case of a Keplerian accretion disc, we have $\omega(r) \propto r^{-3/2}$ and hence $\partial\omega/\partial r < 0$ and also $G_{\text{vis}} < 0$. This shows that with our definition, G_{vis} *denotes the torque exerted by the slower outer ring on the faster inner ring*: the slower ring attempts to hold back the faster inner ring as it glides past it. The torque exerted by the inner ring on the outer ring is just $-G_{\text{vis}}$.

Worked Example 3.1

Determine the viscous torque for the case of Keplerian motion.

Solution

For Keplerian motion the angular speed is given by Equation 3.3. So the radius derivative is

$$\frac{\partial\omega}{\partial r} = (GM)^{1/2}\left(-\tfrac{3}{2}r^{-5/2}\right). \tag{3.9}$$

Inserting this into Equation 3.8 gives

$$G_{\text{vis}} = 2\pi r \, \nu_{\text{vis}} \, \Sigma r^2 (GM)^{1/2} \left(-\tfrac{3}{2} r^{-5/2}\right).$$

Hence

$$G_{\text{vis}} = -3\pi \, \nu_{\text{vis}} \, \Sigma (GMr)^{1/2}. \tag{3.10}$$

Exercise 3.3   Verify that the right-hand side of Equation 3.8 has the units of a torque.   ■

Viscous dissipation

The presence of a viscous shearing flow in the accretion disc also implies the local generation of heat, as there is friction between adjacent disc annuli. The amount of heat generated in the plasma flow by friction, per unit time and unit area, is given by

$$D(r) = \tfrac{1}{2}\nu_{\text{vis}} \, \Sigma \left(r \frac{\partial \omega}{\partial r}\right)^2. \tag{3.11}$$

The quantity $D(r)$ is the rate, per unit surface area, at which the mechanical energy of the rotational motion of the plasma is converted into heat due to viscosity. It is called the **viscous dissipation** rate.

● What is the SI unit of D?

○ From the right-hand side of Equation 3.11 we have

$$\text{m} \times \text{m}\,\text{s}^{-1} \times \text{kg}\,\text{m}^{-2} \times \text{s}^{-2} = \text{kg}\,\text{s}^{-3}.$$

As a rate of converted energy per unit surface area, the quantity D must have the units $\text{J}\,\text{s}^{-1}\,\text{m}^{-2}$, which is indeed $\text{kg}\,\text{s}^{-3}$ as $\text{J} = \text{kg}\,\text{m}^2\,\text{s}^{-2}$.

We can see why this dissipation must occur as follows.

Consider three adjacent gas rings that rotate with their local Keplerian angular speed (Equation 3.3). We label the rings as A, B and C, starting from the inner ring (Figure 3.7). The inner edge of ring B is at radius r and the outer edge is at radius $r + \Delta r$. A rotates faster than B and hence tries to spin-up (increase the speed of) ring B, i.e. there is a positive torque $|G_{\text{vis}}(r)|$ on B from A. Conversely, ring C rotates slower than B and tries to spin-down B, i.e. there is a negative torque $-|G_{\text{vis}}(r + \Delta r)|$ on B from C. Because of the radial dependence of the viscous torque G_{vis}, the sum of these two torques is non-zero. In other words, there is a net torque on ring B (in this case trying to increase its spin). This involves work, or, to be precise, a rate of working. To find an expression for the rate of working, we think of an analogy involving linear motion. If a sledge is pulled horizontally in a straight line, a constant pulling force of magnitude F is exerted to overcome friction with the ground and maintain

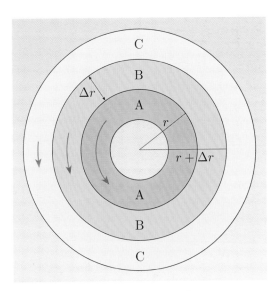

Figure 3.7   Torque balance for an accretion disc ring.

a constant speed v. The pulling power is $\mathrm{d}W/\mathrm{d}t = F \times v$ (where W denotes work). Likewise, in the case of rotational motion the rate of working is given by $\mathrm{d}W/\mathrm{d}t = G_{\mathrm{vis}} \times \omega$. The magnitude of torque G_{vis} takes the place of the magnitude of the force F, and the angular speed ω takes the place of the translational speed v.

Some fraction of the rate of working that is associated with the net torque deposits mechanical energy into the disc in the form of heat. In fact, it turns out that viscous dissipation deposits energy into the ring at a rate $G_{\mathrm{vis}}(r) \times \Delta\omega$, where $\Delta\omega$ is the *difference* between the angular speeds at the outer and inner edges of the gas ring. (This reasoning is sufficient for the purposes of this book; a more thorough justification can be found in the literature.) Making use of the first-order expansion, the difference $\Delta\omega$ is approximately

$$\Delta\omega \approx \frac{\partial\omega}{\partial r}\Delta r$$

if the width Δr of the ring is small. So in the disc ring, viscous dissipation occurs at a rate $G_{\mathrm{vis}}(r) \times \Delta\omega = G_{\mathrm{vis}} \times \Delta r\,(\partial\omega/\partial r)$. We normalize this now to the surface area of the disc ring. Each of the upper and lower sides of the disc ring has an area of $2\pi r\,\Delta r$ ('circumference $\times$ width'). Hence the dissipation rate per unit surface area is

$$D = \frac{G_{\mathrm{vis}}\,\Delta r\,(\partial\omega/\partial r)}{2 \times 2\pi r\,\Delta r} = G_{\mathrm{vis}}\,\frac{\partial\omega/\partial r}{4\pi r}.$$

Using Equation 3.8 this becomes

$$D = 2\pi r\,\nu_{\mathrm{vis}}\,\Sigma r^2\,\frac{\partial\omega}{\partial r} \times \frac{\partial\omega/\partial r}{4\pi r},$$

which simplifies to Equation 3.11.

For Keplerian motion the viscous dissipation rate is

$$D(r) = \tfrac{9}{8}\nu_{\mathrm{vis}}\,\Sigma\frac{GM}{r^3}. \tag{3.12}$$

Exercise 3.4   Show how Equation 3.12 results from Equation 3.11.   ■

3.3   Conservation laws

We now proceed to study the radial structure of geometrically thin accretion discs quantitatively. To this end we rewrite two fundamental conservation laws of physics — the conservation of mass and the conservation of angular momentum — in terms of vertically integrated, or averaged, quantities (variables).

The two resulting expressions are **partial differential equations**. In general, the physical quantities describing the disc, such as the surface density Σ, the viscosity ν_{vis} and the radial drift velocity v_r, depend on both the time t and the radial distance r from the centre of the disc. These quantities are functions of two independent variables, r and t, e.g. $\Sigma = \Sigma(r,t)$. Equations involving these functions include partial derivatives with respect to these variables. Recall that the partial derivative $\partial\Sigma/\partial r$ denotes the rate of change of Σ with distance r for a fixed t, i.e. this is the radial surface density gradient in the disc at a given time (see also Figure 3.5).

● Describe the meaning of the partial derivative $\partial \Sigma / \partial t$.

○ This denotes the rate of change of Σ with time, at a fixed distance r. Therefore this is the rate at which a disc ring at distance r gains or loses mass.

In principle, the same comments apply to the angular speed ω, except that the time-dependence is almost always assumed to vanish (i.e. $\partial \omega / \partial t = 0$).

3.3.1 Conservation of mass

For a geometrically thin accretion disc, the conservation of mass can be written as

$$r\frac{\partial \Sigma}{\partial t} + \frac{\partial}{\partial r}(r v_r \Sigma) = 0. \tag{3.13}$$

To arrive at this expression we consider a narrow disc ring (annulus) between radii r and $r + \Delta r$ (see Figure 3.8), where $\Delta r \ll r$. The mass stored in the ring can be calculated as 'surface density × surface area'. The surface area of the disc ring is just $2\pi r \, \Delta r$, i.e. 'circumference × width'. Hence the mass M_a in the disc annulus is $2\pi r \, \Delta r \times \Sigma$ and the rate of change of M_a is

$$\frac{\partial M_a}{\partial t} = \frac{\partial(2\pi r \, \Delta r \, \Sigma)}{\partial t}.$$

As 2π, Δr and r do not depend on time, this simplifies to

$$\frac{\partial M_a}{\partial t} = 2\pi r \, \Delta r \, \frac{\partial \Sigma}{\partial t}. \tag{3.14}$$

Now, the rate of change of the ring mass is also given as the difference between the mass inflow and outflow rates. To calculate the net inflow or outflow of mass, we first define another quite useful quantity. The *local* **mass accretion rate** $\dot{M}(r,t)$ is the amount of mass that flows per unit time through the boundary at radius r between adjacent disc annuli. By convention, $\dot{M}$ is positive if the mass flows inwards, towards smaller radii, and negative if the mass flows to larger radii. We can relate $\dot{M}$ to the radial drift velocity v_r, which is defined to be positive if the material drifts outwards. Assume that the disc gas drifts inwards with velocity $(-v_r) > 0$. Then in a short time interval Δt, all the mass $\delta M < M_a$ that is originally in a narrow sub-annulus of width $\delta r = (-v_r) \Delta t < \Delta r$ will be able to cross the boundary surface at r (see Figure 3.8). Hence the local mass accretion rate is

$$\dot{M} = \frac{\delta M}{\Delta t} = \frac{2\pi r \, \delta r \, \Sigma}{\Delta t} = 2\pi r \frac{\delta r}{\Delta t} \Sigma = 2\pi r(-v_r)\Sigma.$$

Here we have calculated the mass in the sub-annulus in the same way as above for the full disc ring. Hence we obtain

$$\dot{M}(r,t) = -2\pi r v_r \Sigma. \tag{3.15}$$

We are now in a position to work out the net change ΔM_a of the mass M_a of the disc ring between radii r and $r + \Delta r$ in the time interval Δt. To this end we simply take the difference between the local mass flow rates at $r + \Delta r$ and at r, and multiply it by Δt. Remembering the definition of a partial derivative and the

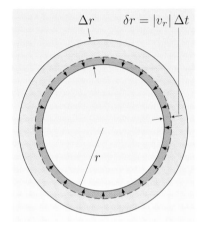

Figure 3.8 Mass crossing the inner boundary of the disc ring at radius r due to a radial drift with speed v_r.

approximation by first-order expansion, we see that the net flow (in minus out) can be written as

$$\Delta M_a = \left(\dot{M}(r+\Delta r, t) - \dot{M}(r,t)\right) \times \Delta t \simeq \Delta r \, \frac{\partial \dot{M}}{\partial r} \times \Delta t.$$

Using Equation 3.15 this becomes

$$\frac{\Delta M_a}{\Delta t} = \Delta r \, \frac{\partial \dot{M}}{\partial r} = -\Delta r \, \frac{\partial(2\pi r v_r \Sigma)}{\partial r},$$

so

$$\frac{\Delta M_a}{\Delta t} = -2\pi \, \Delta r \, \frac{\partial(r v_r \Sigma)}{\partial r}. \tag{3.16}$$

For small time intervals Δt, the expression $\Delta M_a/\Delta t$ in Equation 3.16 becomes the derivative $\partial M_a/\partial t$. Equating the right-hand side of Equation 3.14 with the right-hand side of Equation 3.16, and dividing by $2\pi \, \Delta r$, finally reproduces Equation 3.13. The derivation is exact in the limit $\Delta r \to 0$.

3.3.2   Conservation of angular momentum

The conservation of angular momentum in an accretion disc can be expressed as

$$r\frac{\partial}{\partial t}(\Sigma r^2 \omega) + \frac{\partial}{\partial r}(r v_r \Sigma r^2 \omega) = \frac{1}{2\pi}\frac{\partial G_{\text{vis}}}{\partial r}. \tag{3.17}$$

As we shall see, the derivation of this equation is analogous to the one for mass conservation, with the exception that an additional term occurs due to the presence of a torque.

● Consider the first two terms in Equation 3.17, and compare them to the two terms in Equation 3.13 describing mass conservation. Are there any similarities?

○ The terms in Equation 3.17 are obtained by replacing Σ in Equation 3.13 with $\Sigma r^2 \omega$.

This is not a coincidence. As $r^2\omega$ is the specific angular momentum of the disc material, i.e. the angular momentum per unit mass, $\Sigma r^2 \omega$ is the angular momentum of the disc material per unit surface area. The two terms represent the change of the angular momentum of the disc ring due to an imbalance between incoming angular momentum brought in with the mass flowing into the ring, and outgoing angular momentum carried away by the mass flowing out of the ring.

Exercise 3.5   Following the example of Subsection 3.3.1 on the conservation of mass, derive the first two terms of Equation 3.17 explicitly.   ■

The additional term $(1/2\pi)(\partial G_{\text{vis}}/\partial r)$ on the right-hand side of Equation 3.17 arises because there is also a net viscous torque acting on the disc ring, as we have discussed in Subsection 3.2.2. By analogy to Newton's second law — 'force equals rate of change of linear momentum' — the net torque contributes a rate of change of angular momentum. The torque term constitutes a source or sink of angular momentum. The net viscous torque G_{vis} on the disc ring is just

$$G_{\text{vis}}(r+\Delta r) - G_{\text{vis}}(r) \simeq \Delta r \, \frac{\partial G_{\text{vis}}}{\partial r}.$$

As Equation 3.17 is obtained from the angular momentum balance equation for the disc ring by division by $2\pi\,\Delta r$ (see the last step in Subsection 3.3.1), the final source term in Equation 3.17 is

$$\frac{1}{2\pi\,\Delta r}\,\Delta r\,\frac{\partial G_{\mathrm{vis}}}{\partial r} = \frac{1}{2\pi}\,\frac{\partial G_{\mathrm{vis}}}{\partial r},$$

as required.

3.4 Radial structure of steady-state discs

We shall now turn to a highly significant simplification: we consider *steady-state* accretion, i.e. *time-independent* accretion. This assumption does *not* imply that the accreting plasma is *at rest*; rather, it means that the disc appears the same at all times. The plasma in the disc will still circle the central object on Keplerian orbits, while slowly drifting inwards. For the accretion disc model, we require that none of the quantities describing the steady-state disc depends explicitly on time, so all partial derivatives of the form $\partial/\partial t$ will vanish. This greatly simplifies Equations 3.13 and 3.17 that govern the radial structure of the disc; they will no longer be partial differential equations at all — they reduce to ordinary differential equations, with r as the only independent variable. We shall learn a great deal about discs by just considering steady-state accretion. In fact, we shall even understand time-dependent discs (in Chapter 4) largely by insights that we gain from these steady-state discs.

The assumption of steady-state accretion allows one to *integrate* the accretion disc equations, i.e. to find solutions of these equations. In the case of the conservation of mass we know the integral already: it is the mass accretion rate $\dot{M}$ that we have defined in Equation 3.15 above, as the *local* mass accretion rate through the disc. In steady-state discs the radial mass flow rate has to be the same everywhere in the disc, at all times — otherwise mass would pile up or deplete in certain disc rings, in clear conflict with the assumption of a steady state. This constant mass accretion rate must also equal the rate at which mass is fed into the disc from an external mass reservoir (e.g. the mass transfer rate from a donor star).

The integral of the angular momentum equation (Equation 3.17) is a little more involved. For steady-state discs the term with $\partial/\partial t$ vanishes, so we are left with

$$\frac{\partial}{\partial r}(rv_r\Sigma r^2\omega) = \frac{1}{2\pi}\,\frac{\partial G_{\mathrm{vis}}}{\partial r}. \tag{3.18}$$

Both sides of this equation are derivatives with respect to r, so we can apply the inverse operation, an integration over r, on both sides. Quite generally, for any function $f(r)$, the identity

$$\int \frac{\partial f}{\partial r}(r)\,\mathrm{d}r = f(r) + C \tag{3.19}$$

holds, with C being an arbitrary constant. The ambiguity expressed by the integration constant C appears as we did not specify the integration boundaries: for any C we have

$$\frac{\partial}{\partial r}(f(r) + C) = \frac{\partial f}{\partial r}(r).$$

Integrating Equation 3.18 by applying the rule expressed in Equation 3.19 gives

$$r v_r \Sigma r^2 \omega = \frac{G_{\text{vis}}}{2\pi} + \frac{C}{2\pi}, \qquad (3.20)$$

where we have combined the two integration constants into one new constant $C/(2\pi)$. This is purely for convenience — as long as there is a constant in the equation, it is not important what form it has.

Using Equation 3.15 we can rewrite the left-hand side of Equation 3.20 as

$$r v_r \Sigma r^2 \omega = -\frac{\dot{M}}{2\pi} r^2 \omega.$$

Making the additional assumption that the angular speed is Keplerian (Equation 3.3), this becomes

$$-\frac{\dot{M}}{2\pi} r^2 \left(\frac{GM}{r^3} \right)^{1/2} = -\frac{\dot{M}}{2\pi} (GMr)^{1/2},$$

and hence Equation 3.20 finally reads

$$-\dot{M}(GMr)^{1/2} = G_{\text{vis}} + C. \qquad (3.21)$$

Clearly, to make use of this integrated equation we need to know what the integration constant C is. As Equation 3.21 is valid everywhere in the disc, we can determine the value of C at any disc annulus we like. A clever choice is a point where $\partial \omega / \partial r = 0$, because this implies that the torque vanishes, $G_{\text{vis}} = 0$ (see Equation 3.8).

To determine where $\partial \omega / \partial r = 0$, we make a short detour and introduce the boundary layer.

3.4.1   The boundary layer

The **boundary layer** is the innermost region of the accretion flow where the angular speed of the accreting plasma can deviate significantly from the Keplerian value.

Usually the accretor rotates with an angular speed ω_1 well below the Keplerian value that corresponds to the surface of this object, $\omega_K(R_1) = (GM/R_1^3)^{1/2}$ (Equation 3.3 with $r = R_1$, the accretor radius). Conversely, we know that at large radii r, far away from the accretor, the angular speed of the material in the disc is indeed just the Kepler rate $\omega_K(r)$. Clearly we can expect that the angular speed ω of the disc plasma does not suddenly (discontinuously) drop from ω_K to the smaller ω_1. Rather, we expect a smooth transition as shown in Figure 3.9. Hence ω must start to deviate from the Kepler rate once the radius is smaller than some value $r = R_1 + b$. The region of the accretion flow between the surface of the accretor and the radius $r = R_1 + b$ is the boundary layer. In other words, b is the radial width of the boundary layer.

On Keplerian orbits, the centrifugal force just balances the gravitational force that pulls the disc material inwards towards the central accretor. This balance breaks down in the boundary layer: the centrifugal force is much smaller than the gravitational force. Instead, gravity is roughly balanced by a radial pressure gradient. (See also the related arguments leading to Equation 3.33 below.)

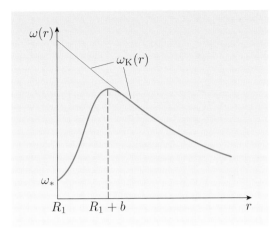

Figure 3.9 The angular speed near the inner edge of an accretion disc around an object with a surface angular speed ω_1 that is smaller than the Keplerian value given in Equation 3.3.

It turns out that this requires that the boundary layer is not very extended in the radial direction — this allows for a larger pressure gradient. More specifically, the width b is small compared to both the radius of the accreting object and the height of the disc just outside the boundary layer.

We now return to the discussion of how to determine the integration constant C in Equation 3.21. We have just identified the outer edge of the boundary layer as the point where $\partial\omega/\partial r \approx 0$. The angular speed itself is still approximately Keplerian at this point, i.e. given by the rate defined in Equation 3.3.

Hence evaluating Equation 3.21 at $r = R_1 + b \simeq R_1$, where we have $G_{\rm vis}(R_1 + b) = 0$ and $\omega(R_1 + b) \simeq (GM/R_1^3)^{1/2}$, finally gives the integration constant as

$$C = -\dot{M}(GMR_1)^{1/2}. \tag{3.22}$$

3.4.2 The steady-state surface density

Using the value for C from Equation 3.22 in Equation 3.21, we now derive a relation that describes the surface mass density profile of a steady-state disc. To achieve this, we make use of the expression for the viscous torque for Keplerian discs, Equation 3.10,

$$G_{\rm vis} = -3\pi\,\nu_{\rm vis}\,\Sigma(GMr)^{1/2}.$$

Inserting this and the expression for C in Equation 3.22, Equation 3.21 gives

$$-\dot{M}(GMr)^{1/2} = -3\pi\,\nu_{\rm vis}\,\Sigma(GMr)^{1/2} - \dot{M}(GMR_1)^{1/2}.$$

Dividing both sides by GM and solving for $\nu_{\rm vis}\,\Sigma$ gives

$$-3\pi\,\nu_{\rm vis}\,\Sigma r^{1/2} = -\dot{M}r^{1/2} + \dot{M}R_1^{1/2},$$

or finally

$$\nu_{\rm vis}\,\Sigma = \frac{\dot{M}}{3\pi}\left[1 - \left(\frac{R_1}{r}\right)^{1/2}\right]. \tag{3.23}$$

This is a very important and useful expression, and we shall make abundant use of it below.

● Describe the radial dependence in Equation 3.23 far away from the central star, and close to the surface of it.

○ Far away from the accreting object, i.e. for $r \gg R_1$, the term $(R_1/r)^{1/2}$ is small compared with 1 and can be neglected. This means that the product $\nu_{vis}\,\Sigma$ of viscosity and surface density is constant throughout the outer disc. This is no longer true close to the surface of the accreting object. When r approaches R_1, the product $\nu_{vis}\,\Sigma$ approaches 0. (See also Figure 3.10.)

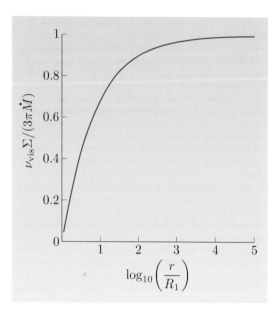

Figure 3.10 $\nu_{vis}\,\Sigma$ as a function of r in a steady-state disc.

Equation 3.23 shows that over a large region of a steady-state accretion disc $\nu_{vis}\,\Sigma = \dot{M}/3\pi = $ constant, so the surface density is inversely proportional to the viscosity. Whatever physical mechanism provides the viscosity, in a steady-state disc with a given mass accretion rate, the disc will adjust its local structure (surface density, temperature) such that Equation 3.23 is fulfilled.

We now make use of Equation 3.23 to calculate how the disc luminosity varies with distance from the central star.

3.4.3   The accretion disc luminosity

We shall derive an expression for the disc luminosity by determining the luminosity of a disc annulus between radii r_1 and r_2, and then summing the contributions of all such annuli in the disc. The key ingredient for calculating the luminosity of a disc annulus is the local viscous dissipation rate $D(r)$. This is the rate at which energy is deposited locally, due to the action of a viscous torque, into the disc plasma. We make the simplifying assumption that the disc radiates this energy at the same rate that viscous dissipation generates it. Although the dissipation rate generally depends on the viscosity, it is possible to eliminate the viscosity using the steady-state relation Equation 3.23.

As we have seen in Equation 3.12, the viscous dissipation rate (energy deposited into the disc plasma per unit time per unit surface area) in a Keplerian disc is

$$D(r) = \tfrac{9}{8}\nu_{vis}\,\Sigma\,\frac{GM}{r^3}.$$

Inserting the steady-state relation Equation 3.23 for $\nu_{\text{vis}} \Sigma$ gives

$$D(r) = \frac{9}{8} \frac{\dot{M}}{3\pi} \left[1 - \left(\frac{R_1}{r} \right)^{1/2} \right] \frac{GM}{r^3},$$

which simplifies to

$$D(r) = \frac{3GM\dot{M}}{8\pi r^3} \left[1 - \left(\frac{R_1}{r} \right)^{1/2} \right]. \tag{3.24}$$

The luminosity (energy generated per unit time) is then given by the integral of $D(r)$ over the surface of the disc ring between radii r_1 and r_2 ($r_1 < r_2$):

$$L(r_1, r_2) = 2 \times \int_{r_1}^{r_2} D(r) \, 2\pi r \, \mathrm{d}r.$$

Here the azimuthal part of the surface integral has already been carried out, giving the factor 2π in the integrand. The factor 2 in front of the integral accounts for the fact that the disc has two faces. Inserting D and collecting all constants in front of the integral gives

$$L(r_1, r_2) = \frac{3GM\dot{M}}{2} \int_{r_1}^{r_2} \left[1 - \left(\frac{R_1}{r} \right)^{1/2} \right] \frac{1}{r^2} \, \mathrm{d}r. \tag{3.25}$$

> **Worked Example 3.2**
>
> Make the substitution $y = R_1/r$ and carry out the integral in Equation 3.25.
>
> **Solution**
>
> As
>
> $$\frac{\mathrm{d}y}{\mathrm{d}r} = -\frac{R_1}{r^2},$$
>
> we can replace $\mathrm{d}r/r^2$ with $-\mathrm{d}y/R_1$ and obtain
>
> $$L(r_1, r_2) = -\frac{3GM\dot{M}}{2R_1} \int_{y_1}^{y_2} [1 - y^{1/2}] \, \mathrm{d}y, \tag{3.26}$$
>
> where $y_1 = R_1/r_1$ and $y_2 = R_1/r_2$. The integral can be carried out as follows:
>
> $$\int_{y_1}^{y_2} [1 - y^{1/2}] \, \mathrm{d}y = \int_{y_1}^{y_2} \mathrm{d}y - \int_{y_1}^{y_2} y^{1/2} \, \mathrm{d}y$$
>
> $$= y_2 - y_1 - \left[\frac{2}{3} y^{3/2} \right]_{y_1}^{y_2}$$
>
> $$= y_2 - y_1 - \frac{2}{3} y_2^{3/2} + \frac{2}{3} y_1^{3/2}$$
>
> $$= - \left[y_1 \left(1 - \frac{2}{3} y_1^{1/2} \right) - y_2 \left(1 - \frac{2}{3} y_2^{1/2} \right) \right].$$
>
> Inserting this for the integral in Equation 3.26, and re-substituting R_1/r_1 for y_1, and R_1/r_2 for y_2, gives
>
> $$L(r_1, r_2) = -\frac{3GM\dot{M}}{2R_1} \times (-1)$$
>
> $$\times \left\{ \frac{R_1}{r_1} \left[1 - \frac{2}{3} \left(\frac{R_1}{r_1} \right)^{1/2} \right] - \frac{R_1}{r_2} \left[1 - \frac{2}{3} \left(\frac{R_1}{r_2} \right)^{1/2} \right] \right\}.$$

Cancelling the leading R_1 finally gives an expression for the luminosity of an accretion disc annulus:

$$L(r_1, r_2) = \frac{3GM\dot{M}}{2} \left\{ \frac{1}{r_1} \left[1 - \frac{2}{3} \left(\frac{R_1}{r_1} \right)^{1/2} \right] - \frac{1}{r_2} \left[1 - \frac{2}{3} \left(\frac{R_1}{r_2} \right)^{1/2} \right] \right\}.$$

(3.27)

Exercise 3.6   Use Equation 3.27 to calculate the luminosity of the whole disc. Set $r_1 = R_1$ and let $r_2 \to \infty$.   ■

This exercise confirms the statement made in Equation 1.8 in Subsection 1.1.2: the integral luminosity of a geometrically thin, optically thick steady-state accretion disc is just half of the accretion luminosity L_{acc} as defined in Equation 1.3 in Subsection 1.1.1. The other half of the accretion luminosity is in fact radiated by the boundary layer. As the boundary layer is so much smaller than the disc but nonetheless has the same luminosity, it inevitably must be hotter than the disc.

● Why is this?

○ It is because the radiant flux (i.e. the luminosity per unit area) is proportional to the temperature to the fourth power. This is the Stefan–Boltzmann law (see Equation 1.22 in Subsection 1.4.1).

3.4.4   The accretion disc temperature profile

The results derived above allow us now to revisit the calculation of the emitted spectrum of a steady-state accretion disc presented in Subsection 1.4.1. *In equilibrium* the flux emerging from a disc annulus must equal the rate at which viscous dissipation deposits energy into this disc ring. We have just used this concept to calculate the luminosity of the disc ring. If the disc is also optically thick (i.e. opaque), we expect the flux emerging from the disc ring to be that of a black body, characterized by the surface temperature — or, to be precise, the effective temperature $T_{eff}(r)$ — of the disc ring. We can equate the viscous dissipation rate for a steady-state disc, Equation 3.24, with the flux $\sigma T_{eff}(r)^4$ emerging from a black body emitter (Equation 1.22). Solving for $T_{eff}(r)$ gives us the temperature profile of a steady-state, optically thick, geometrically thin accretion disc as

$$T_{eff}^4(r) = \frac{3GM\dot{M}}{8\pi\sigma r^3} \left[1 - \left(\frac{R_1}{r} \right)^{1/2} \right].$$

(3.28)

For $r \gg R_1$ this expression is the same as Equation 1.20 in Subsection 1.4.1.

Exercise 3.7   By following the steps below, show that $T_{eff}(r)$ as given by Equation 3.28 attains a maximum value of $0.488T_*$ at $r = (49/36)R_1$, where T_* is defined as

$$T_* = \left(\frac{3GM\dot{M}}{8\pi\sigma R_1^3} \right)^{1/4}.$$

Start from Equation 3.28, and: (a) substitute in T_*; (b) put $r/R_1 = x$ and $(T_{\mathrm{eff}}/T_*)^4 = y$; (c) find $\mathrm{d}y/\mathrm{d}x$ and set $\mathrm{d}y/\mathrm{d}x = 0$ to find the value of x where y is maximal; (d) find y at the maximum; (e) find T_{eff} at the maximum.

Exercise 3.8   Why is it not possible to discover the magnitude of viscosity (the value of α) by just observing a steady-state disc? ∎

3.5 The vertical disc structure

So far we have been mostly concerned with the radial structure of the disc. The simplifying assumption that the disc is geometrically thin made it possible and sensible to work with vertically integrated quantities. We are now considering the vertical disc structure in more detail, to establish that our earlier assumptions were justified.

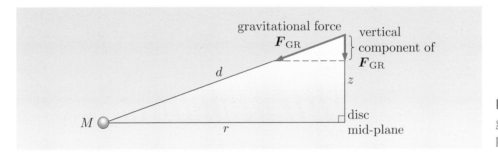

Figure 3.11   Component of gravity perpendicular to the disc plane.

As before, we consider the disc in cylindrical coordinates (r, ϕ, z), where the plane $z = 0$ is the disc mid-plane, and the accreting body is at the origin. At a point $|z| > 0$ above the disc mid-plane, the gravitational acceleration towards the central accreting object has a component in the negative z-direction, i.e. towards the plane of the disc (see Figure 3.11). The magnitude of the total gravitational acceleration g at this point is

$$g = \frac{GM}{d^2} = \frac{GM}{r^2 + z^2},$$

where the distance d to the central body of mass M is

$$d = \sqrt{r^2 + z^2}.$$

From simple trigonometry we have for the vertical component g_z of the gravitational acceleration (see Figure 3.11)

$$\frac{-g_z}{g} = \frac{z}{d}.$$

The minus sign expresses the fact that the acceleration is in the negative z-direction. Therefore we obtain

$$g_z = -\frac{GMz}{d^3} \simeq -\frac{GMz}{r^3}, \tag{3.29}$$

where the last equality holds if $z \ll r$.

In equilibrium the disc will adjust its vertical structure in such a way that the gravitational acceleration g_z is exactly balanced by a vertical pressure gradient, so that the fluid layers are stabilized against gravitational collapse. If this is the case, the accretion disc is said to be in **hydrostatic equilibrium**. To see how this arises, consider a small fixed fluid volume in the disc, with base area A, parallel to the disc mid-plane, and height Δz in the z-direction.

Equation of state

A plasma or gas is characterized by the usual state variables pressure P, density ρ and temperature T. They are related by an **equation of state**, which in many cases is the familiar perfect gas law

$$\frac{P}{\rho} = \frac{k}{\overline{m}}T \simeq c_s^2,$$

(3.30)

where k is the Boltzmann constant, and $\overline{m}$ is the mean mass of the constituent particles (electrons, ions or molecules). The **isothermal sound speed** c_s in a perfect gas is given by

$$c_s^2 = \frac{P}{\rho} = \frac{k}{\overline{m}}T,$$

(3.31)

which can be written as

$$c_s \approx 10^4\,\mathrm{m\,s^{-1}} \left(\frac{T}{10^4\,\mathrm{K}}\right)^{1/2}.$$

(3.32)

Now the mass of the fluid element is obtained as density × volume, $m = \rho A\,\Delta z$. In the presence of a vertical gradient of the gas pressure P there will be a pressure difference ΔP between the top and bottom areas of the fluid element, which in turn gives rise to a net force, $\Delta P \times A$, or net acceleration

$$g_{\text{pressure}} = -\frac{\Delta P\,A}{m} = -\frac{\Delta P\,A}{\rho A\,\Delta z} = -\frac{1}{\rho}\frac{\Delta P}{\Delta z}$$

on the element. In the limit of a very small fluid element this becomes

$$g_{\text{pressure}} = -\frac{1}{\rho}\frac{\partial P}{\partial z}.$$

(3.33)

Hydrostatic balance is achieved if $g_{\text{pressure}} + g_z = 0$. The fact that a pressure gradient implies a corresponding acceleration in a fluid, as expressed in Equation 3.33, is a general result (see also Subsection 3.4.1).

● Is there a similar force balance for disc plasma in the radial direction?

○ In the radial direction the gravitational force is balanced by the centrifugal force, as the disc plasma executes Kepler orbits.

If the typical vertical extent of the disc is H, then the pressure acceleration term becomes

$$g_{\text{pressure}} \simeq -\frac{1}{\rho}\frac{P(z=H) - P(z=0)}{H}$$

$$\simeq \frac{1}{\rho}\frac{P(z=0)}{H} = \frac{c_s^2}{H},$$

where in the first step we used $P(z = H) \ll P(z = 0)$ and in the last step we used $P = \rho c_{\mathrm{s}}^2$ to replace the pressure in the disc mid-plane and cancelled ρ. In hydrostatic equilibrium, the pressure term and gravitational term (Equation 3.29, for $z = H$) must add up to 0, i.e.

$$\frac{c_{\mathrm{s}}^2}{H} - \frac{GMH}{r^3} = 0$$

or

$$H^2 = \frac{c_{\mathrm{s}}^2}{GM/r^3}.$$

This gives for the vertical scale height

$$H \simeq \frac{c_{\mathrm{s}}}{\omega_{\mathrm{K}}}, \qquad (3.34)$$

where ω_{K} is the Keplerian angular speed (Equation 3.3). It is also useful to rearrange Equation 3.34 to read

$$\frac{H}{r} \simeq \frac{c_{\mathrm{s}}}{v_{\mathrm{K}}}, \qquad (3.35)$$

where we introduced the Keplerian speed (which is the azimuthal speed of the disc plasma) from Equation 1.5. This is an important result:

> the disc scale height relates to the distance from the rotational axis as the sound speed relates to the Keplerian speed.

In other words, the accretion disc is flat ($H \ll r$) if the azimuthal motion is highly supersonic. The next exercise shows that this is indeed the case for accretion discs observed in cataclysmic variables.

Exercise 3.9 Calculate the disc opening angle δ (given by $\tan \delta = H/r$) for a typical disc in a cataclysmic variable system. Assume that the accreting white dwarf has mass $1\,\mathrm{M}_\odot$, and that the temperature at the outer disc radius $r_{\mathrm{o}} = 0.5\,\mathrm{R}_\odot$ is $10^4\,\mathrm{K}$. ∎

For a given radial distance r, the disc scale height varies as

$$H \propto c_{\mathrm{s}} \propto T^{1/2}, \qquad (3.36)$$

(see Equations 3.35 and 3.31) where T is the disc temperature in the one-zone model, i.e. the mid-plane temperature of the disc, at radius r. Implicit in the above considerations is the assumption that the vertical disc structure is isothermal, so that in a given disc ring the gas temperature is the same for all heights z. Clearly this cannot be the case exactly. Without a temperature gradient in the vertical direction, the disc could not cool by radiation. But it turns out that in realistic disc models the temperature does not change a great deal from mid-plane to surface, perhaps only by a factor of a few, unlike in the radial direction where the temperature changes by orders of magnitude. So the assumption of an isothermal vertical disc layer is quite acceptable.

3.6   Shakura–Sunyaev discs

A quantitative model of a steady-state accretion disc emerges from the conservation equations that we derived above once we also specify the functions that describe the property of a cosmic plasma in general: the equation of state, the opacity and the viscosity. These functions are collectively known as the **input physics**.

A popular choice is the perfect gas law, **Kramers' opacity**, and the α-viscosity introduced in Equation 3.6. Discs subject to this form of viscosity are sometimes called **Shakura–Sunyaev discs**, in honour of Nikolai Shakura and Rashid Sunyaev who pioneered the study of α-**discs** in 1973.

Here we are not quoting the full set of expressions describing Shakura–Sunyaev discs; rather, we just focus on the one for the surface density profile $\Sigma(r)$:

$$\Sigma(r) = 52\,\alpha^{-5/4} \left(\frac{\dot{M}}{10^{13}\,\mathrm{kg\,s^{-1}}} \right)^{7/10} \left(\frac{M}{\mathrm{M_\odot}} \right)^{1/4} \left(\frac{r}{10^8\,\mathrm{m}} \right)^{-3/4}$$

$$\times \left[1 - \left(\frac{R_1}{r} \right)^{1/2} \right]^{14/5}\ \mathrm{kg\,m^{-2}}. \tag{3.37}$$

This expression holds if radiation pressure is negligible, and the disc is treated as isothermal in the direction perpendicular to the mid-plane.

Rosseland mean opacity

Opacity measures the opaqueness of matter against radiation. Specifically, the quantity κ is the absorbing cross-section per unit mass of the absorber, for example in $\mathrm{m^2\,g^{-1}}$. (See also the box entitled 'Cross-section, mean free path and optical depth' in Section 6.2.) The cross-section — and hence the opacity — is a complicated function of the frequency of radiation. In the study of the stellar interior, and also in the context of the vertical structure of optically thick discs, we are not interested in this detailed frequency dependence. We work with a suitable average of the opacity, the Rosseland mean, usually denoted by κ_{R}. The Rosseland mean allows one to calculate the effect of energy transport by radiation in local thermodynamic equilibrium in the simple radiative diffusion approximation (Equation 3.40).

Kramers' opacity

Kramers' law (or Kramers' opacity) describes the Rosseland mean opacity of a plasma with density ρ and temperature T as a simple power law, $\kappa_{\mathrm{R}} = \kappa_0 \rho T^{-3.5}$, with $\kappa_0 = 5 \times 10^{20}\,\mathrm{m^5\,K^{3.5}\,kg^{-2}}$. This fit is a good approximation to the actual opacity when free–free and, to some extent, bound–free interactions dominate.

It is fairly straightforward to see how the key dependence $\Sigma \propto M^{1/4} r^{-3/4}$ in Equation 3.37 comes about for $r \gg R_1$.

To this end we first express the α-viscosity in terms of the disc (mid-plane) temperature T (or a suitable average of the temperature profile perpendicular to the disc plane). Using Equations 3.6 and 3.34 we find

$$\nu_{\mathrm{vis}} \propto H \times c_{\mathrm{s}} \propto \frac{c_{\mathrm{s}}}{\omega_{\mathrm{K}}} \times c_{\mathrm{s}} \propto c_{\mathrm{s}}^2 \times (GM/r^3)^{-1/2}$$

and hence, with $c_{\mathrm{s}} \propto T^{1/2}$ (Equation 3.32),

$$\nu_{\mathrm{vis}} \propto T \times M^{-1/2} \times r^{3/2}. \tag{3.38}$$

For Shakura–Sunyaev discs, the disc mid-plane temperature turns out to be proportional to the surface temperature (effective temperature). Therefore from Equation 3.28 we have $T \propto r^{-3/4} M^{1/4}$ (still assuming $r \gg R_1$). This gives

$$\nu_{\mathrm{vis}} \propto T M^{-1/2} r^{3/2} \propto M^{-1/4} r^{3/4}.$$

According to Equation 3.23 we also have $\Sigma \propto 1/\nu_{\mathrm{vis}}$ in a steady-state disc, so

$$\Sigma \propto \frac{1}{\nu_{\mathrm{vis}}} \propto M^{1/4} r^{-3/4}, \tag{3.39}$$

which is what we wished to show.

Vertical energy transport

We consider the accretion disc surface density profiles further as they will play a crucial role in our interpretation of the dwarf nova and soft X-ray transient phenomena in the next chapter. To calculate the vertical structure of the Shakura–Sunyaev disc, we made the implicit assumption that the energy transport in the z-direction (perpendicular to the disc mid-plane) is via radiation (the disc is said to be *radiative*), i.e. through the slow diffusion of photons to the surface. This process is controlled by the opacity κ_{R}. In particular, in order to maintain a fixed energy flux F, the required vertical temperature gradient $\partial T(z)/\partial z$ is proportional to κ_{R}:

$$F(z) = -\frac{4\sigma}{3\kappa_{\mathrm{R}}\rho} \frac{\partial T(z)^4}{\partial z}. \tag{3.40}$$

It is useful to express this radiative diffusion equation in a form that is appropriate for the one-zone model where integrated quantities are used so that the vertical structure does not have to be considered in detail. We approximate the gradient

$$\frac{\partial T(z)^4}{\partial z} \simeq \frac{T(H)^4 - T(0)^4}{H},$$

and note that $T(0)^4 \gg T(H)^4$ (this is true even if the mid-plane temperature is only a few times larger than the surface temperature), so that

$$\frac{\partial T(z)^4}{\partial z} \simeq -\frac{T(0)^4}{H}.$$

Thus the radiative flux $F(H)$ emerging from the surface is

$$F(H) \simeq \frac{4\sigma T^4}{3\kappa_{\mathrm{R}}\rho H}, \tag{3.41}$$

where T denotes the mid-plane temperature $T(0)$, or, in terms of the one-zone model, the characteristic disc temperature. The quantity $\kappa_{\mathrm{R}}\rho H$ in this last expression has a physical meaning. It is the optical depth

$$\tau = \kappa_{\mathrm{R}}\rho H = \kappa_{\mathrm{R}}\Sigma \tag{3.42}$$

of the full disc in the direction perpendicular to the mid-plane.

● Show that the optical depth is dimensionless.

○ The opacity κ_R measures the absorbing cross-section per unit mass of the absorber, so $\kappa_R \rho$ is the absorbing cross-section per unit volume of the absorber, with units $\mathrm{m}^2\,\mathrm{m}^{-3} = \mathrm{m}^{-1}$. Therefore the product of $\kappa_R \rho$ with a length is dimensionless. (See also the box entitled 'Cross-section, mean free path and optical depth' in Section 6.2.)

Energy transport by convection

In convective layers, most of the energy is effectively trapped and transported in rising blobs of gas. These **convective eddies** usually do not exchange energy with the surrounding medium while they travel, so they behave effectively adiabatically. Once they have risen for about a pressure scale height (the height over which pressure drops by a factor e), they finally mix with the surrounding medium, depositing their excess energy at this point. As the energy is literally moving *with* the matter, convection is usually a very effective means of energy transport. For it to work, there must be an upward force or buoyancy on the eddies. This occurs *only* if the drop in temperature in the surrounding medium is *steeper* than the drop in temperature in the eddy itself. For if this is the case, the eddy will be hotter after it has risen a little. Consequently it will have expanded somewhat, so that its density is smaller than in the surrounding medium. This causes the desired upward lift by buoyancy. If effective convection occurs, it sets up the adiabatic temperature gradient (i.e. the temperature gradient under adiabatic conditions) in the disc layers.

Energy transport can also occur in the form of **convection**. Convection occurs when the vertical temperature gradient that the disc would adopt in the absence of convection is too steep, i.e. steeper than the critical temperature gradient under adiabatic conditions. A prime cause for such a steep temperature gradient is an unusually large opacity. The opacity does indeed become particularly large when hydrogen, the dominant species in the accretion disc plasma, starts to recombine, i.e. when the ionization of hydrogen is incomplete. Then bound–free and bound–bound transitions become available, in addition to the free–free transitions that dominate the opacity in a fully ionized gas. The opacity no longer follows Kramers' law.

Hydrogen is partially ionized in the temperature region around 6000–8000 K, so we expect convection to take over the vertical energy transport when the disc temperature reaches this value. Given the standard temperature profile $T \propto r^{-3/4}$ of a disc, this will normally be quite far away from the central accreting object.

An important consequence of the increase of opacity and the onset of convection is that the surface density profile $\Sigma(r)$ will deviate from the standard form for radiative discs. According to Equation 3.37, the surface density decreases as we move away from the centre of the disc, $\Sigma(r) \propto r^{-3/4}$. But when convection sets in, Σ starts to increase again, and remains roughly constant at yet larger distances (see Figure 3.12).

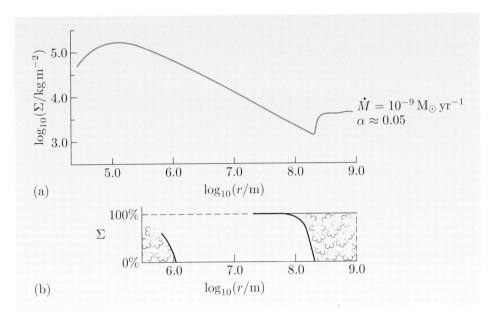

(a)

(b)

Figure 3.12 (a) Surface density profile of a steady-state α-disc (accretor mass $1\,\mathrm{M_\odot}$, mass accretion rate $10^{-9}\,\mathrm{M_\odot\,yr^{-1}}$, $\alpha \approx 0.05$). (b) Mode of vertical energy transport. 'Clouded' regions are convective.

With convection, the disc equations can no longer be solved analytically as for Shakura–Sunyaev discs. We have to rely on numerical calculations to obtain a result like the one shown in Figure 3.12. But it is easy to see why the surface mass density in a convective disc must be larger than in a radiative disc. The magnitude of the temperature gradient established by convection is *smaller* than the magnitude of the temperature gradient in a radiative disc. (Note that the gradient is negative in both cases.) Hence for a given surface temperature (6000 K, say) the convective layer has a *smaller* mid-plane temperature T than the hypothetical radiative layer. The α-viscosity is proportional to the temperature, $\nu_{\mathrm{vis}} \propto T$ (Equation 3.38), and so also smaller. Yet in a steady state the local mass accretion rate must be constant at all radii, and according to the relation $\nu_{\mathrm{vis}}\Sigma \propto \dot{M}$ (Equation 3.23), a smaller viscosity must be compensated for by a larger surface density Σ.

As we shall see in Chapter 4, accretion discs that are large enough that they become convective in their outer parts are *unstable*, i.e. they cannot really exist in the hypothetical steady state that we are assuming here.

Summary of Chapter 3

1. Accretion discs are best described in a cylindrical coordinate system (r, ϕ, z) with the origin at the centre of the accretor. The $z = 0$ plane is the disc mid-plane.

2. Vertically integrated quantities can be used to describe the radial disc structure. The surface density

$$\Sigma(r) = \int_{-\infty}^{+\infty} \rho(r, z)\,\mathrm{d}z \qquad \text{(Eqn 3.1)}$$

is the vertically integrated density ρ. In the one-zone model, the surface density is $\Sigma = H\rho$, where H is the vertical disc scale height.

3. Viscosity is a transport phenomenon and describes the relation between stress and strain: stress = viscosity $\times$ strain. Stress is the force exerted per

91

unit area on a surface. Strain is a measure for the deformation that the stress is causing. For fluid flows, the most important types of stress are pressure, P (force perpendicular to the surface) and shear stress, σ_s (force applied parallel to the surface).

4. The kinematic viscosity due to the turbulent motion of particles or gas blobs with a characteristic speed v_c over a characteristic length λ_c is

$$\nu_{\text{vis}} \simeq \lambda_c \times v_c. \qquad \text{(Eqn 3.5)}$$

5. The phenomenological α-viscosity expresses the disc viscosity in terms of the vertical disc scale height H and the sound speed c_s:

$$\nu_{\text{vis}} = \alpha H c_s. \qquad \text{(Eqn 3.6)}$$

6. A currently very promising candidate for the physical mechanism that provides the viscosity in accretion discs is magneto-hydrodynamic turbulence.

7. The magnitude of the viscous torque in an accretion disc at radius r with local angular speed ω is

$$G_{\text{vis}} = 2\pi r\, \nu_{\text{vis}}\, \Sigma r^2\, \frac{\partial \omega}{\partial r}. \qquad \text{(Eqn 3.8)}$$

8. Viscous dissipation converts mechanical energy of the rotational motion of the disc plasma into heat. For Keplerian discs the conversion rate per unit surface area is

$$D(r) = \tfrac{9}{8}\nu_{\text{vis}}\, \Sigma \frac{GM}{r^3}. \qquad \text{(Eqn 3.12)}$$

9. A good physical model should capture the essential physical ingredients of the process or system, but not dwell on unnecessary details.

10. The radial structure of geometrically thin accretion discs is determined by the conservation of mass and the conservation of angular momentum.

11. The local mass accretion rate

$$\dot{M}(r,t) = -2\pi r v_r \Sigma \qquad \text{(Eqn 3.15)}$$

in the disc is the amount of mass that flows per unit time through the boundary at radius r between adjacent disc annuli. Here v_r is the radial drift velocity.

12. The assumption of steady-state accretion simplifies the integration of the radial disc structure equations. The integral of the equation of mass conservation is the mass accretion rate $\dot{M}$. In a steady-state disc, the mass accretion rate is constant and equal to the rate at which mass is supplied to the disc (e.g. the mass transfer rate from the secondary star).

13. A full integration of the equation describing the conservation of angular momentum requires knowledge of the structure of the boundary layer, the transition zone between the Keplerian accretion disc and the central accreting object. The radial extent of the boundary layer is very small. Pressure forces balance gravity in the radial direction.

14. A steady-state Keplerian disc around a slowly rotating object with radius R_1 will adjust its local structure such that

$$\nu_{\text{vis}} \Sigma = \frac{\dot{M}}{3\pi} \left[1 - \left(\frac{R_1}{r} \right)^{1/2} \right], \qquad \text{(Eqn 3.23)}$$

whatever the physical mechanism of the viscosity.

15. The radial surface temperature profile $T_{\text{eff}}(r)$ of an optically thick, steady-state accretion disc is given by

$$T_{\text{eff}}^4(r) = \frac{3GM\dot{M}}{8\pi\sigma r^3} \left[1 - \left(\frac{R_1}{r} \right)^{1/2} \right]. \qquad \text{(Eqn 3.28)}$$

16. In hydrostatic equilibrium, the vertical component of the gravitational attraction towards the central body must be balanced by the vertical pressure gradient.

17. Accretion discs are geometrically thin if the Keplerian motion is highly supersonic. The disc thickness (or scale height) H is

$$\frac{H}{r} \simeq \frac{c_s}{v_K}. \qquad \text{(Eqn 3.35)}$$

18. The Rosseland mean opacity is a frequency average of the opacity calculated in such a way that energy transport by radiation can be written as a diffusion equation.

19. A Shakura–Sunyaev disc (or α-disc) is a model of a geometrically thin, optically thick steady-state accretion disc with an α-viscosity.

20. Discs become convective when the vertical temperature gradient that the disc would adopt in the absence of convection is too steep. This occurs in particular when the ionization of hydrogen is incomplete, at temperatures of 6000–8000 K.

21. The surface density in radiative discs drops as $r^{-3/4}$. At the point where convection sets in, Σ increases with r, and remains roughly constant at larger radii.

Chapter 4   Accretion disc outbursts

Introduction

So far, we have essentially ignored time-dependent phenomena in the accretion disc itself. This was largely driven by our desire to simplify the complex hydrodynamic problem of the flow of astrophysical fluids. The main simplification was the assumption of time-independence, leading to the theory of steady-state accretion.

Yet many accreting binaries are all but persistently bright. In fact, it was the striking semi-regular brightening known as dwarf nova outbursts that defined the class of cataclysmic variables in the first place. The related soft X-ray transient outbursts in low-mass X-ray binaries are even more spectacular, transforming these systems, for a few weeks, to be among the brightest sources in the X-ray sky. It is clear that the steady-state accretion flows presented in Chapter 3 cannot describe these time-dependent phenomena.

There is also a second, more subtle shortcoming if we were confined to only observe purely steady-state properties. We would fail to discover the origin and magnitude of the most uncertain physical quantity involved in the theory of accretion, namely the viscosity. For, as we have seen in Chapter 3, the viscosity does not appear in the steady-state expression for the temperature and the radiant flux emerging from the disc surface (see, for example, Equation 3.28).

To obtain a physical understanding of the time-dependent phenomena by simple, analytical consideration is much harder than for steady-state accretion. To solve the full problem we would need to write down the disc equations that encapsulate the full time-dependent physics, and then set out to find a solution of these equations — Σ as a function of r and t, say — by numerical methods. Non-linear, higher-order partial differential equations like these disc equations can indeed be solved with sophisticated computer code.

Simply presenting the numerical solution found for time-dependent accretion in a thin disc would not provide deeper insight into the physics of the problem. The calculations do indeed reproduce a time-variability of the disc luminosity that is reminiscent of dwarf nova and soft X-ray transient outbursts, but we want to go beyond this and gain an understanding of *why* such outbursts occur, and identify the physical mechanisms determining the outburst characteristics.

The approach that we take is to investigate the steady-state disc solutions and search for *instabilities*. Are there instabilities — not described by steady-state theory — that might prevent real accretion discs from achieving a steady state in the first place? If so, on what timescale do these instabilities grow, and what is the likely outcome of a system subject to the instability? The advantage of this approach is that we shall be able to describe and understand certain time-dependent phenomena by considering transitions between different steady states. Ultimately we shall discover the limit cycle behaviour of accretion discs that lies at the heart of the observed outburst behaviour.

As a first step towards this goal we first consider viscous diffusion in accretion discs, the process that governs the redistribution of mass (and angular momentum) in the disc, and so the rate of change of the surface density Σ in a non-steady-state situation.

4.1 Viscous diffusion

We recall the conservation equations for mass and angular momentum in accretion discs,

$$r\frac{\partial \Sigma}{\partial t} + \frac{\partial}{\partial r}(rv_r\Sigma) = 0 \qquad \text{(Eqn 3.13)}$$

and

$$r\frac{\partial}{\partial t}(\Sigma r^2\omega) + \frac{\partial}{\partial r}(rv_r\Sigma r^2\omega) = \frac{1}{2\pi}\frac{\partial G_{\text{vis}}}{\partial r}, \qquad \text{(Eqn 3.17)}$$

and note that these do explicitly include terms that describe the rate of change of Σ. In the previous chapter we obtained a steady-state solution by setting $\partial\Sigma/\partial t = 0$, but now we wish to calculate $\partial\Sigma/\partial t$ from known quantities for the case of an evolving disc. For this we need to combine Equations 3.13 and 3.17 to eliminate the other unknown variable, the radial drift velocity v_r. After some straightforward yet lengthy algebra, this gives the desired expression

$$\frac{\partial \Sigma}{\partial t} = \frac{3}{r}\frac{\partial}{\partial r}\left[r^{1/2}\frac{\partial}{\partial r}\left(\nu_{\text{vis}}\Sigma r^{1/2}\right)\right]. \qquad (4.1)$$

The key manipulation needed to obtain this result is to use the product rule to rewrite the second term in the angular momentum equation (Equation 3.17), $\partial(rv_r\Sigma r^2\omega)/\partial r$, as

$$\frac{\partial}{\partial r}(rv_r\Sigma) \times r^2\omega + rv_r\Sigma \times \frac{\partial}{\partial r}(r^2\omega),$$

and substitute

$$\frac{\partial}{\partial r}(rv_r\Sigma) = -\frac{1}{r}\frac{\partial \Sigma}{\partial t} \qquad (4.2)$$

from Equation 3.13.

Equation 4.1 relates the first time derivative of Σ to the second spatial derivative of Σ. This is the characteristic of a diffusion equation. The radial drift of matter in the accretion disc, and hence the rate of change of the local surface density in the disc, can be described as a **viscous diffusion** process. Viscous diffusion is a physical process central to the working of a disc.

To make this clearer, and because the phenomenon of **diffusion** is ubiquitous in nature, we make a short detour to consider diffusion in more detail. Examples of physical processes that constitute diffusion are the random walk of particles and the random walk of photons in an optically thick gas, also known as energy transport by radiation or radiative diffusion (see Section 3.6). Perhaps the best known example is the conduction of heat, such as the loss of heat through the walls of a house. In general terms, diffusion is about the slow spreading of a physical quantity if this quantity is not evenly distributed. The flow rate of the quantity is proportional to its spatial gradient. The tendency to spread is particularly large if the distribution is very uneven.

In the case of a heated house, thermal energy is lost through the walls of the house when it is colder outside than inside. The heat flow rate is proportional to the temperature difference. If it is very cold outside, the heating system has to be turned up to maintain a comfortable room temperature. The greater the temperature gradient between the room and outside, the greater the heat loss through the wall. (The same is true if there are draughty windows or doors, but in this case the heat loss is due to the exchange of warm air for cold air.)

To make this quantitative, we note that the rate of change of heat in a small volume is given by the difference between the rate of heat flowing into this volume and the rate of heat flowing out of it. Suppose that u denotes the amount of heat energy in a unit volume, i.e. the heat energy density (measured in $J\,m^{-3}$), and suppose that j_u denotes the heat flow density, i.e. the amount of heat energy flowing through a unit surface per unit time (measured in $J\,m^2\,s^{-1}$). Then consider the heat flow in the x-direction through a very small cubic volume with side walls, parallel to the $x = 0$ plane, of area A, and width Δx along the x-axis. The inflowing heat energy is just $j_u(x) \times A$, and the outflowing heat energy is $j_u(x + \Delta x) \times A$. Using the first-order expansion this can be written as $(j_u(x) + (\partial j/\partial x) \times \Delta x) \times A$. Therefore the rate of change of the heat energy in the volume $u \times \Delta x \times A$ is the difference between inflow and outflow, so

$$\Delta x \times A \times \frac{\partial u}{\partial t} = j_u(x) \times A - \left(j_u(x) + \frac{\partial j_u}{\partial x} \times \Delta x \right) \times A$$

or

$$\frac{\partial u}{\partial t} = -\frac{\partial}{\partial x} j_u. \tag{4.3}$$

In other words, the rate of change of u is given by the spatial gradient of the flow density. Here we assumed that the heat flow density is positive if the flow is in the positive x-direction. The minus sign signals that if j_u decreases with x ($\partial j_u/\partial x < 0$), this leads to an increase of u ($\partial u/\partial t > 0$) at a given point.

The defining characteristic of a diffusive process is that the flow density itself is proportional to the gradient of the heat energy,

$$j_u \propto -\frac{\partial u}{\partial x}.$$

● Why is there a minus sign?

○ The minus sign is needed because the flow must be in the direction of decreasing heat in order to even out the gradient.

Inserting this into Equation 4.3, we see that the diffusive process gives rise to an equation that relates the *time* derivative of the heat content to the *second* spatial derivative of the heat content,

$$\frac{\partial u}{\partial t} \propto \frac{\partial^2 u}{\partial x^2} \tag{4.4}$$

(see also Figure 4.1). An equation of this form is called a **diffusion equation**, whatever the quantity u stands for. In the example above, u was an energy density (thermal energy per unit volume). Another example is the diffuse mixing of two different types of gas, where u is the mass density of one of the two gas species.

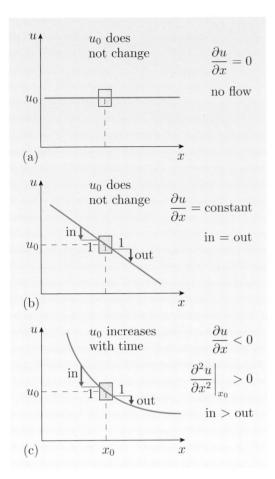

Figure 4.1 The quantity u changes as a result of diffusion if the second spatial derivative of u, i.e. the curvature of $u(x)$, is non-zero.

We can now see that Equation 4.1 for the surface density Σ in the disc is indeed a diffusion equation, as it relates the time derivative of Σ with two spatial derivatives ($\partial/\partial r$ occurs twice) of Σ. The spatial derivative is more complicated than in the case of u above — but this is mainly just because of the cylindrical disc geometry. There is one added complication, however: the viscosity ν_{vis} in Equation 4.1 can be a function of r, Σ and t itself. That is why Equation 4.1 is, in general, a *non-linear* diffusion equation.

For the particularly simple case of a constant viscosity, Equation 4.1 has an analytic solution that describes the viscous evolution of an initially narrow plasma torus. With time, the torus spreads and forms a more extended disc structure. Eventually, almost all of the mass will have accreted to the centre, with all of the angular momentum carried to very large radii by a very small fraction of the mass.

This analytical solution assumes that the disc can spread forever, i.e. the disc is infinite, while the central mass accretes any plasma that arrives at very small radii.

● Is the disc infinite in a binary system?

○ No, the disc has to fit inside the Roche lobe of the accreting star. In a binary system the disc is truncated by tidal effects from the mass donor. The outer disc radius is smaller than the accretor's Roche-lobe radius, perhaps only half as big.

4.2   Hierarchy of timescales

The viscous diffusion or radial drift of the accreting plasma proceeds on a characteristic timescale, the viscous time. This is one of three main characteristic timescales that govern time-dependent phenomena in accretion discs; the other two are the dynamical time and the thermal time. The distinct hierarchy of these timescales is an important element of the stability discussion below, so we shall dwell on it for a while.

Characteristic timescales

Statements to the effect that a certain time or timescale (we use these two terms interchangeably) is *characteristic* of a certain physical effect are quite common in the scientific literature. For instance, the dynamical time is the characteristic time to establish hydrostatic equilibrium. What this really means is that the *actual* time it takes for the process to occur is *of order* of that characteristic time. Depending on the situation, the actual time could easily be half or twice as long as the characteristic timescale. It could also simply mean that it takes the characteristic time for any *significant changes* to occur. An example is the radioactive decay process. The decay rate decreases exponentially with time (Figure 4.2), so strictly speaking the process of decay never finishes. But significant changes — the decay of 50% of the nuclei — occur within the radioactive half-life.

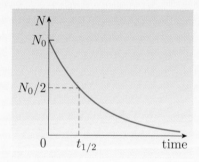

Figure 4.2   Characteristic timescale $t_{1/2}$ for exponential decay.

4.2.1   The dynamical time

The shortest characteristic timescale is the dynamical time, the time it takes disc particles to complete one Kepler orbit around the accreting object:

$$t_{\mathrm{dyn}} = \frac{r}{v_{\mathrm{K}}} = \frac{1}{\omega_{\mathrm{K}}}.$$

For an accretion disc in a cataclysmic variable, for example, the dynamical time is of order minutes, while for an AGN it is $\gtrsim 10^2$–10^4 years. The dynamical time is also the characteristic time for re-establishing hydrostatic equilibrium in the direction perpendicular to the disc plane, should this equilibrium be perturbed. We demonstrate this in the following worked example.

Worked Example 4.1

Perturbations of the hydrodynamical equilibrium imply a pressure imbalance in the affected layers, and such perturbations travel with the local sound speed. Hydrostatic equilibrium is therefore restored on the sound travel time $t_z = H/c_s$ in the z-direction, where H is the disc thickness.

Show that t_z is the same as the dynamical time t_{dyn} defined above.

Solution

The disc thickness is given by

$$\frac{H}{r} \simeq \frac{c_s}{v_K}, \qquad \text{(Eqn 3.35)}$$

which can be rearranged to read $H/c_s = r/v_K$. Therefore

$$t_z = \frac{H}{c_s} \simeq \frac{r}{v_K}, \qquad (4.5)$$

as required.

4.2.2 The thermal time

The (local) thermal time measures how long it takes to generate the thermal energy content of a disc annulus by viscous dissipation at the current rate:

$$t_{th} = \frac{\text{thermal content}}{\text{dissipation rate}}. \qquad (4.6)$$

This is similar in spirit to the Kelvin–Helmholtz time in stars (Equation 2.24), $t_{KH} = GM^2/RL$, which is the time it takes to remove the heat content (thermal energy) $\propto GM^2/R$ of a star with mass M and radius R if the star radiates at luminosity L. The significance of the thermal time is that it indicates the characteristic timescale to re-establish **thermal equilibrium** should this be perturbed. In the state of thermal equilibrium, any local energy losses from the disc, e.g. via radiation, are exactly balanced by the energy that viscous dissipation generates at this disc annulus.

The viscous dissipation rate per unit surface area in Equation 4.6 is the quantity D given by Equation 3.11; we shall use it in the simpler form for Keplerian discs, Equation 3.12. To determine the thermal content, or heat content, of a disc annulus, per unit area, we first note that for a perfect gas with temperature T, the thermal energy per particle is approximately kT. (We are not concerned about factors of order unity here.) If the disc annulus contains N particles, its total heat content is NkT. The number N can be obtained by dividing the total mass of the disc annulus with area A and surface density Σ, $m_{annulus} = \Sigma \times A$, by the mean mass $\overline{m}$ of these particles: $N = m_{annulus}/\overline{m}$. So the heat content per surface area is

$$\frac{NkT}{A} = \frac{m_{annulus}}{\overline{m}} \frac{kT}{A} = \frac{\Sigma A}{\overline{m}} \frac{kT}{A} = \Sigma \frac{kT}{\overline{m}}.$$

But $kT/\overline{m}$ is just the square of the isothermal sound speed c_s^2 (Equation 3.32), so the heat content per unit surface area can be written as Σc_s^2. The thermal time then becomes

$$t_{th} \simeq \frac{\Sigma c_s^2}{D(r)},$$

or, with $D(r)$ from Equation 3.12,

$$t_{th} \simeq \frac{c_s^2}{\nu_{vis}\, GM/r^3}, \tag{4.7}$$

where we again neglected a numerical factor of order unity and assumed $r \gg R_1$. Using the α-viscosity $\nu_{vis} = \alpha H c_s$ (Equation 3.6) in this expression for the thermal time, we can show that

$$t_{th} = \frac{1}{\alpha} t_{dyn}. \tag{4.8}$$

As $\alpha \lesssim 1$, we also have $t_{dyn} \lesssim t_{th}$, i.e. the dynamical time is shorter than the thermal time.

Exercise 4.1   Show that Equation 4.8 follows from Equation 4.7 when the α-viscosity is used.   ∎

4.2.3   The viscous time

The viscous time t_{vis}, sometimes also referred to as the radial drift timescale, is the time it takes the disc fluid to move appreciably in the radial direction. As we have emphasized previously, the dynamics of the disc plasma is dominated by the near-Keplerian azimuthal velocity, but superimposed onto these Kepler orbits is a slow radial drift that, in a steady-state disc, gives rise to a net local mass accretion rate. This is expressed by

$$\dot{M} = -2\pi r v_r \Sigma, \tag{Eqn 3.15}$$

where v_r is the radial drift velocity. The viscous time is simply

$$t_{vis} = \frac{r}{|v_r|}. \tag{4.9}$$

The viscous time can also be expressed as an explicit function of the viscosity. The easiest way to see this is by appealing to the fundamental relation between viscosity and surface density of a steady-state disc. From Equation 3.23 we have, for radii sufficiently far away from the accretor ($r \gg R_1$), $\nu_{vis} \Sigma \simeq \dot{M}/3\pi = $ constant, so from Equation 3.15 we see that

$$|v_r| = \frac{\dot{M}}{2\pi r}\frac{1}{\Sigma} = \frac{\dot{M}}{2\pi r}\frac{3\pi \nu_{vis}}{\dot{M}} \simeq \frac{\nu_{vis}}{r}.$$

Therefore the radial drift velocity is given by

$$|v_r| \simeq \frac{\nu_{vis}}{r}, \tag{4.10}$$

and the viscous time becomes

$$t_{vis} \simeq \frac{r^2}{\nu_{vis}}. \tag{4.11}$$

● Using the α-viscosity, verify that $r^2/\nu_{\rm vis}$ has the dimension of time.

○ The α-viscosity is given by $\nu_{\rm vis} = \alpha H c_{\rm s}$ (Equation 3.6). As α is dimensionless, $\nu_{\rm vis}$ has the unit $\rm m \times m\,s^{-1} = m^2\,s^{-1}$. So $r^2/\nu_{\rm vis}$ has the unit $\rm m^2/(m^2\,s^{-1}) = s$.

We can also obtain an estimate of the viscous timescale from a consideration of the viscous diffusion equation (Equation 4.1) itself, without actually solving it rigorously. To this end we first rewrite Equation 4.1 for the case of a constant viscosity:

$$\frac{\partial \Sigma}{\partial t} = \frac{9\nu_{\rm vis}}{2r}\frac{\partial \Sigma}{\partial r} + 3\nu_{\rm vis}\frac{\partial^2 \Sigma}{\partial r^2}. \tag{4.12}$$

Exercise 4.2 Show that Equation 4.1 can be rearranged into Equation 4.12 for $\nu_{\rm vis} = $ constant. (Use the product rule.) ∎

Next we replace the derivatives in Equation 4.12 with *characteristic values*, or *order of magnitude* estimates. In a sense, we shall be using the very definition of a derivative backwards. First we replace $\partial \Sigma/\partial t$ at radius r with the value of Σ at this radius, divided by the viscous time $t_{\rm vis}$ that we seek, i.e. the characteristic time over which Σ changes:

$$\frac{\partial \Sigma}{\partial t} \simeq \frac{\Sigma}{t_{\rm vis}}.$$

Next we replace the spatial gradient $\partial \Sigma/\partial r$ with Σ, divided by a length scale l that characterizes the spatial gradient (see Figure 4.3). Then Equation 4.12 becomes

$$\frac{\Sigma}{t_{\rm vis}} \simeq \frac{9\nu_{\rm vis}}{2r}\frac{\Sigma}{l} + 3\nu_{\rm vis}\frac{\Sigma}{l^2}.$$

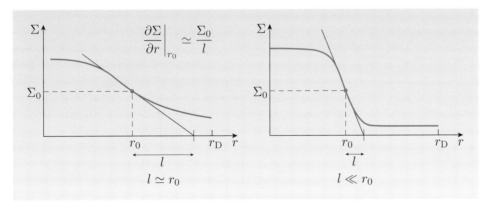

Figure 4.3 Order of magnitude estimates for gradients in differential equations, illustrated for two different surface density profiles in accretion discs. The slope $\partial \Sigma/\partial r$ of the curve at radius r_0 is approximated as Σ_0/l. The outer disc radius is $r_{\rm D}$.

Dropping numerical factors and rearranging, we have approximately

$$\frac{\Sigma}{t_{\rm vis}} \simeq \nu_{\rm vis}\Sigma\left(\frac{1}{rl} + \frac{1}{l^2}\right) = \frac{\nu_{\rm vis}\Sigma}{l^2}\left(1 + \frac{l}{r}\right). \tag{4.13}$$

If $l \ll r$, then the term in brackets is close to unity, and Equation 4.13 can be rearranged to give $t_{\rm vis} \simeq l^2/\nu_{\rm vis}$. If, on the other hand, we have $l \simeq r$, we find from Equation 4.13 that

$$\frac{\Sigma}{t_{\rm vis}} \simeq \frac{\nu_{\rm vis}\Sigma}{r^2}\left(1 + \frac{r}{r}\right),$$

and hence the above result $t_{vis} \simeq r^2/\nu_{vis}$ (again dropping the numerical factor 2). This shows that density enhancements involving sharp gradients, with $l \ll r$, diffuse more quickly than smoother density enhancements.

The viscous time is a rather important quantity in the context of time-dependent accretion. A classical example is the actual formation of an accretion disc in a binary star with Roche-lobe overflow. When the donor star fills its lobe for the first time, the mass transfer stream emanating from the L_1 point free-falls onto the compact accretor along a trajectory that self-intersects as a result of the conservation of angular momentum. Collisions and dissipation cause the plasma stream to settle into a narrow torus in the orbital plane, centred on the accretor, at the so-called **circularization radius** r_c. The torus then slowly spreads in radial direction — matter diffuses inwards (and outwards) — with a characteristic timescale r_c^2/ν_{vis}.

Exercise 4.3   Estimate how quickly an accretion disc forms in a compact binary with a $0.6\,M_\odot$ accretor and $r_c \simeq 0.2\,R_\odot$. Use the α-viscosity with $T \simeq 10^4$ K and $\alpha = 0.3$.   ■

As before we can relate the viscous time to the dynamical time by using the α-viscosity:

$$t_{vis} \simeq \frac{r^2}{\nu_{vis}} \simeq \frac{r^2}{\alpha H c_s} = \frac{1}{\alpha}\frac{r}{H}\frac{r}{v_K}\frac{v_K}{c_s}.$$

In this last expression we recognize r/v_K as t_{dyn}, and v_K/c_s as r/H, so that

$$t_{vis} \simeq \frac{1}{\alpha}\left(\frac{r}{H}\right)^2 t_{dyn}. \tag{4.14}$$

Summary of hierarchy of timescales

Pulling the expressions for all three of the characteristic timescales together shows us a clear hierarchy,

$$t_{dyn} \simeq \alpha\, t_{th} \simeq \alpha\left(\frac{H}{r}\right)^2 t_{vis}, \tag{4.15}$$

which demonstrates that for $\alpha < 1$, the dynamical time is shorter than the thermal time, while both are much shorter than the viscous time (since $H/r \ll 1$).

- ● What are typical values of the three timescales at the outer edge of a Shakura–Sunyaev disc in a binary system with an orbital period of a few hours? Assume $\alpha = 0.1$ and $H/r = 0.01$.

- ○ The dynamical time is the Kepler period of disc particles. According to Kepler's law, the period scales with the distance from the central object as $P_{orb} \propto r^{3/2}$. Hence the dynamical time in the outer disc is somewhat shorter than the binary period, as the disc is smaller than the binary itself. Let's say $t_{dyn} \simeq 1$ h. Then the thermal time is 10 h, while the viscous time is much longer. With $H/r = 0.01$ we have $t_{vis} \simeq 10\,\text{h}/0.01^2 = 10^5$ h, i.e. more than 10 years.

4.2.4 Thermal instability

As we shall see shortly, this hierarchy of timescales is important in analysing a non-equilibrium situation where locally, in a particular disc annulus, the rate of disc heating is out of step with the rate of disc cooling. Clearly, in the case of such an imbalance, the disc must react, e.g. by becoming hotter when heating dominates over cooling. More specifically, the disc will heat up locally on the characteristic timescale associated with a perturbation of thermal equilibrium — the thermal time. If as a result of the disc's immediate reaction to a small perturbation the heating rate is even more out of step with the cooling rate, the perturbation will grow and the disc will continue to heat up, on a thermal time. In this case it is said that the disc is subject to a thermal instability, and the growth time of the instability is the thermal time. (See also the box entitled 'Stability, instability and stability analysis' in Section 2.6.)

As the dynamical time is short compared to this growth time, the disc adjusts its vertical structure effectively instantaneously, and is in hydrostatic equilibrium at all times during the growth of such a thermal instability. But despite this readjustment, the local surface density does not change.

- Why is this?
- ○ The only way that Σ could change is through viscous transport of disc mass from or to neighbouring disc annuli. But this occurs on the viscous timescale, which we have shown is long compared to the thermal time.

4.3 Disc instabilities

We are now in a position to generalize our model description of steady-state, optically thick, geometrically thin accretion discs to viscously evolving discs. We have seen that the dynamical and thermal time are both much shorter than the viscous time. Therefore we make the simplifying assumption that the viscously evolving disc is always in hydrostatic *and* thermal equilibrium, but we shall relax the assumption of thermal equilibrium later.

4.3.1 Viscous instability

We begin by investigating if the steady-state accretion disc models that we presented in Chapter 3 are viscously stable. To this end we apply a *small, local* perturbation to the equilibrium disc, in the form of the addition of a small amount of extra mass in a given disc annulus with negligible width. Therefore the initial steady-state surface density Σ_0 at this annulus increases by a small amount $\Delta\Sigma \ll \Sigma_0$. This set-up is the basis for employing what is known as a *local linear stability analysis*. If the disc is locally viscously stable, the excess mass will diffuse or drift radially away from the disc ring and gradually restore the original surface density Σ_0. If, on the other hand, the disc is viscously unstable, the perturbation grows, causing the affected disc ring to accumulate mass, while neighbouring disc rings are depleted in mass. The disc is breaking up into detached rings!

These two cases are illustrated in Figure 4.4.

$$t_1 < t_2 < t_3$$

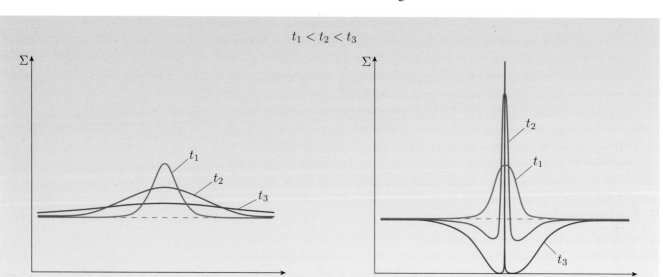

Figure 4.4   Behaviour of a surface density perturbation for (left) viscously stable and (right) viscously unstable discs.

It is useful to restrict the stability analysis to small perturbations $\Delta\Sigma$, as then the consequent change of a physical quantity that depends on Σ can be approximated as $f(\Sigma_0 + \Delta\Sigma) \simeq f(\Sigma_0) + (\partial f/\partial\Sigma) \times \Delta\Sigma$. This is why the stability analysis is *linear*. As we shall see, the sign of the term $\partial f/\partial\Sigma$ ultimately determines if the system is stable or unstable.

In the *local* stability analysis we focus on a very small section of the disc with a very small radial extent, so that $\partial f/\partial\Sigma$ can be treated as constant over the corresponding range of radii.

Exercise 4.4   Re-state the main assumptions for a local linear stability analysis.   ■

Carrying out the full local, linear stability analysis for a steady-state disc by using the viscous diffusion equation for Σ (Equation 4.1) shows the following criterion for viscous stability. A disc is locally, at radius r, viscously stable if

$$\frac{\partial[\nu_{\mathrm{vis}}(r)\,\Sigma(r)]}{\partial\Sigma} > 0. \tag{4.16}$$

This is equivalent to

$$\frac{\partial\dot{M}(r)}{\partial\Sigma} > 0 \tag{4.17}$$

or

$$\frac{\partial T(r)}{\partial\Sigma} > 0. \tag{4.18}$$

● Why is the stability criterion given in Equation 4.16 equivalent to Equation 4.17?

○ From the steady-state surface density relation (Equation 3.23) in Subsection 3.4.2 we see that $\nu_{\text{vis}} \Sigma \propto \dot{M}$ if $r = \text{constant}$, or $\nu_{\text{vis}} \Sigma = k_1 \dot{M}$ with $k_1 = \text{constant}$. Therefore

$$\frac{\partial \dot{M}}{\partial \Sigma} = \frac{1}{k_1} \frac{\partial (k_1 \dot{M})}{\partial \Sigma} = \frac{1}{k_1} \frac{\partial [\nu_{\text{vis}}(r) \, \Sigma(r)]}{\partial \Sigma} > 0$$

as Equation 4.16 holds and $k_1 > 0$. (A similar argument applies for Equation 4.18.)

It is important to note that the derivatives in Equations 4.16–4.18 have to be taken at a fixed disc annulus with radius r. They measure how the equilibrium values of T or $\dot{M}$ or $\nu_{\text{vis}} \Sigma$ change when Σ is varied in such a way that the disc, locally and globally, remains in hydrostatic and thermal equilibrium. In practice, to determine these derivatives we need to compare, at a fixed radius r, steady-state disc models with different mass accretion rates.

The second form of the stability criterion (Equation 4.17) is intuitively clear. Consider a disc where locally, at a certain disc radius r_0, $\partial \dot{M} / \partial \Sigma < 0$. This implies that small changes of the local mass accretion rate $\dot{M}$ and the surface mass density Σ are related as $\Delta \dot{M} \propto -\Delta \Sigma$. A perturbation that locally increases the surface density ($\Delta \Sigma > 0$) causes a corresponding decrease of the local mass accretion rate ($\Delta \dot{M} \propto -\Delta \Sigma < 0$). But the mass supply rate from adjacent disc annuli still has the original higher value. Hence mass accumulates locally, and the surface density Σ increases further in the annulus at r_0, signalling instability.

4.3.2 Thermal–viscous instability

We have now established criteria for the local viscous instability of accretion discs. There is one more piece of information to be taken in before we can investigate systematically if and where an accretion disc might indeed be unstable.

A disc that is subject to a viscous instability will not be able to remain in thermal equilibrium. Rather, a thermal instability will occur as well, allowing the disc to evolve into a different state that is viscously stable.

To see this, we consider the local balance between heating and cooling quantitatively. For consistency with the one-zone model approach of much of Chapter 3, we express this in terms of vertically integrated terms, i.e. we compare the heating per unit surface area and the cooling per unit surface area in a disc annulus with radius r. The former is just the viscous heating $D(r)$ that we established in Equation 3.12, and for $r = \text{constant}$ it scales as

$$D(r) \propto \nu_{\text{vis}} \Sigma.$$

As $\nu_{\text{vis}} \propto T$ (Equation 3.38), this becomes $D \propto T\Sigma$.

The rate of energy lost from the disc annulus due to radiation, per unit disc area, is the flux F that we established in Equation 3.41:

$$F(H) \simeq \frac{4\sigma T^4}{3\kappa_{\text{R}} \rho H}.$$

For $r = \text{constant}$ this scales as $F \propto T^8 \Sigma^{-2}$.

Exercise 4.5   Starting from Equations 3.41 and 3.36, using Kramers' opacity (see the box entitled 'Kramers' opacity' in Section 3.6), show that $F \propto T^8 \Sigma^{-2}$. ∎

In equilibrium, in a steady-state disc, heating equals cooling, $D = F$. Consider now the case where a viscous perturbation increases the surface density by a small amount $\Delta\Sigma > 0$, while the disc (mid-plane) temperature remains constant. Then the heating rate $D \propto T\Sigma$ increases slightly, while the cooling rate $F \propto T^8 \Sigma^{-2}$ decreases slightly. In other words, heating dominates over cooling, and the temperature of the disc annulus will increase.

However, if the disc is viscously unstable, $\partial(\nu_{\mathrm{vis}}\Sigma)/\partial\Sigma < 0$. For a perturbation $\Delta\Sigma > 0$, this implies $\Delta\nu_{\mathrm{vis}} < 0$, so a new steady-state configuration would need a *smaller* temperature and would have to *cool*. Instead, the disc annulus heats up, as we have just shown, and moves away from the local steady-state configuration, until it settles at a new, different steady state, with a higher temperature and mass accretion rate.

Conversely, if the disc is viscously stable, $\partial(\nu_{\mathrm{vis}}\Sigma)/\partial\Sigma > 0$, so a perturbation $\Delta\Sigma > 0$ implies $\Delta\nu_{\mathrm{vis}} > 0$, and the new steady-state configuration is hotter, and this can indeed be reached as the disc is in fact heating up.

We can summarize this also in a slightly different way, as follows. The viscous instability as we studied it in the previous section does not occur, since the assumption of thermal equilibrium that we have made to define it breaks down. Instead, a thermal instability occurs, thus allowing the disc to evolve on a thermal timescale into another, viscously stable state.

If a disc violates the stability criterion, Equation 4.16, the disc is said to be subject to a *thermal–viscous instability*.

4.4   Dwarf novae and soft X-ray transients

Before we apply our stability considerations to the α-disc model developed in Section 3.6, in the hope of finding a mechanism that would give rise to outburst behaviour as observed in dwarf novae and soft X-ray transients, we review here the observed key features of these phenomena.

Both types of systems undergo dramatic outbursts whereby the luminosity of the binary rises by many orders of magnitude. The outburst is followed by a prolonged phase of quiescence, before another outburst may occur.

About half of the known cataclysmic variables with determined orbital period belong to the class of **dwarf novae**. The system SS Cygni is the best studied example. Observations by amateur astronomers document the sequence of outbursts over a long period of time. This long-term light curve illustrates the typical behaviour of dwarf novae (Figure 4.5). The semi-regular increase in brightness is most pronounced in the optical and ultraviolet spectral range. The outbursts have an amplitude of 2–5 magnitudes, last for a few days and recur within weeks or months. Neither the shape of the outburst light curve nor the recurrence time is strictly periodic.

There is a bewildering variety of shapes of dwarf nova light curves, which prompted the definition of numerous dwarf nova subclasses, on the basis of transient features in the light curve such as standstills (i.e. the light curve is stuck at an intermediate level on its return to quiescence) or the existence of superoutbursts (occasional outbursts that are slightly brighter and last longer than normal outbursts; they are also accompanied by so-called superhumps). Dwarf novae in outburst have spectral properties very similar to those of the

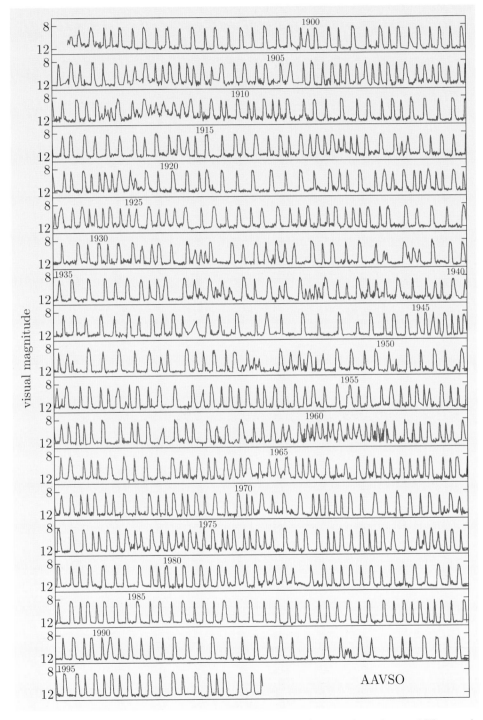

Figure 4.5   The long-term light curve of SS Cygni, spanning almost 100 years! (Constructed from observations made by the American Association of Variable Star Observers. Courtesy John Cannizzo.)

persistently bright **nova-like variables**, i.e. cataclysmic variables that do not experience dwarf nova outbursts. Nova-likes are thought to host bright, steady-state accretion discs, so the implication is that dwarf novae also have discs in a quasi-steady state in outburst. We return to the observational evidence for accretion discs in general in Chapter 5.

Soft X-ray transients are a subclass of low-mass X-ray binaries and display outbursts reminiscent of dwarf nova outbursts, albeit on a much longer timescale and with a much higher outburst amplitude. The light curve morphology is even more varied than for dwarf novae. A fairly common shape, with a fast rise of a few days, followed by a slower exponential decay over a few months, is shown in Figure 4.6. As the name suggests, the outburst is most pronounced in the X-ray regime, with a flux increase in excess of 100 (often much higher), but there is a simultaneous brightening in other wavebands, most notably in the optical range. Many soft X-ray transient outbursts have been observed only once. This implies that the recurrence time is very long — decades or centuries, or longer still. A few systems are seen to recur once a year or every few years.

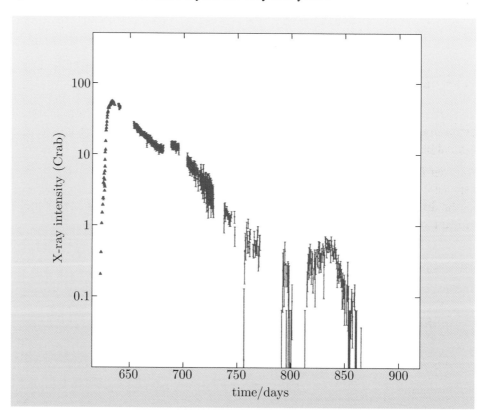

Figure 4.6   The X-ray light curve of the black hole X-ray binary A0620-00 (Nova Monocerotis 1975, also known as V616 Monocerotis), a soft X-ray transient. The X-ray intensity is given in multiples of the Crab nebula intensity. The time axis gives days since Julian Day JD 2442 000.

The shorter timescales of dwarf nova outbursts make them the ideal laboratory for the study of disc outbursts. They allow us to compare theoretical models with observations in some detail. This is why, in the following, we focus on dwarf nova outbursts.

The idealized light curve of dwarf novae (Figure 4.7) consists of four distinct phases: the rise to outburst, the outburst itself, the transition into quiescence, and the quiescent phase. Each phase has a corresponding characteristic timescale; the rise and decline phases are very short, the outburst time is intermediate, and the quiescent time is long.

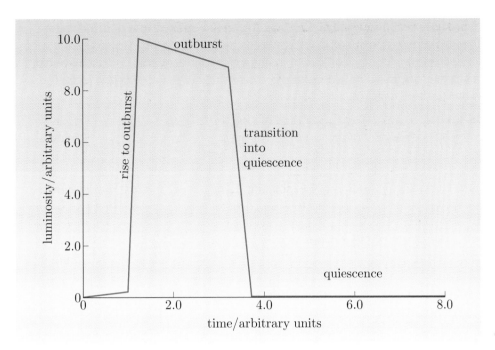

Figure 4.7 An idealized dwarf nova light curve.

4.5 Limit cycles and disc outbursts

We shall now investigate if a stability analysis of Shakura–Sunyaev discs, the idealized models of steady-state accretion discs, can shed light on the origin of the cyclical behaviour observed in dwarf nova systems.

Consider Figure 4.8, which depicts the surface density Σ as a function of distance r from the accretor's centre. This is similar to Figure 3.12 in Section 3.6, but the disc parameters chosen for the figure here are tuned so that the step at the critical radius where convection starts to dominate is emphasized.

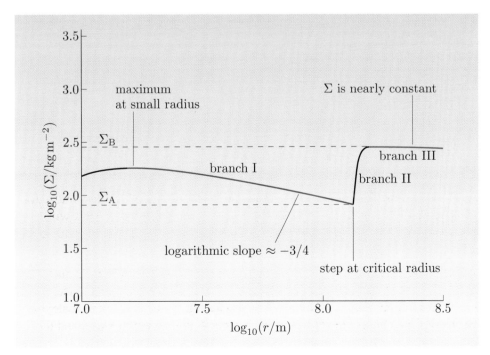

Figure 4.8 The surface density profile of a Shakura–Sunyaev disc, with Σ as a function of r.

As we have seen in Section 3.6, depending on the radial extent of the accretion disc, the surface density profile has various distinct features. At small radii the surface density Σ rises to a local maximum. Beyond the peak, Σ decreases with increasing radius, with essentially a constant slope of $-3/4$. We denote this part of the curve as branch I.

● Why has branch I the constant slope $-3/4$?

○ Note that both axes are logarithmic. From Equation 3.37 we see that $\Sigma \propto r^{-3/4}$ for $r \gg R_1$ in radiative discs. Hence $\log_{10}\Sigma \propto (-3/4) \times \log_{10}r$.

Branch II is the steeply rising part between surface densities Σ_A and Σ_B ($\Sigma_A < \Sigma_B$) at a critical radius r_c, and branch III is the fairly flat part for large radii. The three branches and the two surface densities are clearly labelled in the figure. The critical radius marks the transition between the inner disc regions where the energy transport in the vertical direction, away from the disc mid-plane, is predominantly by radiative diffusion, and the outer disc regions where this energy transport is predominantly by convection. At this point hydrogen, the predominant species in the disc plasma, is only partially ionized. Closer in to the centre ($r \lesssim r_c$) hydrogen is fully ionized, while further out (at $r \gtrsim r_c$) it will be neutral. Figure 4.9 illustrates how the surface density profile of a Shakura–Sunyaev disc varies with the mass accretion rate $\dot{M}$.

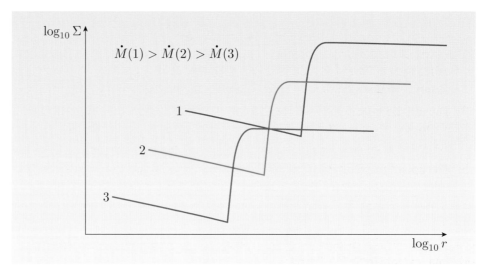

Figure 4.9   The surface density profile for three different values of the mass accretion rate.

In a steady-state disc the local mass accretion rate is the same everywhere in the disc, at all radii. As $\dot{M}$ increases, the overall surface density increases (see Equation 3.37) and crucially the critical radius moves to larger radii in the disc. This becomes clear from the temperature profile

$$T_{\text{eff}}^4(r) = \frac{3GM\dot{M}}{8\pi\sigma r^3}\left[1 - \left(\frac{R_1}{r}\right)^{1/2}\right]. \qquad \text{(Eqn 3.28)}$$

With increasing $\dot{M}$ the disc becomes hotter at all radii, so that the point where the disc temperature equals the critical temperature — signalling the transition between fully ionized and neutral hydrogen — moves away from the accretor.

4.5.1 Generating the S-curve

As we have seen in Equation 4.17, the local stability of the disc at a fixed radius — which we denote by, say, $r = r_S$ — is determined by the sign of the derivative $\partial \dot{M} / \partial \Sigma$, i.e. by the slope of the relation $\dot{M}(\Sigma)$ at this radius r_S. To construct the $\dot{M}(\Sigma)$ relation for a representative radius consider again the generic shape of the surface density profile in Figure 4.8. Note in particular the three branches I, II and III, and the surface densities Σ_A and Σ_B. We have pointed out above that the surface density profile moves up and branch II moves to the right in the Σ versus r diagram as the mass accretion rate increases. This is illustrated in Figure 4.9.

We choose now a fixed reference radius r_S, shown by the *red line* in each of panels 1–5 of Figure 4.10 (overleaf). If we plot a number of surface density profiles for different values of $\dot{M}$, and read off the corresponding surface density Σ at r_S, i.e. where the red line intersects the profile, we can generate the relation $\dot{M}(\Sigma)$ at r_S. Starting with low mass accretion rates, we see that Σ keeps increasing with $\dot{M}$ as long as the red line intersects the surface density profile in branch III. However, as soon as the red line intersects branch II (i.e. when $\Sigma = \Sigma_B$), the intersection moves to lower values of Σ with increasing $\dot{M}$, because branch II is so steep. Once Σ_A is reached, the intersection moves on to branch I, and a further increase of $\dot{M}$ leads again to an increasing value of Σ at the intersection.

The resulting curve in the $\dot{M}$ vs Σ diagram (Figure 4.10, panel 6) therefore has a characteristic 'S' shape with three branches:

1. The lower branch is at small mass accretion rates and extends from small values of the surface density up to Σ_B. As $\dot{M} \propto T^4$, this branch is also called the *cool branch*, or *low-viscosity branch*. The slope $\partial \dot{M} / \partial \Sigma$ is positive, so discs on the cool branch are viscously stable at r_S.

2. The middle branch is at intermediate mass accretion rates and extends from Σ_A to Σ_B. Its slope $\partial \dot{M} / \partial \Sigma$ is negative, hence the disc is locally, at r_S, viscously unstable. Unsurprisingly we call this branch the *unstable branch*.

3. The upper branch, for large mass accretion rates, extends from Σ_A to large values of Σ, and again has a positive slope $\partial \dot{M} / \partial \Sigma$ and is therefore viscously stable at r_S. We call this the *hot branch*, or *high-viscosity branch*.

The $\dot{M}$ vs Σ curve that we have constructed is called the **local S-curve** — *local* because it refers to a specific disc ring at a particular radius (here r_S). Clearly, the horizontal extent (in the Σ-direction) of the unstable branch of the S-curve is larger if the transition between the radiative parts of the disc (branch I) and the predominantly convective part of the disc (branch III) involves a large change in surface density, i.e. if the step in the Σ vs r curve is particularly steep. The vertical height of the step increases if the viscosity parameter α_c in the convective, cool regions of the disc is smaller than the viscosity parameter α_h in the radiative, hot regions of the disc.

● Why is this the case?

○ In a steady-state disc we have $\nu_{vis} \Sigma \propto \dot{M}$, and this is independent of r. So in regions where α is smaller, the viscosity $\nu_{vis} \propto \alpha$ is also smaller, and hence Σ must be larger so that the product $\nu_{vis} \Sigma$ doesn't change. (So far we have always assumed that $\alpha = $ constant, but it is quite plausible that the viscosity mechanism in these very different regimes of density and temperature gives rise to different α parameters.)

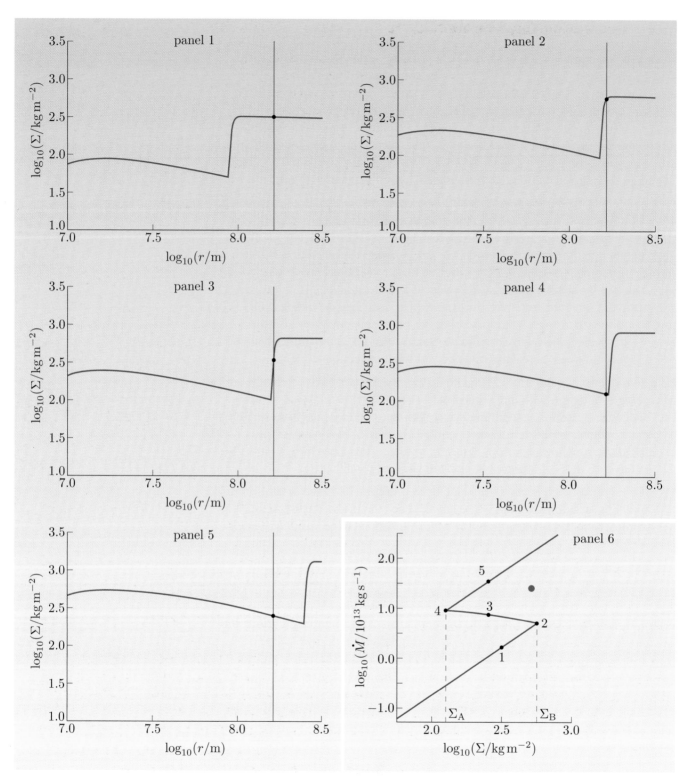

Figure 4.10   The surface density profile for five different values of the mass accretion rate (increasing from panel 1 to 5). Panel 6: The resulting S-curve in the $\dot{M}$ versus Σ diagram.

4.5.2 Local instability and limit cycle

The S-curve in panel 6 of Figure 4.10 was derived from steady-state accretion disc models. Any point *on* the S-curve therefore represents a state of the disc ring at r_S where heating is exactly balanced by cooling, and the disc ring is in thermal equilibrium. Heating is caused by viscous dissipation, while cooling is by either radiation or convection.

Away from the S-curve, the disc ring would be in a state that does not correspond to an equilibrium, steady-state disc. In this case the ring would either cool down if the cooling rate exceeds the heating rate, or heat up if the heating rate exceeds the cooling rate, until thermal equilibrium is established.

More specifically, away from the S-curve the ring state will change on a thermal time, i.e. on a timescale that is much faster than the viscous time on which the surface density of the disc ring might change. If the time evolution of the disc ring state is traced in the $\dot{M}$ vs Σ diagram, then the evolutionary track will be vertically downwards (indicating cooling) if the ring is hotter than the equilibrium state (above the S-curve), or vertically upwards if it is cooler than the equilibrium state (below the S-curve). If the ring state is in between the two stable branches in the surface density range between Σ_A and Σ_B, then the evolution is upwards towards the hot branch if it starts out hotter than the unstable branch, or downwards to the cool, stable branch if it starts out cooler than the unstable branch.

● The green dot in panel 6 of Figure 4.10 indicates a disc ring that is not in thermal equilibrium. In which direction will the disc ring state evolve in the $\dot{M}$ vs Σ diagram?

○ It will evolve vertically upwards, towards the hot branch, as it is cooler than the equilibrium state at the same surface density.

Even if the ring state has reached the S-curve and is then in thermal equilibrium, it will still not necessarily be in a steady state. For this to be the case, the local mass accretion rate needs to be equal to the mass accretion rate elsewhere in the disc, and therefore in practice equal to the mass supply rate into the disc, e.g. from a companion star. This rate is fixed by external processes, and the disc needs to adjust its structure such that the local mass accretion rate equals this value. The disc ring will evolve *viscously* along the S-curve towards this global mass accretion rate. Once this is reached, the disc ring is in a steady state, and remains at a fixed position on the S-curve *if* this fixed position is on a stable branch of the S-curve. Referring to the labels in Figure 4.11 (overleaf), the disc ring will find a steady state on the hot branch if the externally fixed mass accretion rate $\dot{M}$ is larger than $\dot{M}_D$, and on the cool branch if $\dot{M}$ is smaller than $\dot{M}_B$. If, however, the global mass accretion rate falls in between these values, $\dot{M}_D < \dot{M} < \dot{M}_B$, then a stable state doesn't exist! In this case the disc ring will undergo a **limit cycle** evolution.

This comes about because the disc ring perpetually attempts to reach a steady state where there is thermal equilibrium and where the local mass accretion rate would equal the external mass feeding rate of the disc. This steady state is on the unstable branch of the S-curve, so the disc ring cannot settle at this point. The slightest random perturbation would cause the ring to move away from it. Instead, the disc ring is caught in what is called a limit cycle, and effectively circles around the unstable point.

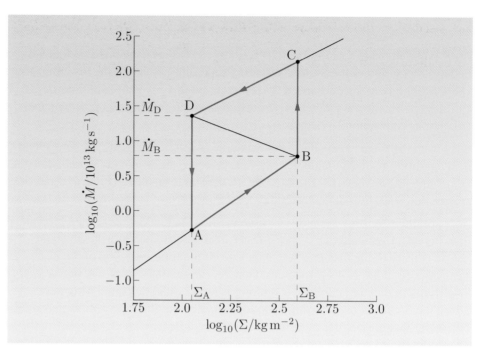

Figure 4.11   Similar S-curve as in Figure 4.10 (panel 6) with limit cycle evolution indicated.

As an example (Figure 4.11), assume that the disc ring starts out with a large $\dot{M}$ and large Σ somewhere on the hot branch, i.e. the disc ring is hot and luminous. Hydrogen is ionized and the opacity is low. The viscosity is large, and the local mass accretion rate is larger than the global mass accretion rate. Therefore the disc ring evolves on the viscous timescale to smaller surface densities — it loses mass. The ring evolves along the S-curve towards point D.

Once the ring state reaches point D, at surface density Σ_A, viscous evolution would lead on to the unstable branch. To move along the unstable branch, the surface density would have to increase, even though the local mass accretion rate is still larger than the global rate. This is of course impossible — all the disc ring can do at this point is to decrease in mass, i.e. its surface density can only decrease. The disc ring avoids this dilemma by cooling rapidly, on a thermal time. It therefore leaves the S-curve and moves at roughly constant surface density towards a smaller local mass accretion rate until it reaches the cool branch of the S-curve at point A.

On the cool branch the disc ring is cool and dim. Hydrogen is neutral and the opacity is high. The viscosity is small, and the local mass accretion rate is smaller than the global mass accretion rate. Therefore the disc ring evolves on the viscous timescale to larger surface densities — it gains mass.

Once the ring state reaches point B, at surface density Σ_B, viscous evolution would again lead on to the unstable branch. As before, this is not possible, and the disc ring makes a rapid transition back into the hot state at roughly constant surface density on a thermal time.

The amount of radiation emitted by the disc ring is proportional to $T_{\text{eff}}^4 \propto \dot{M}$, so the light curve of the disc ring while it is evolving through the limit cycle mirrors the time evolution of the local mass accretion rate. For the limit cycle that we have just discussed, this is shown schematically in Figure 4.12. The resulting light curve is indeed reminiscent of the dwarf nova and soft X-ray transient light curves that we discussed above.

> The dwarf nova phenomenon and soft X-ray transient outbursts can be understood as the limit cycle evolution of an unstable accretion disc that alternates between a hot, high-viscosity state and a cool, low-viscosity state.

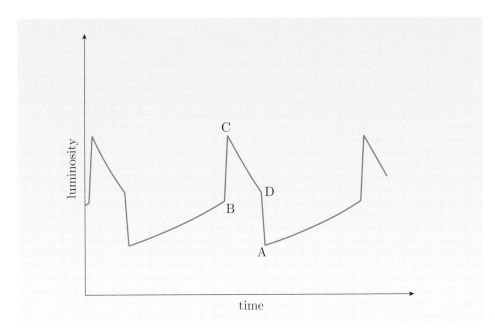

Figure 4.12 Emitted flux versus time for a limit cycle evolution.

However, there is still one piece missing to complete the jigsaw of accretion disc outbursts. So far, we have shown that if a disc ring is unstable, then it undergoes a limit cycle evolution, but if the rest of the disc were to remain stable and persistently bright throughout this limit cycle evolution, then the emission of the stable steady-state disc would completely drown out the luminosity variations of the disc ring. But as we shall see in the next subsection, the accretion disc as a whole does take part in the limit cycle evolution, if only one disc ring is locally viscously unstable.

4.5.3 Global instability

An accretion disc that is locally viscously and thermally unstable can indeed become globally unstable. If a disc ring is locally unstable and undergoes, say, a transition from the cool branch to the hot branch, it will inevitably cause the whole disc (or large parts of the disc) to follow suit. The whole disc will make a transition into a state with high temperature, high viscosity and large local mass accretion rate. This is caused by a heating wave that is launched at the point of instability, and that travels through the disc very quickly, on a fraction of the viscous time. Likewise, if at some point in the disc a transition from the hot state into the cool state is triggered, a cooling wave is launched. The cooling wave travels quickly, on a fraction of the viscous time, through the whole disc (or large parts of the disc), causing the whole disc to go into a cool, low-viscosity state with a small local mass accretion rate.

Figure 4.13 shows the surface density profile $\Sigma(r)$ of the whole disc at different times and illustrates schematically the four main phases of this global disc limit cycle — outburst, transition into quiescence, quiescence, and transition into outburst. The blue and red vertical lines in the figure indicate those two radii, r_1 and r_2, in the disc where the transitions into quiescence (r_2, red line) and into outburst (r_1, blue line) will be triggered. The figure contains two additional straight lines, labelled Σ_A and Σ_B. They represent the critical surface densities Σ_A and Σ_B of the local S-curve and show how Σ_A and Σ_B vary with radius.

● Why are Σ_A and Σ_B increasing with r?

○ At a given r, the values Σ_A and Σ_B denote the lower and upper bounds for the surface density in the steeply rising branch II of the surface density profile of a Shakura–Sunyaev disc (see Figure 4.8). This corresponds to the temperature where hydrogen is partially ionized, i.e. the disc temperature is 6000–8000 K. For small radii, such a low temperature is reached for small values of $\dot{M}$ (see Equation 3.28), and hence correspondingly small values of $\Sigma \propto \dot{M}/\nu_{\mathrm{vis}}$ (Equation 3.23).

In Figure 4.13(a) the disc is in a quasi-steady state, with a very high mass accretion rate everywhere — higher, in fact, than the mass supply rate into the disc. The surface density is larger than Σ_A everywhere, hence the whole disc is on the hot branch. The disc is luminous and hot, and has a large viscosity. This is the outburst state of the disc. As the disc loses mass at a higher rate to the accretor than it gains from the external mass reservoir, the total disc mass is decreasing. The surface density decreases everywhere. The outburst ends when the surface density at r_2 (red line) reaches Σ_A, the upper knee of the S-curve at r_2 (Figure 4.13b).

At this point the disc ring at r_2 makes a transition to the cool state, and launches a cooling wave that propagates through the whole disc. This triggers dramatic changes in the surface density profile (Figure 4.13c). Eventually the whole disc ends up on the cool branch — the disc is dim and cool, and has a low viscosity everywhere. This is the quiescent state of the disc (Figure 4.13d). In quiescence the local mass accretion rate in the disc is everywhere smaller than the mass supply rate into the disc, hence the disc accumulates mass. The surface density increases everywhere, but not necessarily everywhere at the same rate. The quiescence ends when the surface density at radius r_1 (blue line) reaches the critical surface density Σ_B, the lower knee of the S-curve at r_1 (Figure 4.13e). Note that this occurs before the surface density at r_2 (red line) reaches its own knee.

At this point the disc ring at r_1 makes a transition to the hot state, and launches a heating wave that propagates through the whole disc (Figure 4.13f). There is again a dramatic change in the surface density profile, and the disc ends up in the outburst state, as before. The cycle is closed and begins anew (Figure 4.13g).

4.5.4   Link to observations

Exercise 4.6   The accretion disc in a cataclysmic variable is globally viscously stable if hydrogen is fully ionized everywhere in the disc. This is the case when the temperature at the outer disc edge exceeds $T_H \simeq 6000$ K.

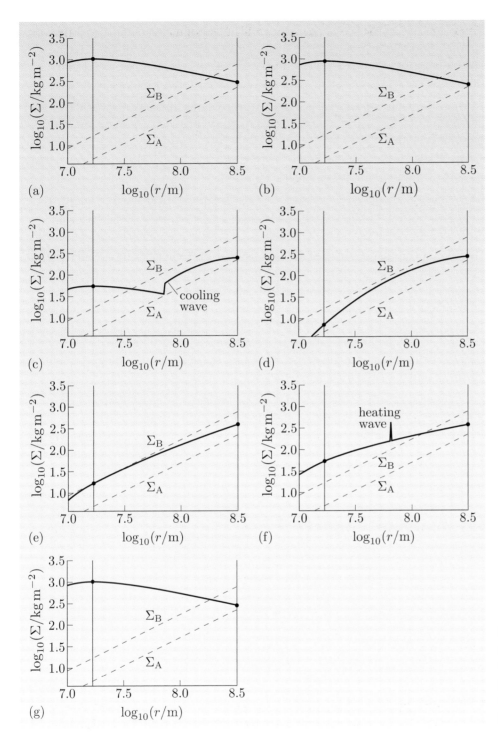

Figure 4.13 The limit cycle evolution affects the whole disc. This is a schematic representation of the evolution of the surface density profile throughout a disc outburst cycle.

(a) Derive an expression for the smallest value of the mass transfer rate $\dot{M}$ in the system so that the disc is still viscously stable, as a function of the orbital period P_{orb}. Assume a mass ratio close to unity, and set the outer disc radius r_{D} as half of the Roche-lobe radius of the accretor.

(b) Determine this limiting value of $\dot{M}$ (in $M_{\odot}\,\mathrm{yr}^{-1}$) for $P_{\mathrm{orb}} = 3\,\mathrm{h}$. ∎

The outburst phase of dwarf novae and soft X-ray transients belongs to the viscous evolution on the hot branch. In outburst, the disc is in a quasi-steady state, with a constant local mass accretion rate at all radii, but this rate is larger than the external mass supply rate (e.g. the mass transfer rate from the donor star). Therefore the disc surface density, and hence the disc mass, decreases slowly. Otherwise the disc is effectively indistinguishable from a steady-state α-disc.

The transition into quiescence is triggered when some disc ring reaches the critical surface density Σ_A and is communicated by a cooling wave. In quiescence, the disc builds up mass on a viscous time. The surface density and temperature are low at all radii; large regions of the disc are likely to be optically thin, so the disc appears very different from a Shakura–Sunyaev model.

An outburst is triggered when a disc ring reaches the critical surface density Σ_B. The local transition to the hot branch triggers a heating wave that causes the whole disc to become hot and luminous.

The duration of the quiescent phase is a measure of the viscous time, and therefore the viscosity, of the disc in quiescence. In cataclysmic variables, the observed ratio of the time spent in outburst and in quiescence implies that the viscosity parameter is smaller in the cold state than in the hot state. Any physical model for the mechanism that is responsible for the viscosity must reproduce this fact. The observed rich phenomenology of dwarf nova outburst light curves and characteristics can be explained by differences in the regions of the discs that take part in an outburst, and by where in the disc the instability is triggered first.

Disc outbursts in soft X-ray transients are additionally affected by the irradiation of the disc from the hot inner disc and the hot accreting object in X-rays, heating the disc and thus suppressing the launch of a cooling wave. In many cases the disc remains in the hot state until most or all of its mass has been accreted onto the neutron star or black hole, explaining why the duration of the outburst and quiescence is so much longer in soft X-ray transients than in dwarf novae.

Summary of Chapter 4

1. Viscosity causes viscous diffusion of the accretion disc surface density. The equation describing viscous diffusion is

$$\frac{\partial \Sigma}{\partial t} = \frac{3}{r} \frac{\partial}{\partial r} \left[r^{1/2} \frac{\partial}{\partial r} \left(\nu_{\rm vis} \Sigma r^{1/2} \right) \right].$$
(Eqn 4.1)

2. An equation of the form

$$\frac{\partial u}{\partial t} \propto \frac{\partial^2 u}{\partial x^2}$$
(Eqn 4.4)

is called a diffusion equation. It describes flow effects that even out spatial gradients. The flow strength scales with the steepness of the spatial gradient.

3. Density enhancements involving sharp gradients diffuse more quickly than smoother density enhancements. The viscous time, the characteristic timescale of the radial drift of matter in the disc, is given by

$$t_{\rm vis} \simeq \frac{r^2}{\nu_{\rm vis}},$$
(Eqn 4.11)

and the radial drift velocity is

$$|v_r| \simeq \frac{\nu_{\rm vis}}{r},$$
(Eqn 4.10)

where ν_{vis} is the viscosity.

4. Hydrostatic equilibrium is established on the vertical sound crossing time $t_z = H/c_{\mathrm{s}}$. The timescale t_z is of order the dynamical time $t_{\mathrm{dyn}} = \omega_{\mathrm{K}}^{-1}$.

5. Thermal equilibrium is established on the thermal timescale t_{th}. In thermal equilibrium, the local cooling by radiation and convection is exactly balanced by the local viscous heating.

6. There is a hierarchy of timescales

$$t_{\mathrm{dyn}} \simeq \alpha\, t_{\mathrm{th}} \simeq \alpha \left(\frac{H}{r} \right)^2 t_{\mathrm{vis}}; \qquad \text{(Eqn 4.15)}$$

i.e. for viscosity parameter $\alpha < 1$, the dynamical timescale is shorter than the thermal timescale, while both are much shorter than the viscous timescale t_{vis}.

7. A steady-state disc in thermal equilibrium is viscously stable if an accidental density perturbation decays away on the viscous time. If the disc is viscously unstable, the perturbation grows and causes the disc to break up into detached rings.

8. The criterion for viscous stability can be written as

$$\frac{\partial [\nu_{\mathrm{vis}}(r)\, \Sigma(r)]}{\partial \Sigma} > 0 \quad \text{or} \quad \frac{\partial \dot{M}(r)}{\partial \Sigma} > 0 \quad \text{or} \quad \frac{\partial T(r)}{\partial \Sigma} > 0$$

(Equations 4.16–4.18).

9. A disc subject to a viscous instability will also develop a thermal instability.

10. Local linear stability analysis considers the reaction of systems to localized, small perturbations.

11. Geometrically thin, optically thick, Keplerian steady-state accretion discs are viscously unstable in the region where hydrogen is partially ionized, i.e. at temperatures around $T_{\mathrm{H}} \simeq 6000\text{–}8000\,\mathrm{K}$.

12. Dwarf nova and soft X-ray transient outbursts can be understood as the result of this thermal–viscous instability. The disc is unstable if at some point in the disc the temperature falls below T_{H}.

13. An accretion disc ring that is subject to a local thermal–viscous instability undergoes a limit cycle evolution, alternating between a hot, high-viscosity state and a cool, low-viscosity state. In the hot state the disc ring loses mass on the viscous time. The transition to the cool state occurs on the thermal time. In the cool state the disc ring gains mass on the viscous time. The transition to the hot state occurs on the thermal time.

14. The local instability launches heating and cooling waves that propagate through the disc, making the disc subject to a global instability. The disc alternates between a hot, bright state with a large mass accretion rate and large viscosity, and a cool, dim state with small accretion rate and low viscosity.

15. In the hot state the accretion disc is in a quasi-steady state. The disc slowly loses mass as the mass accretion rate through the disc is larger than the mass supply rate from the external reservoir. The hot state duration is a fraction of the viscous time in the hot state.

16. In the cool state the disc accumulates mass and slowly evolves on the viscous time of the cold state. The disc may be optically thin, so that the Shakura–Sunyaev description is not an accurate model.

17. The viscosity in the cool state must be much smaller than in the hot state ($\alpha_{\text{cold}} \ll \alpha_{\text{hot}}$).

Chapter 5 Indirect imaging of accreting systems

Introduction

At this point we have achieved a fairly detailed theoretical understanding of accretion discs and the accretion process in cataclysmic variables (CVs), X-ray binaries and, by analogy, active galactic nuclei (AGN). We followed well-founded physical principles and conservation laws to arrive at a model for an accretion disc, i.e. a qualitative and quantitative description of it. But how do we know how good this model is? How accurately does it describe the accretion flow in actual systems out there in the Universe? The only way to find out, and the ultimate test for *any* theoretical model of any physical system, is to check it against observations. Is there any evidence in support of our general picture? Are there any observations in conflict with the model?

In this chapter we consider observational signatures of accreting systems. A particular focus will be various observational techniques that are used to form 'images' of accreting systems indirectly. All compact binary stars and AGN appear as point-like sources of radiation. How then is it possible to 'image' such a system?

The key is that we are able to use *time*-varying flux or spectral information from these sources to infer *spatially*-varying distributions of emitting or absorbing material. In a compact binary star, the brightness and/or spectrum of the object will vary as the two stars orbit each other and present different aspects to our line of sight. Moreover, the orbital periods of compact binaries are often conveniently short (a few hours) to permit study of a complete orbit in a single observing session. The techniques of eclipse mapping and Doppler tomography, that we will discuss below, illustrate this process. In AGN we have the advantage that different regions of interest are sufficiently separated spatially that light-travel time between different components is measurable and can be used to infer the geometry and properties of these structures. The technique of reverberation mapping, discussed later, illustrates the application of such a method to active galaxies.

5.1 Compact binaries as laboratories to study accretion

5.1.1 CVs or X-ray binaries?

In order to study accretion discs observationally, we first need to identify a type of object where the radiation that we observe is emitted primarily from the accretion disc. High-mass X-ray binaries turn out to be unsuitable for this, for two reasons. First, in those systems the dominant sources of radiation are regions very close to the compact object itself (in X-rays) or the donor star (in optical/ultraviolet light), rather than the accretion disc. Second, depending on the nature of the high-mass donor, mass transfer is via a wind rather than Roche-lobe overflow, so that any resulting accretion disc is much less well-defined and smaller than in semi-detached systems. By contrast, in low-mass X-ray binaries and cataclysmic variables, the accretion disc is often the dominant source of radiation, and hence more amenable to study. There is one further complication in low-mass X-ray binaries, namely that a lot of the X-ray emission from the compact object and inner disc is absorbed by the outer disc and the donor star, and subsequently re-emitted in the optical/ultraviolet part of the spectrum, dominating the emission. We are therefore left with cataclysmic variables as a likely class of objects that will enable us to study the accretion flow directly.

Exercise 5.1   Based on what has been discussed in this book so far, summarize the reasons why CVs are likely to be the ideal host systems for the study of accretion discs.

Exercise 5.2   Most known CVs are fairly close, with typical distances of a few hundred parsecs. Here we consider the reason for this.

(a) Calculate the apparent visual magnitude of a typical CV located at a distance of $1000\,\mathrm{pc}$ from the Sun, assuming that the light is dominated by accretion luminosity.

Hint 1: Calculate the accretion luminosity if the white dwarf has mass $1\,\mathrm{M_\odot}$ and radius $R = 8.7 \times 10^6\,\mathrm{m}$, and the mass accretion rate is $10^{-9}\,\mathrm{M_\odot\,yr^{-1}}$.

Hint 2: Calculate the absolute visual magnitude by assuming that the relation between the luminosity and the visual magnitude for this CV is the same as for the Sun. (The absolute visual magnitude of the Sun is $M_V = 4.83$.)

Hint 3: Calculate the apparent visual magnitude, using the distance modulus.

(b) Why could this explain the fact that most CVs are fairly close?   ■

5.1.2   Continuum emission from accretion discs

As we have found in Section 1.4, the continuum emission of an idealized optically thick accretion disc is similar to that of a black body emitter, but the spectrum includes a characteristic power-law branch where $F_\nu \propto \nu^{1/3}$. Would the observation of such a spectral feature not constitute clear evidence for the presence of an accretion disc? Nature, however, is never this straightforward. We have already noted in Figure 1.20 that the extent and therefore prominence of the $F_\nu \propto \nu^{1/3}$ branch hinges on the ratio of outer to inner disc radius.

Exercise 5.3   Make an order of magnitude estimate of the ratio of outer disc radius to inner disc radius, $r_{\mathrm{out}}/r_{\mathrm{in}}$, for a disc in a cataclysmic variable, a neutron star low-mass X-ray binary, and AGN ($10^8\,\mathrm{M_\odot}$).   ■

Comparing the results of Exercise 5.3 with Figure 1.20 shows that discs around white dwarfs and AGN will not display an obvious power-law branch, while discs in LMXBs should. Yet we have just mentioned that in X-ray binaries the outer disc is significantly affected by self-irradiation from the hot inner disc and the central accretor, altering the spectral energy distribution.

Similarly, the thermal continuum emission of AGN is difficult to extract from the actual AGN spectra, as they are contaminated by the host galaxy, added to by prominent emission line regions, and subject to absorption in the interstellar and intergalactic medium. Some researchers believe that the so-called **big blue bump** seen in AGN spectra in the ultraviolet band, at wavelengths short of $400\,\text{nm}$, and the likely high-energy tail of this feature in soft X-rays, represents disc emission. However, this is disputed by others, and nowhere does it have a $\propto \nu^{1/3}$ branch.

So, if the continuum emission is not a good place to look for the existence of accretion discs, then perhaps the line emission will be?

5.1.3 Emission lines from accretion discs

Almost all CVs show strong hydrogen and helium emission lines. Moreover, these emission lines exhibit a periodically changing Doppler shift. The period and magnitude of this changing Doppler shift indicate clearly that the lines emanate from close to the compact object, as it orbits the centre of mass of the system.

Worked Example 5.1

Consider a typical CV with a $0.8\,\text{M}_\odot$ white dwarf orbiting a $0.4\,\text{M}_\odot$ donor star in a circular orbit with an orbital period of $4.0\,\text{h}$.

(a) What would be the maximum orbital speed observed from material in the vicinity of the white dwarf, assuming that the system is viewed edge-on?

(b) What would be the corresponding maximum redshift or blueshift of the hydrogen Hα emission line whose rest wavelength is $656\,\text{nm}$? (See also the box entitled 'Ions and spectral lines' in Section 5.4.)

Solution

(a) Kepler's third law may be written as

$$(a_1 + a_2)^3 / P_{\text{orb}}^2 = G(M_1 + M_2)/4\pi^2,$$

where M_1 and M_2 are the masses of the two stars, a_1 and a_2 are the distances of each star from the centre of mass, and P_{orb} is the orbital period (see Equation 2.3). In this case

$$a_1 + a_2 =$$
$$\left(\frac{6.67 \times 10^{-11}\,\text{N}\,\text{m}^2\,\text{kg}^{-2} \times (0.8 + 0.4) \times 1.99 \times 10^{30}\,\text{kg} \times (4.0 \times 3600\,\text{s})^2}{4\pi^2} \right)^{1/3}.$$

So the separation between the two stars is $a_1 + a_2 = 9.42 \times 10^8\,\text{m}$.

Now, we also know that the mass ratio of the two stars is $q = M_2/M_1 = a_1/a_2$. So in this case $a_1/a_2 = 0.4/0.8 = 0.5$, or $a_2 = 2.0a_1$.

Combining these two results, the distance of the white dwarf from the centre of mass is

$$a_1 = 9.42 \times 10^8\,\text{m}/3.0 = 3.14 \times 10^8\,\text{m}.$$

The speed of the white dwarf in its orbit is therefore $v_1 = 2\pi a_1/P_{\text{orb}}$, and in this case

$$v_1 = (2\pi \times 3.14 \times 10^8\,\text{m})/(4.0 \times 3600\,\text{s})$$
$$= 1.37 \times 10^5\,\text{m}\,\text{s}^{-1} = 137\,\text{km}\,\text{s}^{-1}.$$

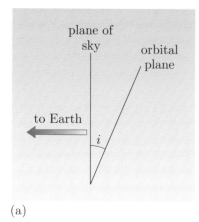

(a)

$i = 0°$

$i = 65°$

(b)

Figure 5.1   (a) The inclination i of a binary orbit is the angle between the orbital plane and the plane of the sky. (b) A binary system viewed at two different inclination angles, $i = 0°$ and $i = 65°$.

This is the maximum speed that would be observed from material in the vicinity of the white dwarf as it orbits around the centre of mass of the system.

(b) The Doppler shift $\Delta\lambda$ of the wavelength is related to the line-of-sight speed $v_\parallel$ by

$$\frac{\Delta\lambda}{\lambda_{\text{em}}} = \frac{v_\parallel}{c}, \tag{5.1}$$

where λ_{em} is the rest wavelength. So in this case the maximum wavelength shift of the Hα line would be

$$\Delta\lambda = (137/3.00 \times 10^5) \times 656\,\text{nm} = 0.3\,\text{nm}.$$

In systems that are viewed close to edge-on, i.e. with a large inclination i (see Figure 5.1), the emission lines are seen to be double-peaked, and significantly broadened (see Figure 5.2 for an example). This structure is exactly what is expected for emission lines that originate in a rotating disc of gas. At any instant we see emission from all parts of the accretion disc, and different regions of the disc will have different rotational velocities with respect to our line of sight, as illustrated in Figure 5.3. The curves superimposed onto the disc in the upper panel of the figure connect points in the disc with the same line-of-sight velocity of the disc plasma (as seen from Earth). Disc plasma within a section of the disc that is bounded by two such lines (such as the shaded areas) has a line-of-sight velocity in between the two velocity values that the two curves correspond to. This disc plasma contributes to the emission in the corresponding velocity bin, shaded in the same colour in the lower panel of the figure. The shape of the resulting Doppler-broadened spectral line emitted by the disc depends on the geometric size of the area that corresponds to a velocity bin, and on the intensity of line emission of the corresponding disc area, which in turn is a sensitive function of temperature and density.

Hence the overall emission line will comprise many different components with a range of Doppler shifts, according to where on the disc the emission originates. Note that the velocity due to motion within the accretion disc is in addition to the velocity of the disc and white dwarf as they orbit the centre of mass of the system, calculated in Worked Example 5.1. The whole of the broad accretion disc emission line, such as the one shown in the lower panel of Figure 5.3, will shift periodically in velocity space (i.e. wavelength) due to the binary orbital motion.

Exercise 5.4   In the upper panel of Figure 5.3, consider a Cartesian coordinate system such that the x-axis is the line that connects the two stellar centres, while the y-axis is parallel (but in the opposite direction) to the line of sight (we assume that $i = 90°$). Explain qualitatively why the pattern on the disc in Figure 5.3 is mirror-symmetric with respect to the x- and y-axes.

Exercise 5.5   (a) Using the coordinate system defined in the previous exercise, determine the function $f(x)$ that defines a curve of constant line-of-sight velocity by $y = f(x)$. Consider the upper right quadrant ($x > 0$, $y > 0$) only.
(i) First, determine the line-of-sight velocity $v_\parallel$ in terms of the azimuth ϕ, accretor mass M, and distance r from the accretor.

(ii) Convert the polar coordinates (r, ϕ) into Cartesian coordinates (x, y), and solve the resulting expression for y.

(b) Determine where this curve intersects the x-axis, i.e. calculate x_0 if $y(x_0) = 0$.

(c) Is there a more direct way to derive the result obtained in part (b)? ■

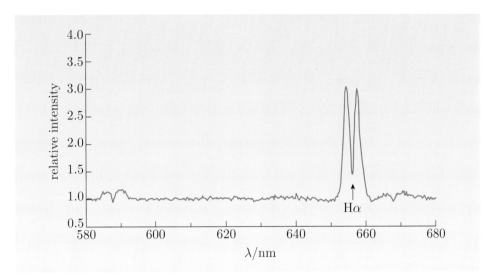

Figure 5.2 The optical spectrum of the cataclysmic variable WZ Sagittae in the wavelength range 580 nm to 680 nm. Wavelength is plotted along the x-axis, and the intensity of the light at this wavelength is plotted along the y-axis. The prominent double-peaked feature labelled 'Hα' is the Hα emission line.

The evidence for the presence of an accretion disc is even stronger in eclipsing systems, where the more extended secondary star moves in front of the white dwarf and disc and blocks out its light.

● What is the inclination angle of eclipsing systems?

○ The line of sight must be close to the orbital plane, so the inclination angle is not much less than 90°.

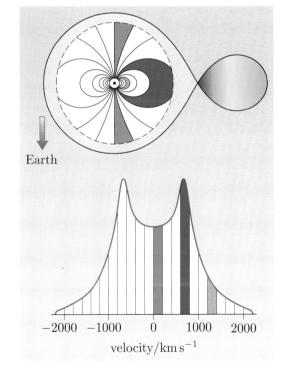

Figure 5.3 The origin of a double-peaked emission line. The upper panel gives a schematic view of the accretion disc residing inside the accretor's Roche lobe at binary phase 0.25 (see Figure 5.4 (overleaf) for a definition of binary phase). The arrow shows the direction to the observer. There is a pattern of curves superimposed onto the disc. The curves connect points in the disc with the same line-of-sight velocity of the disc plasma (as seen from Earth). Disc plasma within a section of the disc that is bounded by two such lines (such as the shaded areas) has a line-of-sight velocity in between the two velocity values that the two curves correspond to. The lower panel shows the profile of a Doppler-broadened spectral line emitted by the disc shown in the upper panel. The spectral line is shown as a function of 'velocity'. This can be translated into the more familiar wavelength scale λ using Equation 5.1. The rest wavelength λ_{em} corresponds to the origin of the velocity axis. Emission in the different velocity bins arises from the regions in the disc that are marked in the same colour.

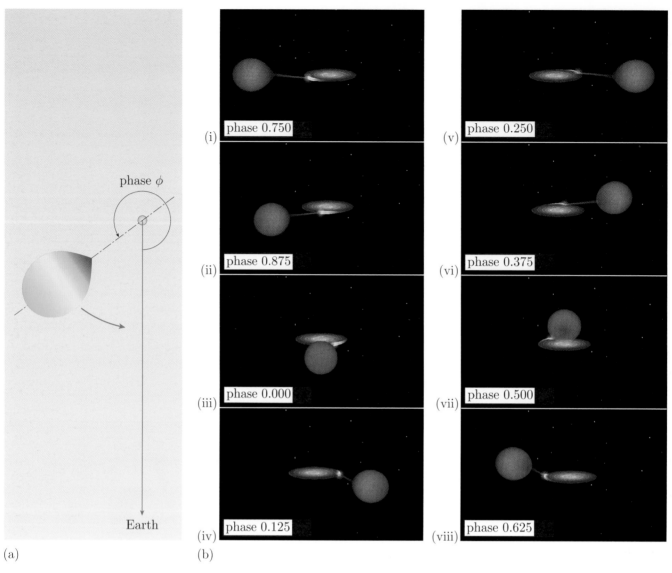

Figure 5.4 (a) The binary phase is the angle, as seen from the primary, between the direction to the secondary star when it is closest to the observer and the instantaneous direction of the secondary star. The binary phase increases in the direction of orbital motion and is expressed as a fraction of $360°$. (b) The binary as seen by the observer for different binary phases, at an inclination of $i = 77°$.

If this is the case, the double-peaked emission lines are seen to lose their blueshifted peaks at the beginning of the eclipse. As the eclipse progresses, the blueshifted peak reappears, as the redshifted peak disappears. Then, finally, the redshifted peak reappears, as the eclipse ends (see Figure 5.5). Observations such as this also show that the material in the accretion disc is orbiting the white dwarf in the same direction as the two stars orbit around their common centre of mass. From the measured velocities of the material in these accretion discs, and assuming the material to be in Keplerian motion, it can be deduced that the accretion disc extends from close to the white dwarf out to almost the edge of its Roche lobe.

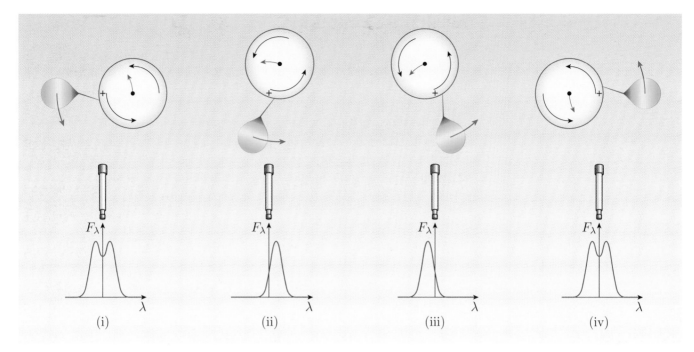

Figure 5.5 The eclipse of a double-peaked emission line in a cataclysmic variable. The side of the disc that is moving towards us is eclipsed first, causing the blueshifted peak of the emission line to disappear. As the eclipse progresses, this side of the disc reappears, and the side of the disc that is receding from us disappears. As a result, the blueshifted peak reappears, and the redshifted peak disappears. Finally, near the end of the eclipse, the receding side of the disc reappears, and so does the redshifted peak of the emission line. (Overall shifts of the emission line due to motion around the centre of mass of the system are not shown.)

Exercise 5.6 (a) Calculate the maximum Doppler shift of the hydrogen Hα line (relative to its laboratory wavelength of 656 nm) emitted from the outer edge of an accretion disc with radius 3.0×10^8 m (i.e. a little less than $0.5\,\mathrm{R}_\odot$) around a white dwarf with mass $0.8\,\mathrm{M}_\odot$.

(b) What is the maximum Doppler shift at the inner edge of the disc (at a radius of 7×10^6 m)?

(Ignore the motion of the accretion disc in its orbit around the centre of mass of the system.) ∎

It is possible to learn a great deal about the nature of a binary star by analysing its orbital light curve — the apparent variability of the emitted radiation as it completes one orbital revolution. The different viewing angle onto the various luminous components in the system as a function of orbital phase (see definition in Figure 5.4) causes the time variability of the observed flux. The light curve can be measured in integral (i.e. white) light, in certain colours (i.e. in restricted but still broad wavelength bands), or in narrow wavelength bands centred on certain spectral lines.

5.2 Eclipse mapping

As we have just seen, the light curve of a binary star becomes particularly interesting when the system is eclipsing. In an eclipsing CV, we can therefore hope to measure the surface brightness distribution across the accretion disc as it is progressively eclipsed by the donor star. However, there is still the problem of additional sources of light in the system, such as the donor star itself, or a **hot spot** where the accretion stream from the donor strikes the outer edge of the disc. The contribution from the donor star may be minimized if we choose to observe short period CVs (e.g. orbital periods of 100 minutes or less) as these will have lower mass (and hence fainter) secondary stars — less than about $0.2\,\mathrm{M_\odot}$.

● How does this upper limit on the donor mass arise?

○ From Equation 2.11 we have $P_{\mathrm{orb}} \simeq 8.8\,\mathrm{h}\,(M_2/\mathrm{M_\odot})$, so $P_{\mathrm{orb}} < 100\,\mathrm{min}$ implies

$$\frac{M_2}{\mathrm{M_\odot}} < \frac{(100\,\mathrm{min}/60\,\mathrm{min})\,\mathrm{h}}{8.8\,\mathrm{h}} = 0.19\,\mathrm{M_\odot} \approx 0.2\,\mathrm{M_\odot}.$$

The hot spot (sometimes also called the bright spot) is the region where the accretion stream from the mass donor star impacts the accretion disc. Kinetic energy of the stream is converted into heat and ultimately radiation, hence the impact region is hotter and more luminous than the rest of the outer disc.

Furthermore, since the hot spot will also present a varying visibility around the orbit, it too produces a characteristic light variation, and this may be disentangled from the light variation due to the rest of the disc. If we choose to observe a system with a high mass transfer rate, the disc will be much brighter than the hot spot, so its contribution may be neglected. Figure 5.6 shows an example of how the overall eclipse light curve of a CV may be decomposed into components that originate from the white dwarf, the hot spot, and the accretion disc.

Exercise 5.7   At what orbital phase is the hot spot brightest?   ■

Eclipse light curves such as the one shown in Figure 5.6 offer the exciting opportunity of generating a map, almost a photograph, of the accretion disc as seen from above the orbital plane. The **eclipse mapping technique** exploits the fact that when the secondary's shadow moves across the disc, it allows us to disentangle the contributions of different luminous parts of the disc from the integral light. Here 'shadow' denotes the disc regions that are hidden from our view because they are eclipsed by the secondary. By comparing the light from the CV when some fraction of it is blocked with the light from the un-eclipsed disc, we can deduce the contribution that the eclipsed region made to the integral light. The shadow effectively probes different zones on the disc and therefore provides additional information needed to construct a two-dimensional map from a point source. Figure 5.7 (overleaf) shows how the probing shadow moves across the disc in a system with inclination $78°$ and mass ratio 0.3.

Key to this method is the fact that shorter wavelength light predominantly comes from inner (hotter) regions of the accretion disc, while longer wavelength light predominantly comes from outer (cooler) regions of the disc (see the temperature profile of accretion discs given by Equation 3.28). Hence we should see a deep, narrow eclipse at short wavelengths, but a shallower, broader eclipse at longer wavelengths. By observing eclipses as a function of wavelength, one can therefore build up a model of the radial temperature distribution of the accretion disc. Observations indicate that a relationship of the form $T(r) \propto r^{-3/4}$ provides a reasonable fit to the data in systems with bright discs.

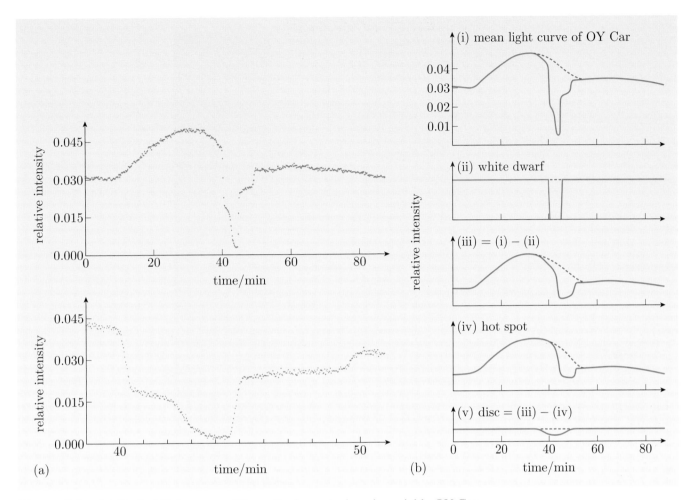

Figure 5.6 (a) Optical light curve of the eclipsing cataclysmic variable OY Car, showing a hot spot hump. (b) Decomposition of the eclipse light curve into its components: white dwarf, hot spot and accretion disc. The dashed curve shows how the light curve would appear in the absence of an eclipse.

Exercise 5.8 How does the length of the donor's shadow in the middle panel of Figure 5.7 depend on the inclination? ■

The eclipse mapping technique is a rather indirect method and requires a good deal of computer simulation to construct an image that is consistent with the observed eclipse light curve. In fact, one usually starts with a guessed model image and calculates how the light curve would look if the model were a perfect representation of the disc. The model light curve from this first guess is then compared to the actually observed light curve. If model and data do not agree, then the constructed model image is modified, and a new model light curve obtained. The correction procedure is repeated until the match between model and real light curve is satisfactory. The quality of fit in this iterative, step-by-step inversion method is determined by a statistical quantity that measures the overall deviation between data and model. Quite often this is the so-called χ^2 **value** ('chi-squared' value).

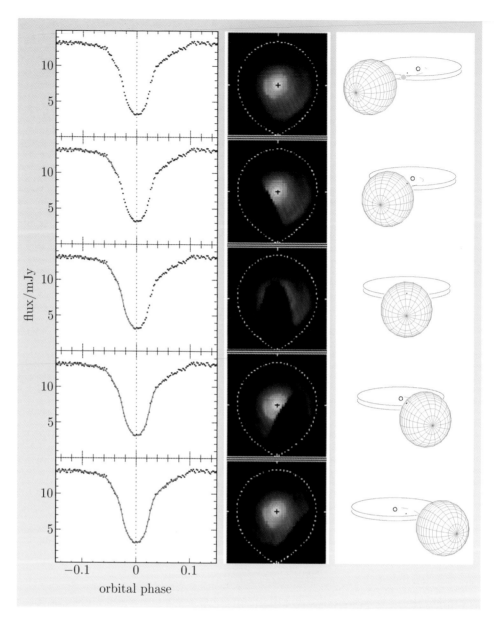

Figure 5.7 Simulated eclipse of an accretion disc, from binary phase -0.08 to binary phase $+0.08$. The right-hand panel shows how the binary star would appear from Earth if we had a powerful enough telescope to resolve it. The middle panel is a view from above the disc plane, with the Roche lobe of the accretor and the shadow of the secondary star indicated. The left-hand panel shows the actually observed light curve (dots) and the model light curve (red line), reconstructed from the assumed brightness distribution in the accretion disc as shown in the middle panel. The red model light curve is drawn only up to the respective binary phase. (Courtesy of Raymundo Baptista.)

χ^2 test

Assume that an observation determines the values of $N + 1$ variables as $x_0, x_1, \ldots, x_N$ (e.g. a light curve with photon counts x_i per time bin Δt_i), while a theoretical model predicts that the values should be $\mu_0, \mu_1, \ldots, \mu_N$.

Then the quantity

$$X^2 = \sum_{i=0}^{N} \frac{(x_i - \mu_i)^2}{\sigma_i} \qquad (5.2)$$

(X^2 is sometimes also written as $X^2 = \sum_{i=0}^{N}(x_i - \mu_i)^2/\mu_i$) is a measure of how well the model reproduces the observed data. Here σ_i is the uncertainty of the observationally determined value x_i.

X^2 is the sum of the squared differences between model and observations, weighted by the corresponding uncertainties. The weighting is such that a given discrepancy between model and measurement for a value with a small uncertainty makes a larger contribution to X^2 than in the case of a large uncertainty.

If the measurement of the $N + 1$ variables is repeated many times, then each time a new X^2 value can be calculated. Due to the statistical nature of the uncertainties, each new value of X^2 will be slightly different. For many repeated measurements, a distribution of X^2 values will emerge. If the number N is large, and for certain constraints on the nature of the measurements, this distribution of X^2 is approximately given by a well-known theoretical function called a χ^2 distribution. (Note that our notation distinguishes between the measured quantity X^2 and the theoretical distribution χ^2, a distinction not usually made in the astronomical literature.) Crucially, the χ^2 distribution has a mean value of N.

Therefore, if a given measurement of the $N + 1$ variables results in a value of $X^2/N \approx 1$, i.e. X^2 is close to the mean value of χ^2, then this indicates that the model is a sensible representation of the observations. Conversely, the model does not fit the data well if the value of X^2/N is much larger than 1, while a value of X^2/N that is much smaller than 1 indicates that the assumed uncertainties σ_i are probably overestimated.

The one-dimensional light curve does not in general uniquely constrain the two-dimensional brightness distribution in the disc. Hence the fitting procedure must impose an additional constraint, the closeness of the model to a default image. The default image represents an educated guess of the final image and should contain all essential features that one would expect to find, e.g. axis-symmetry except for a bright spot, and a radial brightness gradient. The fitting technique finds the model image that is most like the default image but still consistent with the data. The closeness of default and model image is measured by a numerical quantity called the **image entropy** (*not* the entropy of a gas, as used in thermodynamics!), and the fitting involves maximizing this quantity. Similar **maximum-entropy** imaging techniques are widely used in the reconstruction of images from incomplete data.

The image entropy is a quantity defined in the field of information theory, where it is known as information entropy, or information uncertainty. It measures the average uncertainty, or 'surprise', of the outcome of a set of random variables (here the intensity of the pixels in the reconstructed map of the system). Information theory postulates that among all the possible probability distributions (i.e. velocity maps) that are consistent with the assumed (image) constraints, the one with the largest average uncertainty is the actual probability distribution.

5.3 Doppler tomography

A related and perhaps even more powerful indirect imaging technique is **Doppler tomography**. The phase-dependent velocity information in emission lines replaces the shadow as a probe of different zones in the binary. As we have seen above in the context of double-peaked emission lines, the wavelength offset from the laboratory wavelength measures the radial velocity of the emitting material (Equation 5.1).

Consider the emission of a point-like source that is fixed in the co-rotating binary frame, e.g. a certain point on the surface of the Roche-lobe filling donor star. The radial velocity of this point would change sinusoidally with orbital phase, hence the corresponding emission line would trace out a sinusoid — an **S-wave** — when plotted against binary phase (Figure 5.8). This representation is also referred to as a *trailed spectrum*.

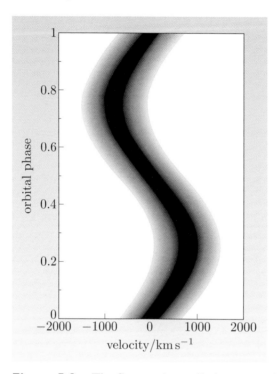

Figure 5.8   The S-wave (or trailed spectrum) from a luminous spot fixed in the binary frame is obtained by plotting the spectrum horizontally, as a function of orbital phase in the vertical direction (wavelength is expressed as velocity, intensity indicated as brightness; the image is a 'negative', i.e. the darkest regions indicate the highest intensity). The centre of the spectral line traces out a sinusoid.

If there is more than one luminous spot in the system, the resulting phase-dependent spectral line is a superposition of many individual S-waves, one for each point. The amplitudes of these individual S-waves, and the binary phase at zero velocity, depend on the location of the corresponding spot in the system. For example, a disc annulus can be thought of as the superposition of luminous points that all trace out an S-wave with the same amplitude, but each with a slightly different phase at zero velocity.

Figure 5.9 shows how certain features of a semi-detached binary as seen in the co-rotating binary frame correspond to features in velocity space, i.e. a coordinate system where the velocity in the y-direction is plotted against the velocity in the x-direction. Worked Example 5.2 explains this figure.

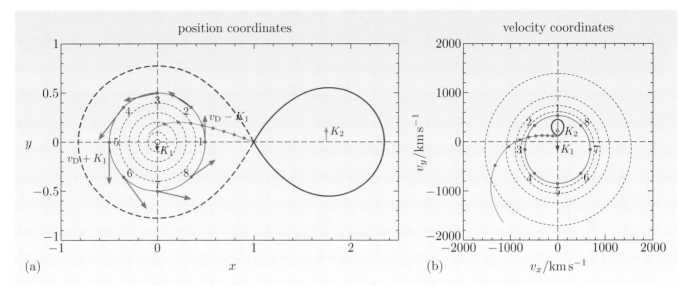

Figure 5.9   Position coordinates and velocity coordinates in a semi-detached binary with accretion disc. Worked Example 5.2 provides a detailed explanation of this figure.

Worked Example 5.2

(a)   Explain how the point labelled '1' in Figure 5.9a relates to the point labelled '1' in Figure 5.9b.

(b)   What is the physical significance of the point with velocity coordinates $(v_x, v_y) = (0, 0)$?

Solution

(a)   Figure 5.9a shows the orbital plane (the xy-plane) of an interacting binary with the familiar Roche lobes. The circles around the accretor (in the left lobe; star 1) indicate the orbital motion of plasma in a Keplerian accretion disc around star 1. The red arrows along the circle with various points labelled by numbers indicate the velocity of the plasma flow, as seen from Earth. The velocity v at each of the labelled points is the sum of two vectors,

$$v = v_D + K_1,$$

where v_D is the velocity of the plasma on its Keplerian disc orbit, i.e. with magnitude (Equation 1.5)

$$|v_D| = v_K = \sqrt{\frac{GM_1}{r}}$$

(where M_1 is the mass of the accretor and r is the radius of the disc orbit) and direction tangential to the circle. The vector K_1 is the velocity of star 1 due to its binary orbital motion and, for the binary phase of 0.25 shown in Figure 5.9, points straight at the observer.

Figure 5.9b shows the same orbital plane, but in velocity coordinates. Consider a certain point in the orbital plane with coordinates (x, y) in

Figure 5.9a. Suppose that the stellar plasma at this point moves with velocity $\boldsymbol{v} = (v_x, v_y)$, where v_x is the x-component and v_y is the y-component of $\boldsymbol{v}$. Then this point will appear as a point in Figure 5.9b at a position given by these velocity components, i.e. at (v_x, v_y).

At point 1, the velocity $\boldsymbol{v}_D$ is in the positive y-direction, i.e. opposite to the direction of $\boldsymbol{K}_1$. Accordingly, the vector sum $\boldsymbol{v} = \boldsymbol{v}_D + \boldsymbol{K}_1$ has a vanishing x-component ($v_x = 0$), while the y-component is just the difference of the magnitudes of $\boldsymbol{v}_D$ and $\boldsymbol{K}_1$, i.e. $v_y = v_D - K_1$. Therefore point 1 reappears in Figure 5.9b at the point with coordinates $(v_x, v_y) = (0, v_D - K_1)$. As $v_D \gg K_1$, we have $v_y > 0$.

(b) The only point that does not move with respect to the observer is the centre of mass. This point would appear at $(v_x, v_y) = (0, 0)$. (This assumes that the binary system as a whole does not have a velocity relative to the observer.)

● Where does the accretor appear on the velocity map?

○ The white dwarf moves in the negative y-direction with speed K_1. So its velocity coordinates are $(v_x, v_y) = (0, -K_1)$. The white dwarf appears in the centre of the circle that connect the points labelled 1 to 8.

In practice, the luminous surfaces in a binary are not point-like but extended. But they can be thought of as a superposition of a large number of luminous spots. The Doppler tomography technique reconstructs the brightness distribution over the surfaces that these points make up from the fine structure in the observed broad S-wave.

This imaging technique is also an inversion method, and maximum-entropy methods are applied abundantly. Usually one constructs an image in velocity space first, and derives a modelled trailed spectrum from it. This is then compared against the observed trailed spectrum. The image in velocity space is modified until the model trailed spectrum satisfactorily fits the observed data. The translation into real space follows only after this is achieved. It is sometimes easier to discuss the results only in velocity space, as this is independent of uncertainties and imperfections of the translation.

The situation becomes somewhat more involved when the emitting gas is moving in the binary frame. Of course, this is the case for accretion disc material on Keplerian orbits around the primary star. It is still possible to construct an image in velocity space, as before, with all the complications and problems attached to it. The translation into real space is even more complicated as it introduces further ambiguity. The same velocity pattern can arise from emitting gas at very different locations in the binary, with very different velocities with respect to the binary frame. If the construction of the velocity map is complicated and involved, the interpretation of the map can be an art!

Exercise 5.9   The velocity map of an accretion disc turns the disc 'inside-out'. Explain this statement.   ■

Medical imaging: the CT scanner

The imaging techniques discussed in this and the preceding section — eclipse mapping and Doppler tomography — are reminiscent of a familiar medical imaging technique: the use of a computed tomography (CT) scanner. A CT scan makes use of the attenuation of X-rays that pass through the patient: different parts of the body are more or less transparent to X-rays, causing a pattern of varying shades on the X-ray image of the body. A CT scan involves the imaging of a succession of two-dimensional slices through the body (*tomos* is Greek for 'slice') to build up three-dimensional information of the body (see Figure 5.10).

Each two-dimensional slice of the CT scan is exposed to a probing X-ray beam that rotates around the body. In this way a 'projection' of the slice, i.e. an edge-on view of the slice in X-rays, can be obtained from different angles. By combining the information contained in the different projections, the full two-dimensional image of the slice can be reconstructed by a computer.

This is in close analogy to the eclipse mapping technique where the probing shadow of the secondary moves across the disc due to the binary motion, and the binary light is registered as a function of orbital phase. The two-dimensional disc structure is reconstructed from the information contained in the light curve.

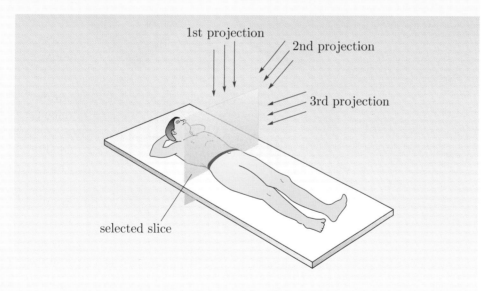

Figure 5.10   The principle of a CT scanner in medical imaging. X-ray beams are passed through a slice of the body in different directions.

A remarkable success story of the Doppler imaging technique came in 1996 with the discovery of a spiral shock pattern in the accretion disc of IP Pegasi (Figures 5.11 and 5.12 overleaf). These spiral shocks are a consequence of the gravitational perturbations from the secondary star on the accretion disc. (See Subsections 7.3.2 and 7.3.3 for a general discussion on shocks in fluids.)

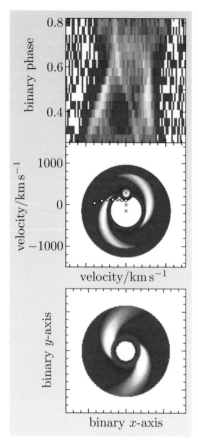

Figure 5.11 A synthetic (model) spiral shock pattern in an accretion disc. *Lower panel*: In the binary frame (position coordinates). *Middle panel*: In velocity space. *Upper panel*: The corresponding S-wave pattern, the so-called trailed spectrum. The crosses in the middle panel denote the centre of the donor star, the centre of mass, and the centre of the white dwarf (from top to bottom). The open circles, connected by a curve, show the location of the mass transfer stream.

In our theoretical considerations in Chapters 3 and 4, we neglected any effects that the secondary might have other than the restriction on the size of the disc imposed by the primary's Roche lobe. Spiral shocks have been predicted by numerical calculations of accretion discs. The shocks are roughly stationary in the co-rotating binary frame.

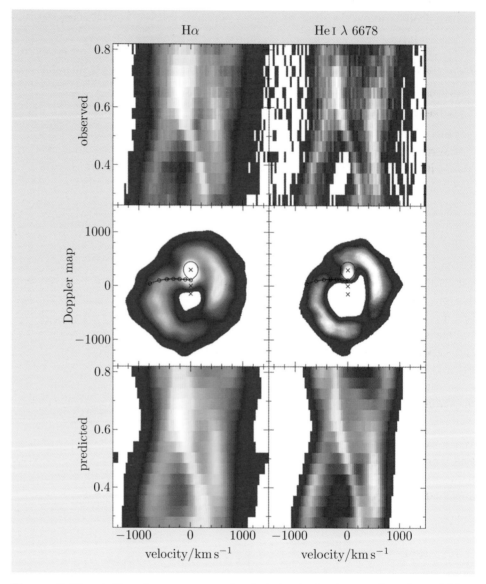

Figure 5.12 A Doppler tomogram of spiral shocks in the cataclysmic variable IP Pegasi. *Left*: Hα emission line. *Right*: He I line at 667.8 nm. The upper panel shows the actually observed trailed spectra (S-wave pattern); the middle panel shows reconstructed images in velocity space. The 'predicted' trailed spectra in the lower panel are obtained from this best fit and serve to check the method.

We now turn to indirect imaging techniques in accreting systems on a much larger scale: the broad-line region of AGN, which is located above and below the AGN accretion disc, and is enclosed by the obscuring dust torus (see Figure 1.13).

5.4 The broad-line region in AGN

Broad emission lines are prominent in the infrared/optical/ultraviolet regions of many AGN spectra, but their widths show substantial differences from one object to another. Broad lines typically have full width at half maximum of around $5000 \, \text{km s}^{-1}$, but even in a single spectrum, different lines originating in different atomic species can have different widths. In addition, many broad lines in AGN spectra can be blended together, such as Lyα λ 1216 and NV $\lambda\lambda$ 1239, 1243 or the so-called *small blue bump* comprising the Balmer continuum and many Fe II lines (see the Box 'Ions and spectral lines' in this section).

● If the width of the broad lines in AGN spectra is due to **thermal broadening**, what temperature hydrogen gas is implied by a line width of $5000 \, \text{km s}^{-1}$?

○ We can equate the average kinetic energy of a hydrogen atom with the thermal energy per particle, $\frac{1}{2} m_{\text{p}} v^2 \simeq \frac{3}{2} kT$ (where m_{p} is the proton mass). So in this case, the implied temperature would be

$$T \simeq \frac{m_{\text{p}} v^2}{3k} = \frac{1.67 \times 10^{-27} \, \text{kg} \times (5.0 \times 10^6 \, \text{m s}^{-1})^2}{3 \times 1.38 \times 10^{-23} \, \text{J K}^{-1}}$$
$$\approx 10^9 \, \text{K}.$$

The estimate of a billion Kelvin for a thermal width of $5000 \, \text{km s}^{-1}$ is a very high temperature. A gas at this temperature would be extremely highly ionized, so that the spectral lines found in the optical region of the spectrum would not be present. In particular, hydrogen would be completely ionized at this temperature, so the Balmer series would not be emitted. In fact, the intensities of the broad lines in AGN spectra indicate gas temperatures of order 10^4 K, which implies a thermal velocity of only around $10 \, \text{km s}^{-1}$ for the atoms. This is too low to account for the broad-line widths, which must instead be due to bulk motions of individual line-emitting clouds, as illustrated by Figure 5.13 (overleaf). The region of the AGN in which these clouds are found is called the **broad-line region** (BLR).

When examined closely, it becomes apparent that the fluxes of broad emission lines vary with time and are correlated with flux variations of the continuum. This indicates that the lines arise in clouds that are optically thick to the ionizing continuum emission emanating from the central source, the accreting compact object.

Ions and spectral lines

Atoms are characterized by discrete electron energy levels, and transitions of electrons between these levels give rise to the emission of spectral lines. As hydrogen is the most abundant chemical element in the Universe, its spectral lines are particularly important in astrophysics. The **Balmer series** consists of the transitions in which the $n = 2$ energy level is the lower level. This series of lines appears in the optical wavelength region, and the **Hα line** (transitions between $n = 2$ and $n = 3$) is often the most prominent line in optical spectra. The highest energy transitions in the hydrogen line spectrum are those to and from the lowest energy level, i.e. transitions to and from $n = 1$. This series of lines is called the **Lyman series**, and the **Lyα line** (transitions between $n = 1$ and $n = 2$) is prominent in ultraviolet spectra.

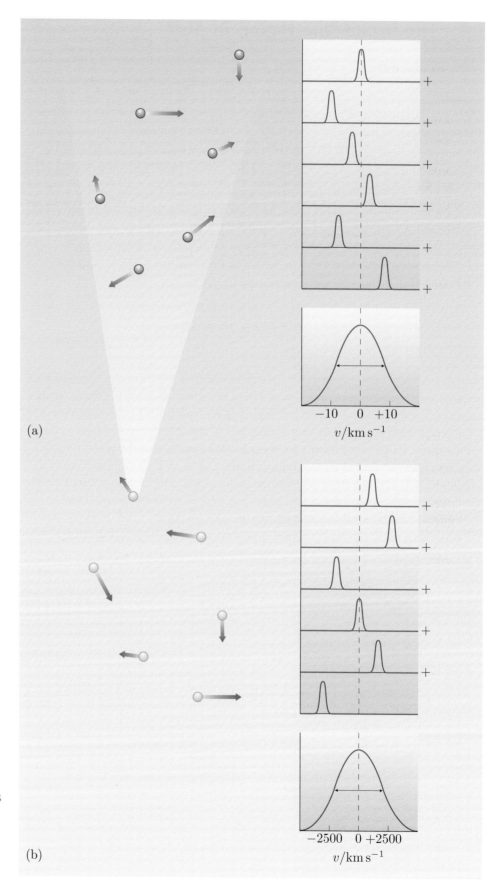

Figure 5.13 (a) Thermal Doppler broadening due to the random motions of gas molecules at a particular temperature can give rise to line widths of a few tens of kilometres per second. (b) Doppler broadening due to the random motions of gas clouds can give rise to much larger Doppler shifts and hence much broader lines. Each cloud in (b) has a narrow-line profile due to the thermal motions of its constituent molecules. The clouds have large velocities, leading to large Doppler shifts and hence a broad-line profile.

Ions and spectral lines (continued)

The **Balmer limit**, which occurs at 364.6 nm, corresponds to transitions between the $n = 2$ level of hydrogen and unbound states, i.e. this is the ionization transition from the $n = 2$ level. Similarly, the **Lyman limit** at 912 nm corresponds to ionization from the ground state ($n = 1$). Consequently, radiation with $\lambda < 912$ nm is known as **ionizing radiation**. Photons with energies just exceeding the Lyman limit are highly prone to absorption by neutral hydrogen gas; consequently, the plane of our own Milky Way galaxy is essentially opaque at wavelengths just short of the Lyman limit.

Generally, astronomers label spectral lines using notation like C IV $\lambda\lambda$ 1548, 1551, which concisely gives an enormous amount of information. Going through this piece by piece:

'C' indicates the chemical element, carbon in this example.

'IV' indicates the ionization state. 'I' indicates the neutral atom, with successive roman numerals indicating successive positively charged ions, hence 'IV' indicates that three electrons have been removed. The C IV ion is C^{3+}; the roman numeral is always one more than the number of positive charges.

'λ' indicates that the following numerals give wavelength. In this case we have '$\lambda\lambda$', which means that the spectral line is a **multiplet**. In the case of a multiplet there are two or more 'fine structure' sub-levels to one or both of the energy levels involved in the transition, so that there are multiple components to the spectral line, corresponding to the various possible energy differences between the initial and final states. This is illustrated in Figure 5.14. In our example, the line is a **doublet**, having two components. **Triplets** and higher multiplets are also possible.

'1548, 1551' gives the wavelength in Å of the two components of the doublet. (10 Å $= 1$ nm, so these wavelengths are 154.8 nm and 155.1 nm.)

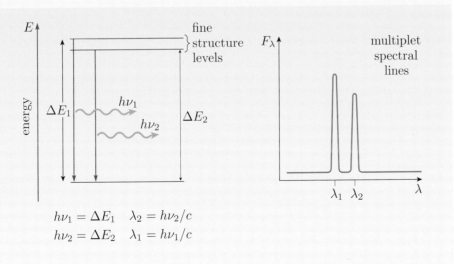

$$hv_1 = \Delta E_1 \qquad \lambda_2 = hv_2/c$$
$$hv_2 = \Delta E_2 \qquad \lambda_1 = hv_1/c$$

Figure 5.14   Fine structure sub-levels in initial or final energy levels lead to a multiplet spectral line.

Ions and spectral lines (concluded)

All spectral lines arise as a result of electronic transitions that are governed by quantum mechanical selection rules. Most common lines correspond to transitions that are **permitted** by the selection rules. For reasons that we shall not explore here, it is also possible (with low probability) for transitions to occur that do not obey all of the selection rules. Such transitions are called **forbidden** or **semi-forbidden lines**, depending on which of the selection rules are violated. The astrophysically important point is that permitted transitions are, in general, much more likely than forbidden or semi-forbidden transitions. This means that if an excited atom has a permitted transition that it can make to a lower level, then it is not likely to make a non-permitted transition. Consequently, the only forbidden and semi-forbidden lines that are observed are those where no permitted transition is available. Furthermore, because non-permitted transitions have a low probability of occurrence, an excited atom will remain in the excited state for a long time before the transition occurs. During this time, a collision with another atom, ion or free electron may occur, and **collisional de-excitation** will result. Hence forbidden and semi-forbidden lines are observed only from low-density regions, where collisions are relatively infrequent.

The notation used to indicate a forbidden line is a pair of square brackets around the chemical element and ionization state, e.g. [O II]. Similarly, a semi-forbidden line is indicated by just the closing bracket, e.g. N IV].

In AGN broad-line spectra, virtually all forbidden lines are subject to **collisional suppression**. This indicates that the electron densities of the broad-line region are high. The absence of the [O III] $\lambda\lambda$ 4363, 4959, 5007 lines in AGN broad-line spectra indicates a lower limit to the electron density of around $10^{14}\,\mathrm{m}^{-3}$, while the only strong semi-forbidden line seen is C III] λ 1909, implying an upper limit of order $10^{17}\,\mathrm{m}^{-3}$.

Despite being able to learn about the bulk properties of the broad-line region from studying the emission lines it produces, line profiles from broad-line regions cannot give us definitive results for the structure and dynamics of broad-line regions. In this section we shall introduce reverberation mapping, another *indirect imaging* tool that can, in principle, allow us to make some progress with this problem. We begin with an Earth-bound analogy, and then examine a particularly simple extragalactic astrophysics example, before considering AGN.

5.4.1   The echo-mapping idea

A submarine's sonar equipment uses sound reflections — echoes — to locate objects underwater. In a similar way, **reverberation mapping** uses **light echoes** to probe the reprocessing structures surrounding the central source in AGN.

In the case of sound echoes, it is easiest to interpret the reflected signal if the initial sound is a short sharp pulse, like a hand clap or the intense pulses (pings) produced by sonar. In this case the reflected signal will also be a short sharp pulse, which will be delayed by the sound travel time difference between the direct and reflected signals.

Figure 5.15a shows the principle used in sonar. By measuring the time delay τ between the clap and its echo, the distance to the cliff d can be determined from $2d = c_s\tau$, where c_s is the sound speed. More generally, a similar method could be used by two people, one providing the sound, and the other measuring the sound travel time delay between the direct and the reflected signal, as shown in Figure 5.15b.

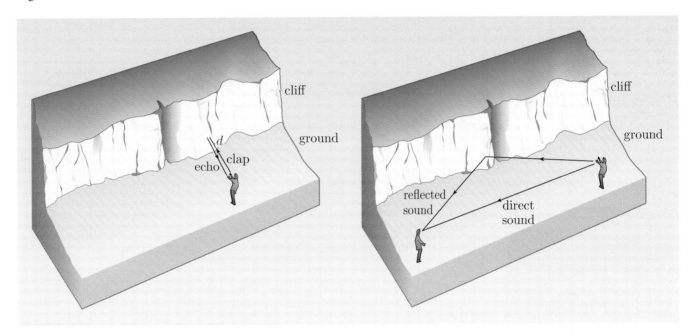

Figure 5.15 (a) A person claps his hands near a cliff. He hears two claps: the first is immediate, the second is delayed by the time taken for sound to travel from him to the cliff, where the sound is reflected, and back. If he measured the time delay, then knowing the speed of sound, he could calculate the distance to the cliff from the sound travel time delay. This is the basis of the method used in sonar. (b) The sound travel time delay between the direct and the reflected sound depends on the difference in the lengths of the two paths travelled.

Light echoes are also easiest to interpret if the initial signal is a short sharp pulse of light. We illustrate how this method works in an astronomical context by applying it to the light generated by a supernova explosion. Compared to the light travel time across even a small galaxy, the time taken for a supernova to brighten and fade is short. Consequently, one of the most impressive applications of reverberation mapping so far is its application to the study of the region surrounding SN 1987A, which is in a nearby satellite galaxy of the Milky Way, called the **Large Magellanic Cloud** (LMC).

● Why this particular supernova?

○ Because this is the closest one that has been discovered in the era of modern astronomy. More distant supernovae have low apparent brightness, and the spatial resolution of optical telescopes is too low to be able to detect their light echoes.

Figure 5.16 illustrates the geometry of the light echoes from SN 1987A. The light echoes shown in Figure 5.17 are caused by light reflected by the interstellar material around the supernova. In principle, the geometry is as drawn in Figure 5.15b, but of course the distance to the Earth is much greater than any of the distances depicted in Figure 5.15b. For any given time delay there is a whole family of points in the LMC that could possibly have caused the reflection. As can be seen in Figure 5.16, this family of points falls on a parabola. In fact, in three dimensions the family of points follows the shape formed by rotating the parabola around the axis provided by the line joining the Earth and the supernova. This three-dimensional surface is called a **paraboloid**. Different values of τ correspond to a series of nested paraboloids, as shown in Figure 5.18.

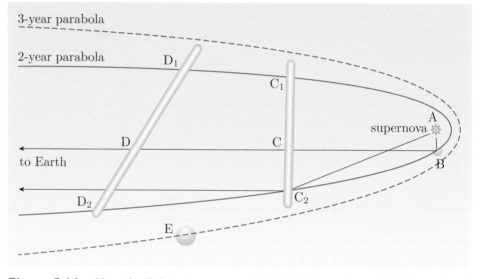

Figure 5.16   How the light echoes around SN 1987A are formed. Light emitted by the exploding supernova travels at speed c in all directions. If this light illuminates clouds of dust, an observer on Earth will see the dust glow. Because the distances between the supernova at A and the dust clouds are large, there is a time delay of months or years between the light that travels directly from the supernova and that which is seen by illumination of dust clouds. The figure shows two parabolas: the inner parabola shows the positions at which the time delay would be 2 years, the outer parabola corresponds to a 3-year time delay. Hence an image taken 2 years after the supernova will reveal bright spots (B, C_1, C_2, D_1, D_2) wherever dust lies on the 2-year parabola; similarly, 3 years after the supernova, the dust lying on the 3-year parabola (at E) is revealed. By watching the continuously changing pattern of reflected light, astronomers have been able to learn about the three-dimensional distribution of dust in the Large Magellanic Cloud.

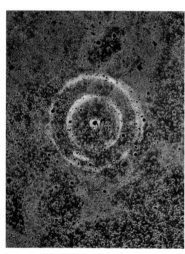

Figure 5.17   The bright rings are light echoes from SN 1987A. The mottled appearance is because an image of the same area of the LMC without the light echo has been subtracted, to remove most of the light due to stars. Because the image qualities of the two images are not identical, mottling is created when one image is subtracted from the other. Each of the bright rings is caused by light reflected by a sheet of dust, as shown in Figure 5.16.

Reflecting material at any point on the paraboloid will cause a light echo to be observed at time τ after the time when the supernova itself was seen. By combining the position on the sky of the light echo with the time τ, we can locate the reflecting material's three-dimensional position in space. Note that this is a completely new piece of information — usually astronomers can definitely measure only the direction to objects, and have almost no direct information on relative distances.

5.4.2 Reverberation mapping applied to AGN

We have seen that light echoes can, in principle, lead to novel information on the relative positions of structures in distant objects. Consequently, it could be beneficial in solving some of the mysteries about the structure of AGN, particularly the nature of the broad-line region.

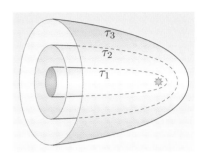

Figure 5.18 The surfaces corresponding to differing time delays τ are nested paraboloids.

● Why is the broad-line region in particular suitable for reverberation mapping?

○ Because it is easy to isolate light that comes from this region (i.e. the broad emission lines), and the emission line fluxes vary with time in a way that is highly correlated with the continuum fluxes, suggesting that they may be powered by absorbing and reprocessing the illuminating continuum.

● Suggest two reasons why the application of reverberation mapping to AGN may be more difficult than its application to SN 1987A.

○ First, the illuminating continuum is not a short sharp pulse of light; instead, it is continuously varying. Second, we can't spatially resolve the broad-line region in the way that we can spatially resolve the dust clouds in the LMC.

The basic method used in reverberation mapping of AGN is illustrated in Figure 5.19.

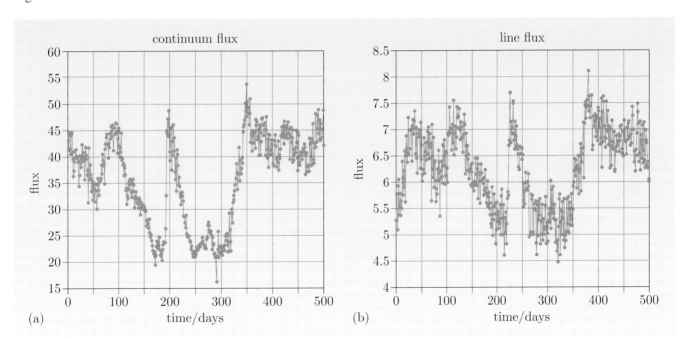

Figure 5.19 The continuum light curve (a) and the light curve of the line emission (b). Just by examining them visually, we can see that they seem to have similar shapes.

The light curves of both continuum light and broad emission lines are observed. The continuum light from the central source illuminates the broad-line region, causing photoionization that powers the line emission. Sharp changes in the luminosity of the central source appear directly in the continuum light curve, and cause changes in the line emission. Consequently, blurred and delayed echoes of the features in the continuum light curve should appear in the emission line light curves. By measuring the typical delay time τ between the direct feature in the continuum light curve and the delayed feature in the light curve for a particular emission line, we can estimate the size of the region in which that particular emission line is produced. The method for measuring the time delay between two light curves is called **cross-correlation**. A cross-correlation analysis reveals whether two light curves have the same features, and the value of the time delay τ between them. The cross-correlation technique multiplies the two light curves together, first shifting one of them in time by an amount τ, and sums over all the points in the light curve. When the features in two functions coincide for a particular shift τ, this will maximize the sum because big numbers will be multiplied by big numbers, and small numbers will be multiplied by small numbers.

Exercise 5.10   (a) Looking at the two light curves in Figure 5.19, what shift exists between the continuum flux and line flux?

(b) Does the emission line flux lag or lead the continuum flux in this case?

(c) What might be an interpretation of this shift?   ■

In order to understand how we might make sense of such time-delay measurements, we start by imagining that the BLR clouds are located in a thin spherical shell of radius r centred on the source of ionizing radiation (i.e. the black hole at the centre of the AGN), as shown in Figure 5.20. Suppose further that the central source emits a sudden burst of continuum radiation, as illustrated in Figure 5.21.

Figure 5.20   A schematic illustration of how light travel time delays will occur, assuming that the BLR may be represented as a thin spherical shell. The paths travelled by the illuminating continuum and the resulting observed emission line photons are shown. By summing the lengths of these two paths, the light travel time delay between the direct continuum and the reprocessed emission line signals can be calculated. The line marked 'isodelay surface' connects regions from which the same time delay would be recorded.

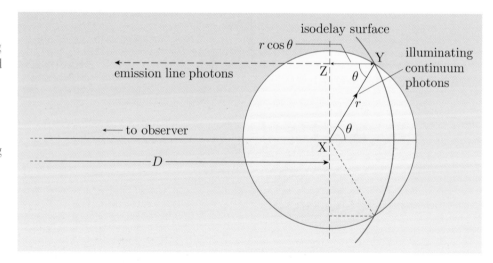

The light travel time delay is $\tau = d/c$, where the distance d is the sum of the two paths XY and YZ shown in Figure 5.20, i.e.

$$d = \mathrm{XY} + \mathrm{YZ} = r + r\cos\theta = (1 + \cos\theta)r.$$

Consequently,

$$\tau = \frac{d}{c} = \frac{(1 + \cos\theta)r}{c}.$$

This is, in fact, the equation of a paraboloid, as indicated in Figure 5.20.

Now, the great advantage of reverberation mapping is that it does not depend on any particular geometry. By determining the response of an emission line to the variability of the continuum flux, it is possible to infer the geometry of the BLR. In order to apply the technique, however, certain assumptions have to made. These include assuming the following.

- The continuum originates in a central source that may be regarded as a point, compared with the size of the BLR.

- The broad-line clouds have a small **filling factor**, and photons travel directly from the central source to the broad-line cloud in which they cause a photoionization.

- The emission line flux responds to the continuum flux. When the continuum flux increases, the emission line flux increases in response.

- The time taken for the continuum light to travel from its source to the BLR is much longer than the time between the ionizing photon being absorbed and an emission line photon being produced in response.

- The time for continuum light to travel from its source to the BLR is much shorter than the time needed for the overall structure of the BLR to change, i.e. the broad-line clouds do not move appreciably in relation to the position of the central source, nor does their ionization structure change appreciably, during the time for a continuum photon to travel between the central source and the BLR.

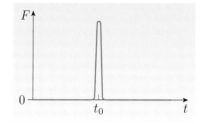

Figure 5.21 A sharp pulse of light would provide the ideal illuminating signal for reverberation mapping.

The filling factor is the proportion of space filled with line-emitting clouds.

Reverberation mapping measurements have now been applied to many AGN. One of the key results is that different emission lines have different time lags with respect to the continuum flux. Lines from highly ionized gases, such as He II, N V and C IV, respond with time lags of only a few days, while lines from lower ionized species, such as the hydrogen Balmer lines, have time lags of up to a few weeks. This clearly indicates an ionization structure within the BLR, with the most highly ionized species being closer to the central source. The overall size of the BLR in the AGN that has been observed is $\sim$10 light-days.

Given this approximate size for the BLR, and the measured velocity dispersion among the broad-line clouds, the approximate mass of the central object in AGN may be estimated from $M \simeq rv^2/G$ (see Subsection 1.3.2, in particular Equation 1.19 and Exercise 1.8).

Summary of Chapter 5

1. Most accreting systems appear as point sources of radiation. Indirect imaging methods are employed to infer *spatially*-varying distributions of emitting or absorbing material from *time*-varying flux or spectral information.

2. Cataclysmic variables are the ideal laboratories for studying accretion discs. The optical and UV emissions are dominated by the disc, and the short orbital periods provide detailed phase-dependent information.

3. The characteristic accretion disc continuum emission of the form $F_\nu \propto \nu^{1/3}$ is difficult to detect in actual accreting systems. In cataclysmic variables and AGN the ratio of outer to inner disc radius is not large enough, while in X-ray binaries the disc emission is altered by X-ray irradiation from the central source.

4. A clear observational signature of accretion discs is double-peaked emission lines. They reflect the Keplerian motion of the plasma in the disc.

5. The light curve of accreting binaries is a powerful diagnostic tool. Some systems show a pronounced orbital hump, caused by the bright (or hot) spot where the mass transfer stream from the secondary impacts the disc.

6. The eclipse mapping technique reconstructs a two-dimensional map of the brightness distribution in the accretion disc from eclipse light curves in different wavebands. The technique exploits the fact that as the secondary's shadow moves across the disc, it covers and uncovers the different luminous parts in the system. The one-dimensional light curve in general does not uniquely constrain the two-dimensional map. The fitting procedure enforces the likeness of the model to a default image, by maximizing a quantity called the image entropy.

7. Doppler tomography is a powerful imaging technique that uses the phase-dependent radial velocity information in emission lines to probe different zones in the binary. Using maximum-entropy techniques, the brightness distribution in the binary is reconstructed from the fine structure in the observed S-wave of spectral lines. The reconstruction gives an image in velocity space.

8. One of the achievements of Doppler tomography has been the discovery of a spiral shock pattern in some accretion discs. The spiral shocks are a consequence of gravitational perturbations of the secondary star on the accretion disc.

9. Reverberation mapping uses light echoes to probe the reprocessing structures surrounding the central source in AGN. This is achieved by observing the response of emission lines to continuum variations.

10. The reverberation mapping technique applied to AGN assumes that: the continuum flux originates in a single compact source; the BLR filling factor is small; the observed optical/ultraviolet continuum flux is related in a simple way to the ionizing continuum flux; and the light travel time across the BLR is the only important timescale in the process.

11. Different emission lines in AGN spectra are seen to have different time lags, such that lines from highly ionized species (such as He II, N V and C IV) have shorter time lags than lines from lower ionization levels (such as the Balmer lines). This is an indication for a radially stratified ionization structure in the BLR.

12. Cross-correlation of continuum and line fluxes for a range of AGN indicate that a typical size for the BLR is around 10 light-days. Assuming a BLR size of 10 light-days and a velocity range of up to $5000 \, \text{km s}^{-1}$ implies a mass for the central object of $M \simeq rv^2/G \simeq 5 \times 10^7 \, \text{M}_\odot$.

Chapter 6 High-energy radiation from relativistic accretors

Introduction

In this chapter, we shall discuss high-energy radiation from accreting systems; by this we mean X-rays and γ-rays, although we shall be mainly discussing X-rays. The X-ray band corresponds to photon energies between a few tens of eV up to a few tens of keV, while the γ-ray range is upwards of 10–100 keV. Indeed, the X-ray Universe is heavily dominated by accreting systems. The X-ray output of our Galaxy, and many others, is largely produced by X-ray binaries, while more than 90% of the X-ray background is contributed by the active nuclei of distant galaxies (AGN). We shall first find out what the Universe looks like with X-ray vision. Next, we shall start small, with the X-ray binaries. We shall discuss the various ways that these systems produce X-rays, and the Eddington limit: how a bright enough star system can defeat gravity! We shall also discuss how we can tell which processes are happening, from the emission spectrum (i.e. colour) and the variations in brightness of the X-rays. Finally, we shall make everything a million times bigger, in order to see how AGN are very like scaled-up X-ray binaries.

6.1 An X-ray view of the Universe

As we saw back in Chapter 1, the X-ray sky looks rather different to the one that we are used to in the optical waveband, being dominated by point sources of strongly varying brightness on short timescales. To see this version of the Universe, we must use X-ray telescopes. Since the Earth's atmosphere is opaque to X-rays, we must move above it to get a good X-ray view.

The first X-ray experiments were sent up on rockets, to work for only a short time before coming back down to the ground again. In 1962 Riccardo Giacconi and his team deployed an experiment to find X-rays from the Moon. They were greatly surprised to detect X-rays from somewhere else entirely! They only succeeded in pinning down the location of this newly discovered strong X-ray source in 1966, and called it Scorpius X-1, or Sco X-1 for short (see the box below). Unsurprisingly, Sco X-1 is an X-ray binary.

A rose by any other name

Optical astronomers have had thousands of years to think of exciting and romantic names for their stars. It helps that they can look up to the heavens and point out Bellatrix and Betelgeuse and the rest. X-ray astronomers are rather more practical (or unimaginative). Most celestial X-ray sources are named after their position, although really bright X-ray sources are named after the constellation that they are in. They can also be identified by their counterparts in other wavebands. For example, Sco X-1 is also known as 4U 1617−15 and V818 Sco, as well as a whole host of other names. The numbers 1617−15 refer to the position of Sco X-1, i.e. right ascension = 16 h 17 min, declination = −15°, while the leading letter/number combination indicates a catalogue or survey of a certain X-ray mission (here, 4U denotes the 4th catalogue of sources discovered by the Uhuru satellite). V818 Sco refers to the donor star, which is the 818th variable star to be classified in the constellation of Scorpius in the visual band.

Modern X-ray telescopes are mounted on satellites that orbit the Earth well above the atmosphere (see Figure 6.1 for an example). They use CCD imaging, but they can also measure the energy and arrival time of each X-ray photon. We shall find out how to use this information later in this chapter. Major X-ray observatories are listed with their main features in the box below.

Figure 6.1 An artist's impression of the XMM-Newton X-ray observatory, built by the European Space Agency with contributions from NASA. On the front (right) we see the entrance aperture of three X-ray telescopes. A mirror module is shown in Figure 6.5.

Some key X-ray observatories through history

Uhuru was the first ever X-ray satellite, a NASA mission that was operational from December 1970 to March 1973. It gave us the first catalogues of bright X-ray sources, as shown by the names of many X-ray sources, e.g. 4U 1624−490.

Einstein was another NASA satellite, operating from 1978 to 1981, and was the first fully-imaging satellite. It resolved X-ray sources in other nearby galaxies, including the Andromeda galaxy and the Magellanic Clouds.

Ginga was a Japanese satellite that was in operation from 1987 to 1991. It revealed the first black hole transients and a 6.4 keV emission line in many active galaxies, due to fluorescent iron (see Subsection 6.7.2).

ROSAT was a German-led satellite with contributions from the USA and the UK, and operated from 1990 to 1999. The name comes from Röntgensatellit, which translates as X-ray satellite. It provided a high resolution all-sky survey of ∼150 000 objects in the 0.1–2 keV band, and was crucial in determining the line-of-sight absorption to many X-ray binaries.

ASCA was a Japanese follow-up to Ginga that operated from 1993 to 2000. Its high spectral resolution in the 1–10 keV band allowed the first detection of broadening in the 6.4 keV line discovered by Ginga, proving that the line originated in the accretion disc (see Subsection 6.7.2).

The **Rossi X-ray Timing Explorer** (**RXTE**) sacrificed imaging capabilities to achieve the highest possible time resolution: 2.6 μs. As a result, RXTE discovered the kHz oscillations from low-mass X-ray binaries (LMXBs) predicted by theory. It is a NASA mission, launched in 1995 into low Earth orbit (altitude of 600 km). The short 90-minute period allows an on-board all-sky monitor to keep track of the brightest X-ray sources in the sky, but means that light curves are often broken up, as the Earth eclipses the target.

Chandra is a NASA mission launched in 1999 and is still currently active. It was optimized for imaging, and has unprecedented spatial resolution of order 1 arcsec or better. The sharp focus and extremely low background noise level enable detection of an X-ray source with as few as 5 X-ray photons. However, this was only achieved by sacrificing sensitivity, as the quality of the mirrors was chosen over quantity.

XMM-Newton is a European-led satellite that was launched a few months after Chandra in 1999, and is still currently active. It has three co-aligned X-ray telescopes, and one optical/ultraviolet telescope. It is the most sensitive X-ray imaging observatory to date (depending on energy range and source spectrum, about five times more sensitive than Chandra), but has poorer spatial resolution than Chandra (5 arcsec rather than 1 arcsec). XMM-Newton is currently the only telescope that can detect sufficient photons for studying the variability and emission from many X-ray binaries outside our Galaxy.

Even though all stars produce X-rays, most are too faint for current detectors. As we have touched upon in Chapter 1, and shall revisit in the next section, a strong source of X-rays requires exceedingly high temperatures, usually more than 10^7 K. Figure 6.2 shows the nearby galaxy NGC 253, as seen in the optical waveband. This image shows the combined emission of around 10^{11} stars.

Figure 6.2 The nearby spiral galaxy NGC 253, known as the Sculptor galaxy (Capella Observatory, Namibia, 16 September 2004). The major axis (apparent length) of this galaxy is around $26'$, nearly the size of the full moon.

The same galaxy in X-rays, using the XMM-Newton observatory, is displayed in Figure 6.3. The white ellipse shows the outline of the galaxy, and contains only about 100 X-ray sources; most of these sources are expected to be X-ray binaries, possibly with a few background AGN. We can also see lots of X-ray sources outside the galaxy; these are distant AGN.

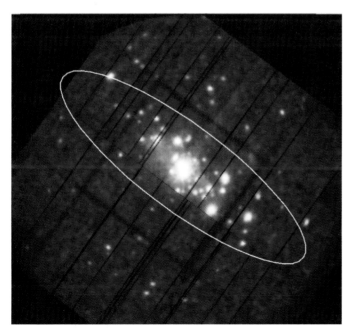

Figure 6.3 A three-colour X-ray image of NGC 253. Red, green and blue filters correspond to 0.3–2.0, 2.0–4.0 and 4.0–10 keV energy ranges. The white ellipse traces the outline of the galaxy.

X-ray units of measurement

In this book, we always use watts as the measure of power. However, in the literature, X-ray luminosities are almost always measured in $\mathrm{erg\,s^{-1}}$. As $1\,\mathrm{erg} = 10^{-7}\,\mathrm{J}$, we have $1\,\mathrm{erg\,s^{-1}} = 10^{-7}\,\mathrm{J\,s^{-1}}$, i.e. $10^{-7}\,\mathrm{W}$. For example, the luminosity of the Sun is $3.83 \times 10^{33}\,\mathrm{erg\,s^{-1}}$.

Unfortunately, classical telescope designs don't work for X-ray astronomers, as X-rays would scatter or go through the mirror rather than being focused. We can only focus X-rays that encounter the mirror at **grazing incidence**; if $0°$ is normal to the surface and $90°$ is along the surface of the mirror, then we need an angle $> 88°$. In 1952 Hans Wolter worked out how to focus X-rays with sets of two mirrors with the same focal point, one being a section of a parabola and the other being a section of a hyperbola; all the X-ray telescopes today use the Wolter Type I arrangement, sketched in Figure 6.4.

Figure 6.4   Cross-section of an effectively cylindrical arrangement of two mirrors used to focus X-rays in X-ray telescopes. The incoming X-rays hit two mirrors at grazing incidence, $\leq 2°$ from the surface of the mirror. The first mirror is formed from a section of a parabola, while the second mirror is a section of a hyperbola. The Wolter Type I mirror system uses many such structures nested inside each other to improve the collecting area. This can be seen in Figure 6.5.

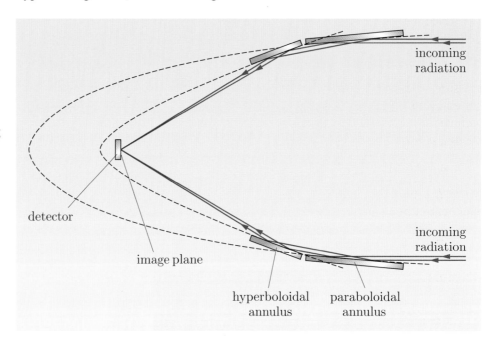

incoming radiation

detector

image plane

hyperboloidal annulus

paraboloidal annulus

incoming radiation

Figure 6.5 shows the testing of one of the mirror modules from the XMM-Newton observatory. It consists of 58 sets of nested paraboloid and hyperboloid gold-coated mirrors, all focused to the same point.

Exercise 6.1   In X-ray astronomy it is customary to describe photons using their energy, rather than frequency or wavelength. We describe the energy in terms of keV, and $1\,\mathrm{keV} = 1.6 \times 10^{-16}\,\mathrm{J}$. What is the frequency of a $1\,\mathrm{keV}$ photon? And its wavelength?   ∎

6.2   The Eddington limit

As we have seen in Chapter 1, the luminosity of an accreting system is proportional to the mass accretion rate. However, a higher luminosity leads to higher **radiation pressure** on the material as it falls in towards the accretor.

Logically, there comes a point where the outward radiation pressure on a particle exactly balances the inward gravitational pull. This is called the **Eddington limit**, or **Eddington luminosity**, and it is an important concept that we shall refer to many times in this and later chapters.

Figure 6.5 Testing one of the mirror modules of the XMM-Newton telescope before launch.

Pressure is defined as the force per unit area, and also as the change in momentum per unit area. In the case of radiation pressure, photons get absorbed, and momentum is transferred from the photon to the absorber. The momentum of a photon, p, is given by

$$p = E/c,$$

where E is the photon energy. The flux F (energy per unit time per unit area) experienced by a particle at a distance r from the X-ray source is simply given by

$$F = \frac{L}{4\pi r^2},$$

where L is the source luminosity. Hence we can define a momentum flux as F/c.

Cross-section, mean free path and optical depth

Here we review how the concepts of cross-section σ, mean free path λ, optical depth τ and opacity κ describe the absorption or scattering of particles, including photons (i.e. including radiation) in a medium consisting of target particles such as atoms, ions or electrons.

Consider particles that propagate in the x-direction in a slab of matter with thickness Δx in this direction, and area ΔA perpendicular to it. The propagating particles may interact via, for example, collisions or absorption with target particles in the slab. The probability for an interaction with one target particle in the slab can be described by an effective cross-section σ, i.e. an area with the dimension m^2. If the number density of target particles in the medium is n, then the slab with volume $\Delta V = \Delta x \, \Delta A$ contains $N = n \times \Delta V$ such targets. The total target cross-section is $\sigma \times N$, while the total target cross-section per unit volume is σn, and the opacity κ, the total target cross-section per unit mass, is

$$\kappa = \sigma n / \rho. \tag{6.1}$$

(Note that $\rho = \Delta M / \Delta V$, where ΔM is the mass of the slab.)

The probability P_{target} for interaction is therefore

$$P_{\text{target}} = \frac{\text{target cross-section}}{\text{cross-section of slab}} = \frac{\sigma N}{\Delta A} = \frac{\sigma n \, \Delta A \, \Delta x}{\Delta A},$$

so

$$P_{\text{target}}(\Delta x) = \sigma n \, \Delta x. \tag{6.2}$$

P_{target} becomes larger than unity if the slab is thick ($\Delta x > 1/\sigma n$), indicating that more than one interaction takes place for a given propagating particle.

We apply this formalism now to radiation, i.e. the propagation of photons through an absorber. If the intensity (or the flux, i.e. power per unit area) of the radiation as it enters the absorbing slab at $x = 0$ is I_0, then the amount $I_0 \times P_{\text{target}}$ is absorbed, while the intensity exiting the absorber at $x = \Delta x$ is $I(x) = I_0 - I_0 \times P_{\text{target}}$, or $\Delta I = I(x) - I_0 = -I_0 \times P_{\text{target}}$. For very small absorber depths $\Delta x \to \mathrm{d}x$, this becomes

$$\frac{\mathrm{d}I}{\mathrm{d}x} = -\sigma n \, I(x). \tag{6.3}$$

If the product $\sigma \times n$ is constant, this differential equation has the simple solution $I(x) = I_0 \exp(-\sigma n x)$.

The quantity

$$\lambda = \frac{1}{\sigma n} \tag{6.4}$$

is called the **mean free path**, because λ is the average distance that a photon/particle will travel before it interacts with a target (i.e. is absorbed or scattered). The **optical depth** τ of a layer with thickness x is defined as

$$\tau = \int_0^x \sigma n \, \mathrm{d}x, \tag{6.5}$$

so that $\mathrm{d}\tau/\mathrm{d}x = \sigma n$ or $\mathrm{d}\tau = \sigma n \, \mathrm{d}x$ (or, equivalently $\mathrm{d}\tau = \kappa \rho \, \mathrm{d}x$). Therefore Equation 6.3 can also be written elegantly as

$$\frac{\mathrm{d}I}{\mathrm{d}\tau} = -I(\tau),$$

which has the simple solution

$$I(\tau) = I_0 \exp(-\tau). \tag{6.6}$$

Consulting Equation 6.2, we see that the optical depth $\tau(\Delta x)$ can also be interpreted as the probability for interaction in a target material with thickness Δx, $P_{\text{target}}(\Delta x) = \tau(\Delta x)$. In particular, as $\Delta x/\lambda = \sigma n \, \Delta x = \tau(\Delta x)$, the optical depth quantifies the average number of interactions that a particle will undergo while travelling a distance Δx in the target medium. For radiation, a medium is called **optically thick** (opaque) if $\tau > 1$, as this indicates that typically a photon cannot escape before it is absorbed. Conversely, if $\tau < 1$, it will be able to escape and the medium is **optically thin** (transparent).

Let us consider a blob of accreted material with area ΔA facing the central accretor, and thickness Δr in the radial direction, made of ionized hydrogen (i.e. protons and electrons) with density ρ. The probability of absorbing a photon is given by the number density of electrons ($n_e = \rho/m_p$) and the Thomson scattering cross-section (σ_T). Using Equation 6.3, the change in flux over the thickness of the blob, $\Delta F/\Delta r$, is

$$\frac{\Delta F}{\Delta r} = \left(\sigma_T \times \frac{\rho}{m_p}\right) \times \left(\frac{L}{4\pi r^2}\right).$$

This results in a change in momentum flux, and therefore change in pressure, $\Delta P/\Delta r$, given by

$$\frac{\Delta P}{\Delta r} = \frac{1}{c} \times \frac{\Delta F}{\Delta r} = \left(\sigma_T \times \frac{\rho}{m_p}\right) \times \left(\frac{L}{4\pi r^2 c}\right). \tag{6.7}$$

The blob therefore experiences a net acceleration in the radial direction, away from the accretor,

$$\frac{1}{\rho}\frac{\Delta P}{\Delta r} = \left(\frac{\sigma_T}{m_p}\right) \times \left(\frac{L}{4\pi r^2 c}\right) \tag{6.8}$$

(see Equation 3.33, with r instead of z). By definition, for $L = L_{Edd}$, this must be balanced by the gravitational acceleration of the blob, GM/r^2, where M is the mass of the accretor. Equating the two acceleration terms and rearranging the equation to get L_{Edd} gives

$$L_{Edd} = \frac{4\pi G M m_p c}{\sigma_T}. \tag{6.9}$$

This assumes that the accreted material consists purely of ionized hydrogen, whereas accretion from evolved donors may include heavier elements such as helium.

● Would you expect L_{Edd} to be higher or lower if the accreted material is heavier than hydrogen?

○ Equation 6.9 shows us that L_{Edd} is proportional to the mean particle mass per free electron of the incoming material, i.e. m_p in the case of hydrogen. Therefore L_{Edd} increases if the accreted material is composed of heavier elements.

Exercise 6.2   Now we shall consider the Eddington limit for primaries of different sizes.

(a) Express the Eddington luminosity in terms of the accretor mass M in solar units. Determine the numerical constant in this expression.

(b) What is L_{Edd} for a $1.4\,M_\odot$ neutron star?

(c) What is L_{Edd} for a $10\,M_\odot$ black hole?   ■

Exercise 6.2 shows that Equation 6.9 can also be written as

$$L_{Edd} = 1.26 \times 10^{31} \left(\frac{M}{M_\odot}\right) \text{W}. \tag{6.10}$$

We shall see an excellent demonstration of the Eddington limit in action later in the chapter, when we discuss the X-ray burst phenomenon seen in low-mass X-ray binaries with neutron star primaries. We shall also see how the Eddington limit can be used to discover black holes, and possibly misused in creating the so-called ultraluminous X-ray sources (ULXs).

6.3 Analyzing X-ray spectra

Accreting systems can produce X-rays in a host of weird and wonderful ways, resulting in a wide variety of X-ray emission spectra. We shall see how combining the spectral information with variability analysis provides an invaluable diagnostic of the physical processes at play. Furthermore, we shall discuss how we can use this information to pick out new black holes.

Detector effective area

One might think that the spectrum seen by an X-ray observatory is the same as the X-ray spectrum emitted by the source. However, the detector (camera) responds to different energies (i.e. frequencies) in different ways. Figure 6.6 shows the response of the pn camera on board the XMM-Newton observatory to X-rays in the range 0.3–10 keV for a source spectrum with 1 photon $keV^{-1}\,cm^{-2}\,s^{-1}$ at all energies; in this case, a filter was used which cuts out some of the low-energy photons. The variation in the number of collected photons with energy may be thought of as a change in collecting area of the detector, hence the y-axis is expressed in cm^2. We often refer to the sensitivity of an X-ray detector to a particular energy as the **effective area**. For all the many nested mirrors, the peak effective area for the pn camera is only $\sim$1100 cm^2, about the same as a 40 cm-diameter telescope, and this is the most sensitive imaging X-ray telescope to date.

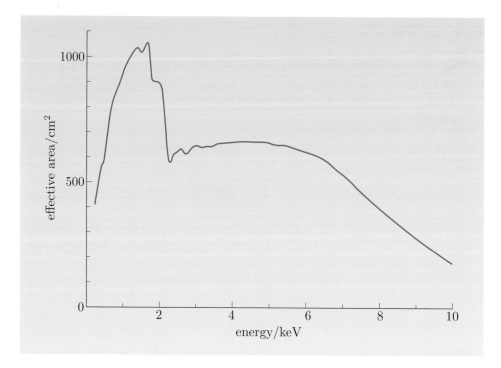

Figure 6.6 The variation in effective area as a function of energy for the pn CCD detector, one of the instruments on board the XMM-Newton X-ray observatory.

6.3.1 Thermal emission

The familiar black body spectrum is an example of thermal emission. Anything can produce thermal X-ray emission, so long as it is hot enough: around 10^7 K. Stellar coronae manage this, so all stars emit thermal X-rays. Indeed, the Sun is the brightest X-ray source in the sky, but this is thanks to its proximity, rather than its strength: the ratio for X-ray to optical output for the Sun is only $\sim 10^{-7}$. We shall see an example of a much stronger X-ray black body when we look at X-ray bursts; these have typical temperatures of around 2×10^7 K, and have a luminosity of 10^5 L$_\odot$.

In Subsection 1.4.2 we have discussed the multi-temperature disc black body spectrum, the thermal emission of an optically thick accretion disc, and this is indeed observed in many X-ray binaries.

6.3.2 Non-thermal emission

Thermal emission often contributes to the X-ray radiation from X-ray binaries, but it is the non-thermal X-ray emission that really sets them apart from other star systems, and produces the bulk of AGN emission. In disc-powered systems, the process responsible for this is **unsaturated inverse Compton scattering**. Despite the name, the concept is quite simple, and it is here that the idea of photons having a radiation temperature (Equation 1.26 in Subsection 1.4.2) is incredibly useful.

It is generally agreed that disc accretion results in a hot, optically thin cloud of electrons with temperatures exceeding 10^9 K (100 keV) in many cases. This is called the **accretion disc corona** (ADC). However, the location of the ADC is not so certain. Some people believe that the accretion disc gets so hot near the primary that it becomes optically thin, producing a corona between the disc and the primary that surrounds the primary. Others believe that the ADC forms co-rotating slabs above and below the accretion disc. In this case, the corona could be evaporated from the surface of the disc. Schematics of both scenarios are given in Figure 6.7. The physical mechanisms that cause the electrons in the ADC to be so hot in the first place are not fully understood — the same applies to the related case of the **solar corona**. Possible heating mechanisms include plasma waves and electric currents induced in the ADC by magnetic fields.

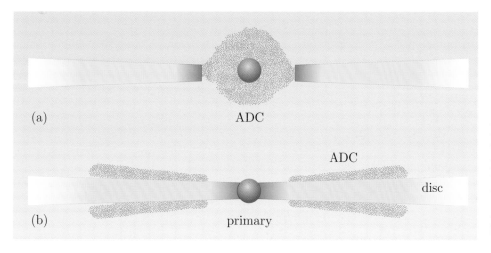

Figure 6.7 Schematics of the two possible ADC scenarios. In the top scenario, the ADC lies between the inner accretion disc and the primary; the input photons are expected to mostly come from the surface of the primary (this obviously wouldn't work for a black hole!). The bottom scenario shows a planar ADC that co-rotates above and below the accretion disc.

The non-thermal emission process starts with thermal photons, from either the disc or the surface of the primary (so long as it is not a black hole). These photons pass through the ADC, interacting with a number of electrons before escaping the corona. The radiation temperatures of these photons ($\lesssim 2\,\text{keV}$) are far lower than those of the electrons in the ADC ($\sim 100\,\text{keV}$ or more in many cases). Therefore interactions between a photon and an electron will result in energy being transferred from the electron to the photon. This process is called **inverse Compton scattering** to distinguish it from normal **Compton scattering**, where the photon gives energy to a lower-energy electron. We shall discuss inverse Compton scattering in more detail later in the book (in Subsection 7.3.4). Irradiating the corona with X-rays from the accretion process cools the corona, rather like adding cold water to a hot bath. Therefore more luminous X-ray sources have cooler coronae. As an example, the system Sco X-1, often the brightest X-ray source in the sky after the Sun, has a corona temperature of $\sim 10\,\text{keV}$, rather than the more typical $\sim 100\,\text{keV}$.

For most low-mass X-ray binaries (LMXBs) the mean number of scatterings that a photon experiences before it exits the ADC is close to 1, which is much less than the number of scatterings required for the photons to reach an energy of kT_e. This is the reason why the process constitutes *unsaturated* inverse Compton scattering; saturated scattering occurs when the mean photon temperature reaches the electron temperature, and the photons thermalize to a black body spectrum with temperature T_e. The distribution of escape times depends on the geometry of the corona, and on the location of the X-ray source with respect to the corona; the details are too complex for this book.

The result is that the emission spectrum of radiation exiting the corona is markedly different from the spectrum of the radiation entering the corona, since the number of times that the photon is scattered determines its energy as it exits the corona. The Comptonized emission spectrum can be approximated by a power law

$$P_\text{E}(E) = E^{-\Gamma_\text{p}} \tag{6.11}$$

for photons with energies lower than kT_e. Here P_E is the number of photons received per unit time per unit area per photon energy interval. The quantity Γ_p is the photon index. For energies higher than kT_e, we find that $P_\text{E}(E)$ is described by the Wien tail of a black body with temperature T_e:

$$P_\text{E}(E) \propto E^2 e^{-E/kT_\text{e}} \tag{6.12}$$

(cf. Equation 1.27 with $E = h\nu$). Figure 6.8 shows an example of a Comptonized emission spectrum. Many X-ray telescopes have energy ranges that are much lower than kT_e, so the emission looks like a simple power law. However, if we see a Wien tail as well as the power law, astronomers can estimate the parameters of the corona.

X-ray spectra are often displayed as photons received $P_\text{E}(E)$ per unit time and unit detector area, per photon energy interval, versus photon energy E. This is related to the more familiar flux density F_E, the energy received per unit time, detector area and photon energy interval, as $F_\text{E} \propto P_\text{E} \times E$. Displaying F_E versus E is in turn equivalent to a spectrum shown as F_ν versus ν, such as the black body intensity in Figure 1.19. If the photon count rate is given as a power law with photon index Γ_p, then the flux density follows the power law $F_E \propto E^{-\Gamma_p+1}$.

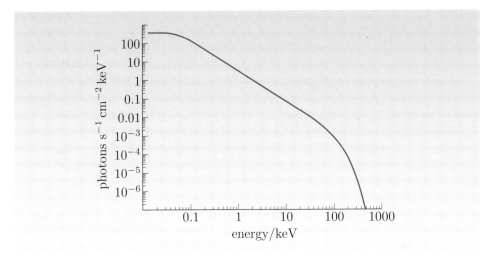

Figure 6.8 Example of a Comptonized spectrum for a spherical corona with $T_e = 50\,\mathrm{keV}$. The x- and y-axes are log-scaled, so the spectrum for energies $< 50\,\mathrm{keV}$ appears as a straight line with gradient $-\Gamma_p$.

6.3.3 X-ray absorption

Some of the X-ray radiation is absorbed by particles in the interstellar medium in between us and the X-ray source. This could be due to **interstellar dust**, or the material in dense **molecular clouds** if the X-ray source is located within or behind the cloud. Furthermore, X-ray binaries may suffer extra absorption from the secondary star, or from material in the accretion disc. For X-ray binaries that we see nearly edge-on, this last type of absorption shows up as regular variations in the X-ray light curve on the orbital period; such behaviour is discussed in Subsection 6.5.2.

The amount of radiation removed by the absorption is energy (i.e. frequency) dependent: low-energy X-rays are absorbed more than high-energy X-rays. The main culprit for this is hydrogen, with other contributions by helium, carbon, oxygen, nitrogen and various heavier elements; the contribution of each element is decided by (i) how much there is (the abundance), and (ii) how likely it is to absorb photons of a given energy. These two factors are combined into an absorbing cross-section. The probability for absorbing photons of a particular energy dramatically increases when that energy corresponds to an ionization state in one of the absorbing elements. We call the resulting jump in cross-section an **absorption edge**.

Figure 6.9 (overleaf) shows an approximation of the absorption cross-section as a function of energy. As hydrogen is the most abundant element, it is customary to express the observed absorption as N_H, the equivalent number of neutral hydrogen atoms per cm^2 in a column between us and the X-ray source that would generate the same amount of absorption. The absorption curve shown in Figure 6.9 assumes **cosmic abundance**, i.e. that the relative abundances of the different elements are the same as are measured for the Solar System.

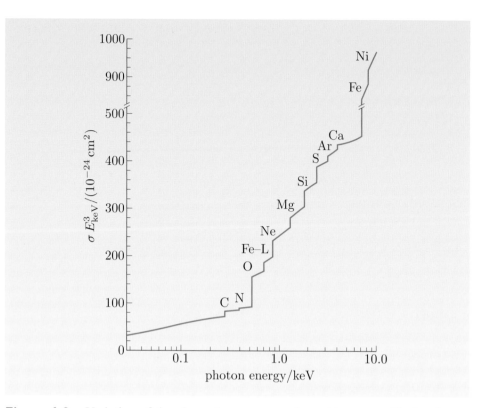

Figure 6.9   Variation of the absorption cross-section with energy. Notice that the y-axis shows the cross-section (σ) multiplied by $(E/1\,\text{keV})^3$ to make it easier to see; for example, the true value of σ at $10\,\text{keV}$ is 1000 times smaller than shown. Each little jump in absorption (absorption edge) is caused by the element shown.

Typical absorptions range from 10^{20} to 10^{22} H atom cm^{-2}. This sounds quite a lot, but an empty pint glass contains around 1.4×10^{22} atoms of air. So if we took all the air in a pint glass, and stretched it into a column with a base of $1\,\text{cm}^2$, but with a length of say $10\,\text{kpc}$, we would reproduce the density of the interstellar medium.

Comparing N_{H} with A_{V}

Surveys of nearby X-ray binaries have allowed us to compare the N_{H}, measured from the X-ray spectra, with the visual extinction A_{V}, measured in magnitudes. This resulted in the empirical equation

$$N_{\text{H}} = (1.79 \pm 0.03) \times A_{\text{V}} \times 10^{21} \text{ H atom cm}^{-2} \tag{6.13}$$

given by Predehl and Schmitt in 1995. For example, the highest line-of-sight N_{H} for known Galactic X-ray binaries is 8×10^{22} atom cm^{-2}. This is equivalent to a visual extinction of ~ 45 magnitudes!

The optical depth, at energy E, towards an X-ray source at distance $x = D$ from the observer at $x = 0$ is

$$\tau(E) = \int_0^D \sigma(E)\, n(x)\, \text{d}x = \sigma(E) \int_0^D n(x)\, \text{d}x$$

(from Equation 6.5, where σ depends on E but not on x, and therefore moves to the front of the integral). Hence

$$\tau(E) = N_H\,\sigma(E). \tag{6.14}$$

The fraction of radiation transmitted through the absorber, $f_{trans}(E)$ $(= I(D)/I(0)$, see Equation 6.6), is then given by

$$f_{trans}(E) = \exp\left[-\tau(E)\right] = \exp\left[-\sigma(E)\,N_H\right]. \tag{6.15}$$

Exercise 6.3   In 1983, Morrison and McCammon produced an analytical approximation for the cross-section shown in Figure 6.9, to give $\sigma(E)$ for energies in the range 0.01–10 keV:

$$\sigma(E) \simeq (c_0 + c_1 \times E + c_2 \times E^2)E^{-3} \times 10^{-24}\,\text{cm}^2,$$

where c_0, c_1 and c_2 vary depending on the energy, and E is measured in keV.

(a) If $N_H = 1.5 \times 10^{22}$ atom cm^{-2}, calculate the fraction of 1 keV photons absorbed ($c_0 = 120.6$, $c_1 = 169.3$ and $c_2 = -47.7$).

(b) For 5 keV photons, $c_0 = 433$, $c_1 = -2.4$ and $c_2 = 0.75$. What is the fraction of 5 keV photons absorbed if $N_H = 1.5 \times 10^{22}$ atom cm^{-2}? ∎

6.4   Time variability in accreting systems

Accretion-powered systems are in general not constant sources of radiation. They display significant variability on very diverse timescales; some changes happen hundreds of times per second, while other changes take days, months or years, such as the soft X-ray transient outbursts discussed in Section 4.4. This variability reveals a wealth of information about the physical processes that drive the emitting systems. We are familiar with the concept of light curves already; however, we shall find out how X-ray light curves are particularly complex. Then we shall introduce another valuable tool for studying variability: the power density spectrum.

6.4.1   Analyzing light curves

Analyzing X-ray light curves is a particularly flexible technique, because we can extract light curves covering many different energy ranges from the same data. This is because the X-ray detectors determine the energy of each photon, where it came from, and the time that it arrived. The X-ray spectra discussed in the previous section give us the distribution of photons as a function of energy (frequency), summed over a time period of our choosing. The light curves that we shall discuss in this section take a different approach, giving us the number of photons as a function of time, summed over a particular energy range. We can choose any energy range that we like, so long as it is within the capabilities of the X-ray observatory. Figure 6.10 shows light curves of the X-ray binary 4U 1624−490 over three different energy ranges (often referred to as energy bands).

Telling the time

The time axis in Figure 6.10 is measured in seconds; the exposures of X-ray observations are usually very long, and it is common to see observations lasting 10 ks (~3 hours), 50 ks (~14 hours) or even 100 ks (~28 hours). This observation is particularly long, lasting ~92 hours, which is almost 4 days!

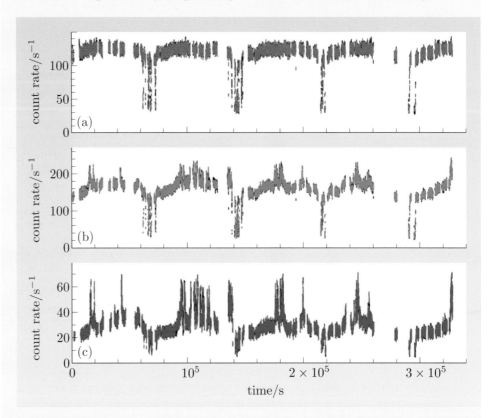

Figure 6.10   Three light curves from the September 1999 observation of the low-mass X-ray binary (LMXB) 4U 1624−490, obtained with the RXTE satellite, in three different energy bands. Panel (a) covers the 2.0–5.0 keV band, panel (b) gives the light curve for the 5.0–10 keV band, and panel (c) shows the 10–25 keV band. These light curves demonstrate the energy-dependent nature of many high-energy phenomena: the low-energy light curve is dominated by deep intensity dips, while the high-energy light curve is characterized by sporadic outbursts of energy called flares; however, there is very little trace of dips at high energy, or of flares at low energy.

Once we have light curves in different energy bands, we can manipulate them in many ways, constructing so-called colours. For example, if we name the different energy bands used in Figure 6.10 Band 1 (2.5–5.0 keV), Band 2 (5.0–10 keV) and Band 3 (10–25 keV), then we might define the intensity ratio Band 2/Band 1 as a *soft colour*, and the ratio Band 3/Band 2 as a *hard colour*. The possibilities are endless, and are chosen to show off the data in the best way possible, or extract the maximum information; if we try one set of colours, and find them to be unhelpful, then we can simply choose different ones. For this reason astronomers must be very careful to define their colours before using them!

A hard look at colour

A more general way of describing an X-ray spectrum, without detailing specific features, is its hardness. We denote high-energy photons as *hard*, and low-energy photons as *soft*; hardness quantifies the relative amounts of hard and soft photons, and is therefore a coarse measure of the shape of the emission spectrum. If we define a colour as the intensity in a high-energy band divided by the intensity in a low-energy band, then we can also refer to it as a hardness ratio.

In Figure 6.11 we take a look at an X-ray observation of Cygnus X-1, the most famous black hole X-ray binary; it shows just how useful X-ray colours can be. Cygnus X-1 is a high-mass X-ray binary, and is powered by disc accretion. However, the disc is fed mainly by the wind from the secondary rather than Roche-lobe overflow. From time to time, blobs of in-falling material cross the line of sight, causing momentary increases in absorption and dips in the intensity. Photo-electric absorption preferentially removes low-energy photons, so the dips are deeper at lower energies than at higher energies.

It is practically impossible to see these dips in light curves of Cygnus X-1, because the accretion process causes large, random-intensity variations called **stochastic noise**; we shall discuss stochastic noise in more detail later in the chapter. However, these dips are easily seen if we look at how the colour varies with time (see Figure 6.11).

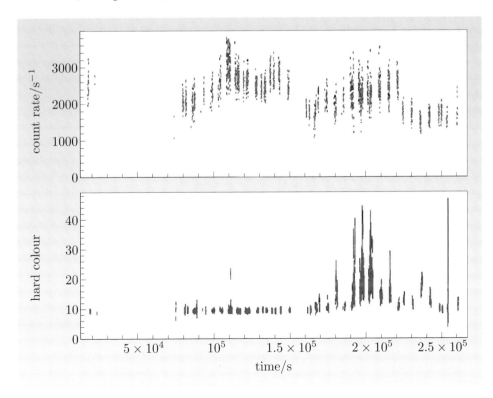

Figure 6.11 This is an example of the usefulness of colours. The top panel shows the 1.5–11.6 keV light curve of Cygnus X-1 from the Ginga observation made in August 1987. For the bottom panel we define a 'hard colour' as the 2.3–11.6 keV intensity divided by the 1.5–2.3 keV intensity. We can see intervals of high absorption (dips) as dramatic increases in the hard colour; however, these dips are totally invisible in the light curve.

6.4.2 Frequency analysis of variations in the light curve

Here we shall discuss one of the most useful diagnostic tools in X-ray astronomy, the power density spectrum. This allows us to break down the variability in a light curve into its component parts. To ease into things, we shall demonstrate the concept by building up light curves using sine waves of different amplitudes and frequencies.

Figure 6.12a shows a simple sine wave

$$I(t) = A \sin(2\pi f t),$$

with frequency f and amplitude $A = 1$. If the intensity of an X-ray source, $I(t)$, varied sinusoidally, then its light curve would look rather like Figure 6.12a. However, the emission might be more complicated than that. Figure 6.12b shows a combination of three sine waves:

$$\sin(2\pi f t) + 0.5 \sin(2\pi(3f)t) - 0.65 \sin(2\pi(4.4f)t).$$

We show a yet more complicated curve in Figure 6.12c. We could carry on adding sine waves with various amplitudes and offsets, to make any light curve we like.

Figure 6.12 Increasingly complicated combinations of various sine curves. Panel (a) shows a simple sine wave with an amplitude of 1 and frequency f. Panel (b) shows a combination of three sine waves: $\sin(2\pi f t) + 0.5 \sin(2\pi(3f)t) - 0.65 \sin(2\pi(4.4f)t)$. Panel (c) shows an even more complicated curve, adding $0.77 \sin(2\pi(f/10)t) + 7 \sin(2\pi(f/10)t + 1.26)$ to the existing curve. If we added more and more sine waves, we could reproduce any light curve we liked.

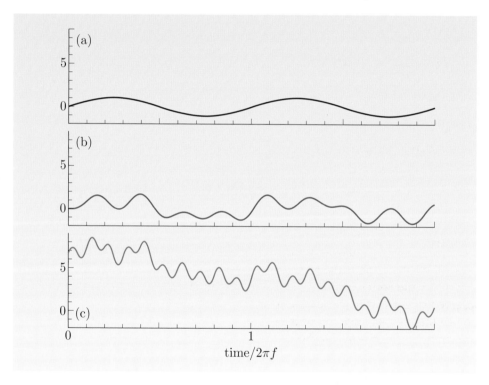

Similarly, we can take any observed light curve and break it down into its component frequencies.

First, we shall consider a light curve consisting of N data points, with a fixed time interval ΔT between readings. ΔT is known as the **time resolution**: we can see any intensity changes that are slower than this, but cannot see any changes that are quicker.

● What is the lowest frequency that we can observe in this light curve?

○ The lowest frequency that we can measure gives one cycle over the whole observation of duration $N \Delta T$, hence $f_{min} = 1/(N \Delta T)$. We may see evidence for variability over longer timescales, but cannot be confident unless we observe a whole cycle.

● The highest frequency, also called the **Nyquist frequency**, is given by $f_{max} = 1/(2 \Delta T)$. Why is this?

○ We need to measure something at least twice to detect any changes. As ΔT is the time resolution, the shortest detectable timescale for variability is $2 \Delta T$. This gives a high-frequency limit of $f_{max} = 1/(2 \Delta T)$.

Now we need to describe this light curve. One way is to give the intensity as a function of time. At a given time t_j, we measure an intensity I_j, where j is an integer between 1 and N. We can also construct this light curve from sine waves with $N/2$ evenly spaced frequencies: $f_k = k/N \Delta T$, where k is an integer between 1 and $N/2$ ($k = N/2$ corresponds to the Nyquist frequency). The challenge is to find the correct contribution of each frequency that enables us to replicate the light curve.

If we create a light curve as a sum of sine waves with frequency f_k, amplitude C_k and relative phase shift ϕ_k, then the intensity I_j at time t_j is given by

$$I_j = \sum_{k=1}^{N/2} C_k \sin(2\pi f_k t_j + \phi_k). \tag{6.16}$$

Using the identity

$$\sin(\alpha + \beta) = \sin\alpha\cos\beta + \cos\alpha\sin\beta, \tag{6.17}$$

this can also be written as

$$I_j = \sum_{k=1}^{N/2} \left[A_k \sin(2\pi f_k t_j) + B_k \cos(2\pi f_k t_j) \right], \tag{6.18}$$

where $A_k = C_k \cos\phi_k$ and $B_k = C_k \sin\phi_k$, and $A_k^2 + B_k^2 = C_k^2$.

The mathematical technique of **Fourier transformation** allows one to determine the coefficients C_k from the measured intensities I_j, and hence the relative weight with which the corresponding sine wave with frequency f_k contributes to the observed signal I_j (the phase ϕ_k remains undetermined). This method is based on the fact that the time integral of the intensity times a sine wave with a given frequency f_k' effectively vanishes, except when this frequency is one contained in the signal.

To see this, consider the simple case where the signal consists of only one sine wave ($A_{k'} = 0$ for any $k' \neq k$, and $B_k = 0$ for all k): $I(t) = A_k \sin(2\pi f_k t)$. Then consider the integral

$$\int I(t) \times \sin(2\pi f_k' t) \, dt = A_k \int \sin(2\pi f_k t) \sin(2\pi f_k' t) \, dt.$$

The time integral is over the whole duration of the measurement, $N \Delta T$.

For the case $k = k'$, the integrand is proportional to $\sin^2(2\pi f_k t)$ and therefore non-negative everywhere. The two sine waves with identical frequency amplify each other, maximizing the integral. For $k \neq k'$, on the other hand, the different frequency sine waves tend to cancel each other out, so that the integrand is a strongly variable function of time that fluctuates between positive and negative values, giving a negligible integral.

In other words, we have

$$A_k \propto \int I(t) \times \sin(2\pi f_k t) \, \mathrm{d}t.$$

If $I(t)$ is instead given by a cosine term, a similar argument shows that

$$B_k \propto \int I(t) \times \cos(2\pi f'_k t) \, \mathrm{d}t.$$

If $I(t)$ is a superposition of sine and cosine terms, as in the sum in Equation 6.18, then the arguments apply separately for each term, and the above equations remain valid for each k.

Finally, we can approximate the integral $\int I(t) \times \sin(2\pi f_k t) \, \mathrm{d}t$ with a discrete sum of N rectangles with width ΔT and height $I_j \sin(2\pi f_k t_j)$, so that

$$\int I(t) \times \sin(2\pi f'_k t) \, \mathrm{d}t \simeq \sum_{j=1}^{N} I_j \sin(2\pi f_k t_j) \, \Delta T.$$

The contribution of the frequency f_k to the light curve is described by its power $P(f_k) \propto C_k^2 = A_k^2 + B_k^2$. Hence

$$P(f_k) \propto \left[\sum_{j=1}^{N} I_j \cos(2\pi f_k t_j) \right]^2 + \left[\sum_{j=1}^{N} I_j \sin(2\pi f_k t_j) \right]^2. \tag{6.19}$$

The function $P(f_k)$ is known as the **power density spectrum** (PDS) and gives us the recipes that tell us the correct power required for each frequency to produce the desired light curve. Figure 6.14 shows the PDS for each example light curve given in Figure 6.12. Unlike these ideal light curves, every real PDS includes a Poisson noise component from random variations in the light curve (see Subsection 6.5.3). Astronomers commonly set C to be $2\,\Delta T/N\overline{I}$, where $\overline{I}$ is the mean intensity, because this gives the Poisson noise a power of 2 for every frequency in the PDS.

Exercise 6.4   We shall now create a power spectrum for a very simple light curve, shown in Figure 6.13.

(a) What are the lowest and highest frequencies in the PDS of this light curve? How many frequencies are there in the PDS?

(b) Using Equation 6.19, calculate the power for each frequency in the PDS, $P(f_k)$. ∎

Much of the variability seen in X-ray binaries results from the accretion of blobs of material at random times.

Figure 6.15 shows the mean PDS of the black hole X-ray binary Cygnus X-1 in different brightness states — the so-called low state (red curve) and high state (blue curve). In different sections of the frequency range we find $P(f_j)$ to be proportional to some power of f_j:

$$P(f_j) \propto f_j^{-\alpha_f}.$$

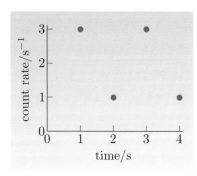

Figure 6.13   A simple light curve.

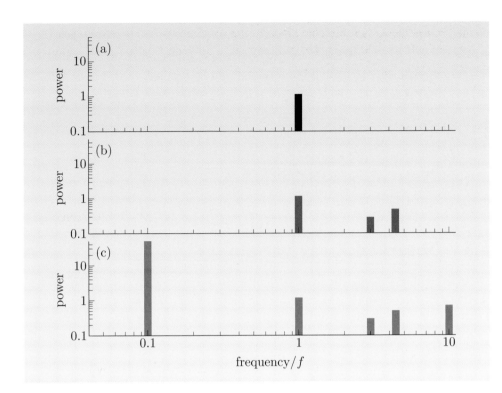

Figure 6.14 Power density spectra for the three light curves shown in Figure 6.12. Each PDS is log-scaled because of the wide range in frequency ($0.1f$–$10f$) and power (0.25–49).

For instance, in the low state, $\alpha_f = 1$ for $f \lesssim 10\,\mathrm{Hz}$; this kind of behaviour is discussed in Subsection 6.5.3. Note that the x- and y-axes are logarithmic, and that the y-axis shows frequency times power, making the power density spectrum appear as a straight line with slope $-\alpha_f + 1$.

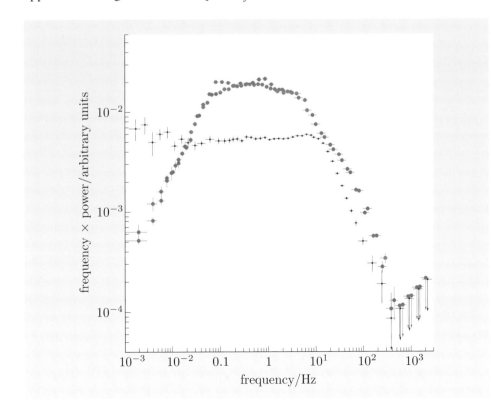

Figure 6.15 Power density spectrum of the black hole X-ray binary Cygnus X-1 in the low state (red) and high state (blue). The y-axis shows frequency times power, hence the gradients in this figure are given by $-\alpha_f + 1$ rather than $-\alpha_f$.

● Why does the low-state power density spectrum below 10 Hz in Figure 6.15 appear as a straight line with gradient $\simeq 0$?

○ As $y = \log_{10}(f \times P)$ we have, using $P = kf^{-\alpha_f}$,

$$y = \log_{10}(f \times P) = \log_{10}(f \times k \times f^{-\alpha_f})$$
$$= \log_{10}(k \times f^{1-\alpha_f})$$
$$= \log_{10} k + (-\alpha_f + 1) \times \log_{10} f.$$

Setting $x = \log_{10} f$ we therefore have $y = \text{constant} + (-\alpha_f + 1) \times x$ for the curve shown in the diagram. In the low state $\alpha_f = 1$, so the slope is $-\alpha_f + 1 = 0$.

6.5 Observed X-ray variability

Equipped with the diagnostic tools used in X-ray astronomy discussed above, we now turn to interpreting the manifold phenomena observed in accretion-powered X-ray sources. We shall start with X-ray binaries, because they are geometrically smaller than AGN, and therefore show variability on much shorter timescales.

6.5.1 X-ray bursts

X-ray bursts are analogous to **classical nova outbursts** in accreting white dwarf systems. The timescales associated with the nova phenomenon are much longer than those of X-ray bursts. The rise to outbursts can take several days, the decline several weeks. Classical novae are not seen to recur but are thought to do so within $\gtrsim 10^4$–10^5 years. The outburst amplitude is $\gtrsim 9$ mag.

X-ray binaries containing accreting neutron stars differ from those with black holes in that there is a physical surface for material to fall onto. Material builds up and can trigger a violent explosion, known as an X-ray burst. Type I X-ray bursts are most common, observed in nearly half of the 200 or so currently known low-mass X-ray binary systems. They involve a sharp rise in luminosity, sometimes reaching the Eddington limit, followed by an exponential decay. A typical Type I burst (see Figure 6.16) lasts 10–100 s, and can recur in a few hours. The much rarer Type II X-ray bursts have been observed in only two X-ray binary systems. They can occur hundreds of times per day, so cannot be driven by the same physical mechanism as the one that drives the Type I bursts. From now on we shall ignore Type II bursts, and refer to Type I X-ray bursts simply as **X-ray bursts**.

X-ray bursts are driven by a thermal instability that exists when nuclear burning is confined to a thin shell wrapped around a degenerate core (Schwarzschild and Härm, 1965). In the case that we are interested in, the whole neutron star is degenerate while the accreted material is burned in a thin layer on the neutron star surface.

Nuclear burning is temperature-sensitive, and will proceed more quickly as the temperature rises. If the energy gain through burning is larger than the energy loss by radiation, as is the case for thin (but not too thin) shells, a **thermonuclear runaway** is triggered, with ever increasing temperature and nuclear burning rate. The result is an X-ray burst.

So, at the start of the burst cycle, accreted material settles on the neutron star surface. This material burns in a layer that is thin enough for the temperature to increase, but also the ratio of surface area to volume is high for the shell, so it radiates heat quickly, like tea that has been spilled in a saucer. Therefore the rise in temperature results in a net energy loss, and equilibrium is restored. As more

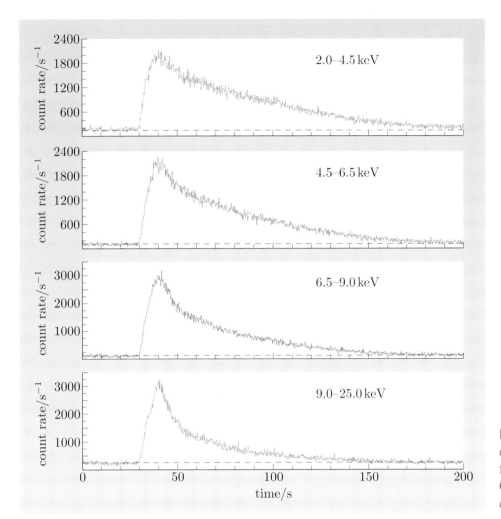

Figure 6.16 RXTE observation of an X-ray burst from the X-ray binary GS 1826−238 seen in four different energy bands.

material builds up, the shell of burning material gets thicker. When the shell is finally thick enough for the rising temperature to result in a net energy gain, the burst is triggered; the material is consumed and the cycle starts again.

The first estimates of X-ray energy spectra from bursts were made by subtracting the non-burst X-ray spectrum from the spectra of the burst at different stages, giving *difference spectra* that describe the burst itself. Burst difference spectra are well represented by a simple black body model, with peak temperatures of around 2.5×10^7 K. The total flux received from a black body source with luminosity L at distance d is $F_{\text{tot}} = L/4\pi d^2$. If the radius of the black body emitter is R_{BB} and its temperature is T, the luminosity is $L = 4\pi R_{\text{BB}}^2 \, \sigma T^4$ (using the Stefan–Boltzmann law, Equation 1.22, to calculate the flux). Hence the detected flux is

$$F_{\text{tot}} = \frac{L}{4\pi d^2} = \frac{R_{\text{BB}}^2 \, \sigma T^4}{d^2}. \tag{6.20}$$

If we determine the flux and temperature of a burst from its X-ray spectrum and know the distance to the X-ray binary, then we can calculate the size of the black body emitting region. X-ray burst spectra are consistent with a black body emitter with a ~10 km radius. This is taken as proof for the neutron star nature of sources that exhibit X-ray bursts.

Some X-ray bursts have properties that give physical proof of the Eddington limit. These bursts are preceded by a small peak some 5–10 s before the main burst, or appear to be double-peaked. These bursts have so much fuel that they reach the Eddington limit, and radiation pressure causes the neutron star's photosphere to expand by up to 2–3 times. The luminosity can't go any higher, so the photosphere cools as it expands, since

$$L = 4\pi R_{BB}^2 \,\sigma T^4 = L_{Edd} = \text{constant},$$

so $T \propto R_{BB}^{-1/2}$. We see the small precursor peak when the temperature of the photosphere drops so low that it no longer emits X-rays. Double-peaked bursts have less radius expansion, and therefore the temperature doesn't drop so far.

The photosphere can contract again only when the total luminosity drops below the Eddington limit. As the photosphere retreats, it heats up, causing the rise of the main burst. The peak of the main burst occurs when the photosphere has shrunk to its original size, and the X-ray luminosity is again very near (but below) the Eddington limit. Therefore radius-expanding bursts can give good estimates of the size of a neutron star.

Some of the low-mass X-ray binaries that exhibit radius-expanding bursts live in globular clusters. Their distances are well known, hence we can measure the peak luminosities of the bursts accurately. Surveys have shown the peak luminosities to be $(3.0 \pm 0.6) \times 10^{31}$ W, or $\sim 10^5 \, L_\odot$. This is a little higher than the Eddington limit that we calculated for a $1.4 \, M_\odot$ neutron star accreting hydrogen, but consistent with the accretion of material that is not pure hydrogen (see Section 6.2, in particular Exercise 6.2).

Exercise 6.5   An X-ray burst is observed from an X-ray binary known to be around 8 kpc away, with a mini-burst seen several seconds before it. The main peak flux is 4.5×10^{-11} W m^{-2}, and the difference spectrum is best described by a black body with a temperature of $kT \sim 2.1$ keV.

(a) What is the peak luminosity of the main burst? Is this what you would expect?

(b) Using the peak luminosity and temperature of the main burst, estimate the radius of the neutron star. ∎

6.5.2   Periodic intensity dips

We briefly mentioned in Subsection 6.3.3 that a small number of high-inclination (almost edge-on) LMXBs can experience absorption that varies over the orbital cycle. The outer edge of the accretion disc is not smooth and uniform, but varies in thickness from region to region; hence some disc material may stray into the line of sight for some time during the orbital cycle, while we may have a clear view of the X-ray source at other times. The main culprit is the hot spot where the accretion stream crashes into the accretion disc (see also Section 5.2); the energy from the impact creates a bulge of hot material that extends well out of the plane of the disc. This is illustrated in Figure 6.17.

For LMXBs with inclinations of ~ 60–$80°$, these dips are energy-dependent, so they appear deeper in low-energy light curves than in high-energy light curves. An example of this is shown in Figure 6.18. These systems are known as the dipping sources, and they are very important for learning about the geometry of the different emission components, as we shall see shortly.

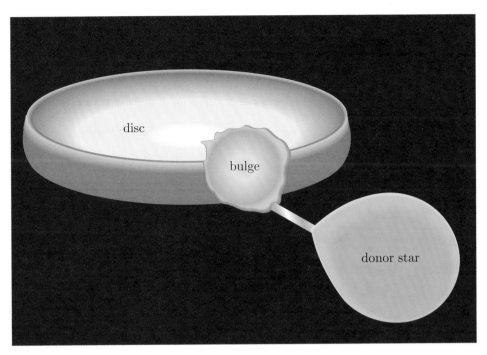

Figure 6.17   Illustration of the vertical structure of an accretion disc. The main feature is the bulge created where accreted matter impacts on the outer disc; however, the height of the outer disc can vary all round the disc.

For LMXBs with exceptionally high inclinations ($\sim$80–90°), the light curve varies continually over the orbital period, and the dips are energy-independent as the optical depth is $\gg 1$ for all X-ray energies. These sources are known as the ADC sources, as the observed emission is thought to come from the corona discussed in Subsection 6.3.2.

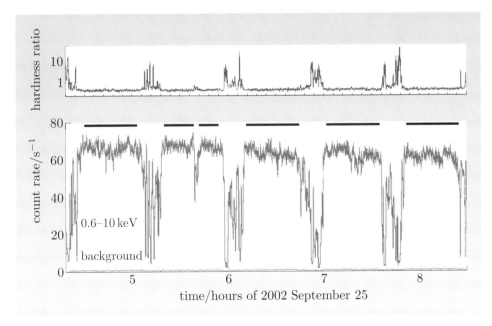

Figure 6.18   Light curve (red) and hardness ratio (blue) for the 2002 XMM-Newton observation of the dipping X-ray binary 4U 1916−053. Here the hardness ratio is defined as the 2.5–10 keV intensity divided by the 0.5–2.5 keV intensity. We see that the hardness ratio increases in the dips, showing that low-energy photons are preferentially removed.

As one might expect, the X-ray spectrum of a dipping source changes quite drastically during dipping, and this evolution is normally due only to the changes in absorption; having said that, it is not unknown for an X-ray burst to occur during a dip! This makes dipping sources an excellent laboratory for testing emission models, as we have to successfully describe every stage in the dip. For example, many bright LMXBs exhibit X-ray spectra composed of two components — one thermal component and one non-thermal component. However, there has been a long history of arguments about where the emission comes from, and studies of spectral evolution during dipping have a useful contribution to make.

A particularly striking example of spectral evolution during dipping came from an observation of 4U 1624−490 with the EXOSAT satellite, on 25 March 1985. One of the dips was found to have a flat bottom; we can see light curves covering the dip in the 1.2–4.7 and 4.7–9.8 keV bands in Figure 6.19. This suggests that one of the emission components was completely absorbed by the disc material that got in the way, while emission from elsewhere was still at least partially visible.

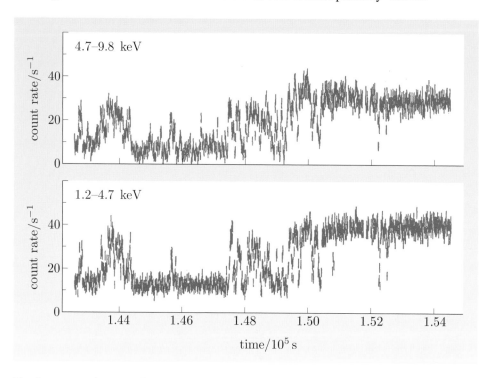

Figure 6.19 Light curves of 4U 1624−490, known as the Big Dipper, in two energy bands. Remarkably the dip is completely flat in the 1.2–4.7 keV band; the fact that the count rate bottoms out at 10 count s^{-1} indicates that at least two emission components are involved, and that one component has been removed completely.

Furthermore, the non-dip and dip emission spectra were strikingly different: the non-dip emission spectrum was well modelled by a black body that contributed ∼70% of the emission plus a power law, while the dip spectrum appeared to be a pure power law (see Figure 6.20).

Since the only difference between the two spectra was increased absorption during the dip, it was realized that the black body and power-law components must come from physically separated regions of the X-ray binary system. This led to a model where the absorption was set to different values for each component. It was revealed that N_H for the black body region dramatically increased from ∼5×10^{22} to at least 3×10^{24} atom cm^{-2} in the dip, while the absorption for the non-thermal component stayed at ∼5×10^{22} atom cm^{-2} throughout. These results show that the black body component comes from a relatively small region; likely candidates are the surface of the neutron star itself, or the innermost regions of the accretion disc. Conversely, the corona, which contributes the power-law component, must extend over a much larger area than the black body. In other dipping sources, the corona can be partly or wholly obscured, but all dipping LMXBs support the idea of a small black body and a large corona.

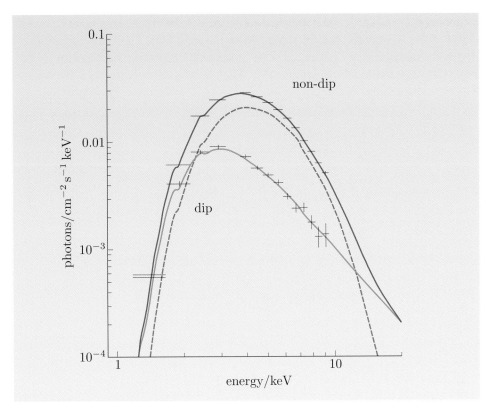

Figure 6.20 Comparison of non-dip (blue) and dip (green) spectra for 4U 1624−490. The non-dip spectrum is the sum of a black body component (red) and a power-law component, while the dip spectrum is a pure power law.

Figure 6.19 demonstrates that the intensity does not drop immediately: it takes a few minutes to go into the dip at the start (the **ingress**), and another few minutes to come out at the end (the **egress**). If the absorbing material in the disc covered the emission region all at once, the edges of the dips would be vertical; hence the absorber must cover the emission regions gradually. We call this *progressive covering*, and a cartoon of this situation is shown in Figure 6.21.

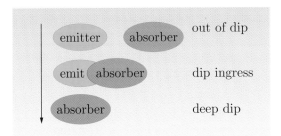

Figure 6.21 Illustration of the progressive covering of an extended emitter by an extended absorber.

We can estimate the size of the corona D_{ADC} from the ingress time ΔT_{ing} if we can work out how fast the absorbing material is moving. Assuming that the extra absorption during dipping is caused by matter on the outer edge of the disc, the velocity of the absorber, v_{abs}, is simply given by

$$v_{\mathrm{abs}} = \frac{2\pi R_{\mathrm{disc}}}{P_{\mathrm{orb}}} = \frac{D_{\mathrm{ADC}}}{\Delta T_{\mathrm{ing}}}, \tag{6.21}$$

where R_{disc} is the radius of the disc, and P_{orb} is the orbital period.

In ADC systems R_{disc} is thought to be 30–50% of the Roche-lobe radius of the neutron star. Measured values of D_{ADC} range over $\sim 10^7$–10^9 m, increasing as the X-ray luminosity increases.

Exercise 6.6   A new LMXB is discovered, which exhibits X-ray bursts as well as dips every 4.4 hours. The dip ingress time is found to be 2000 s. Estimate the size of the corona. *Hint*: To determine the size of the accretion disc, first estimate the donor mass from the orbital period–mass relation, then use the fact that for low-mass main-sequence stars $R_2/R_\odot \simeq M_2/M_\odot$ to calculate the Roche-lobe radius $R_{L,2}$ of the companion star. Then make use of the identity

$$\frac{R_{L,1}}{R_{L,2}} = \frac{f(1/q)}{f(q)} \tag{6.22}$$

(where $q = M_2/M_1$, and f is given by Equation 2.7) to determine the Roche-lobe radius of the neutron star, which in turn gives the disc radius.   ■

Warped discs

The occurrence of X-ray dips in non-eclipsing systems points towards the existence of disc material high above the orbital plane, perhaps as high as $H \simeq 0.2 \times r$ if r is the distance from the neutron star or black hole.

● For comparison, how thick are typical CV accretion discs?

○ The relative disc scale height H/r is of order of the local sound speed divided by the Keplerian speed, i.e. typically less than a few per cent (see also Equation 3.35). This is more than a factor of 10 less than in X-ray binaries.

One possible reason for why the discs may *appear* much thicker is that they could be warped. A **warped accretion disc** is, of course, no longer globally flat, and hence, strictly speaking, no longer a 'disc'. But *locally* it is still flat. We illustrate what we mean by a warped disc with the following recipe of how to make one. Break the initially flat disc up into many narrow disc rings. Then tilt each ring by a small angle δ against the orbital plane, and rotate the tilted ring about the rotation axis of the original disc by an angle ε (Figure 6.22). If ε is the same for each ring, the resulting structure is of course again a flat disc, inclined against the orbital plane by the angle δ. But if ε is a function of the ring radius, e.g. linearly increasing with r, the result is a tilted and twisted disc.

If the angle δ is large enough, the disc might appear rather thick to a distant observer who cannot resolve the detailed, warped structure — even though the initial disc was perfectly thin (Figure 6.23). A physical mechanism that can cause the disc to warp is an instability that arises when the disc is irradiated. The irradiating luminosity will be re-radiated or scattered from the disc surface. The resulting radiation pressure causes a reaction force. If there is a deviation from complete symmetry about the orbital plane, a net torque results, and a warp will form.

Exercise 6.7   In the above recipe for tilted and twisted discs, how large a tilt angle δ is necessary to explain an apparent disc of thickness $H/R = 0.2$, if the underlying disc is infinitesimally thin?   ■

6.5.3   Stochastic variability: organized chaos

Accretion-powered objects are fuelled by the accretion of discrete blobs of matter, hence there is a considerable amount of random variation in their luminosities, in

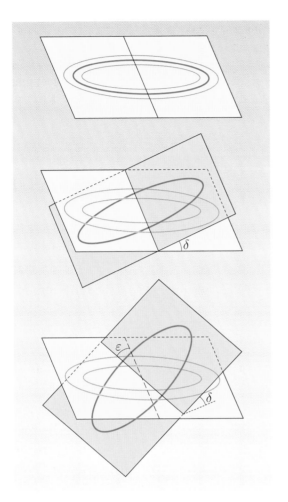

Figure 6.22 Constructing a warped disc: the angles δ and ε.

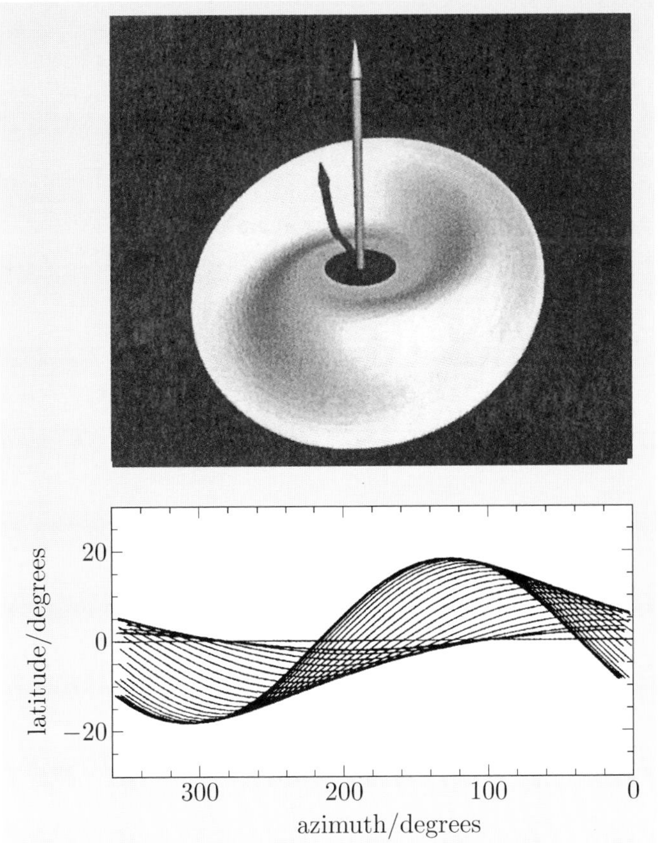

Figure 6.23 The shape of a warped disc (top), and an edge-on view of the disc (bottom).

addition to events such as X-ray bursts or dips. This is known as stochastic variability (in cataclysmic variables seen as *flickering*). However, there is often method behind the madness, and this is revealed in the power density spectra (PDS). The stochastic variability exhibited by X-ray binaries falls into two groups: **broadband noise** and **quasi-periodic oscillations** (QPOs). We shall discuss a particular kind of QPO that is seen during X-ray bursts, as these are fairly well understood. We shall also get an overview of the broadband noise, as it is very useful for characterizing the various types of behaviour in X-ray binaries.

Quasi-periodic oscillations

QPOs are features in the power density spectra of some X-ray binaries that bear similarities to the features seen in a pulsar PDS. Pulsar emission is extremely stable, and so distinctive that it is possible to identify hundreds of pulsars from their light curves alone. The pulse periods are often known to 14–16 significant figures, and they show up as an extremely narrow peak in the PDS of the pulsar (see Figure 6.24).

QPOs are created by stochastic processes, therefore they are not strictly periodic. As a result, they show up in a PDS as a broader peak that covers a range of

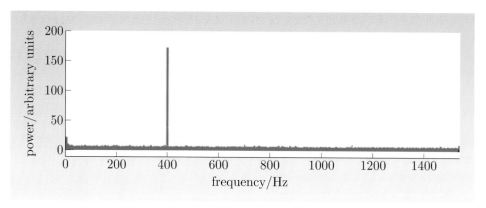

Figure 6.24   Example of a power density spectrum for a pulsar. The pulsation frequency is clearly seen as a spike, indicating a single, coherent modulation period in the light curve.

frequencies. We can define any PDS feature as a QPO if the **full width half maximum** (FWHM) spread in frequencies is less than half the central frequency. Figure 6.25 shows an example QPO exhibited by Sco X-1. Similar QPOs are seen in many Galactic X-ray binaries near the Eddington limit. One possible explanation assumes that the extremely high radiation pressure experienced by the inner disc material sets up some sort of feedback loop or sound waves.

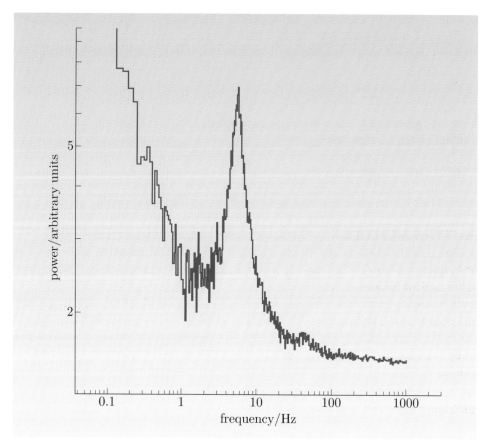

Figure 6.25   The power density spectrum of Sco X-1 during its so-called normal branch phase. The QPO is the large spike centred at ~6 Hz. We can also see strong broadband noise, as discussed in the next subsection.

Exercise 6.8   Figure 6.25 shows a QPO in the PDS of Sco X-1 with a peak of 6.0 Hz; it is unclear what causes it. If the QPO were caused by variations in the disc structure at some distance from the neutron star, what distance would that be? Assume Keplerian orbits and a neutron star mass of $1.4\,\mathrm{M_\odot}$.   ■

Broadband noise

In addition to QPOs, we can often see variability in LMXBs in excess of **Poisson noise** (see the box below) over a wide range of frequencies; this is called broadband noise. This broadband noise takes on different characteristics that depend on the accretion rate. LMXBs with neutron star primaries exhibit very similar broadband noise to LMXBs with black hole primaries; since both types of LMXB accrete from a disc, it is clear that the broadband noise must be a product of disc accretion. Indeed, all known black hole high-mass X-ray binaries (HMXBs) are also disc accreting, and show the same types of broadband noise.

Poisson noise

Even if a light source has a constant intensity, the emission of photons is a discrete process where photons can be produced at any time. Therefore, if we observe a star with a mean intensity of 100 count s^{-1}, then we might see, for example, 97 photons from it in one second, then 113 in the next. In 1838, Siméon-Denis Poisson derived the equation that governs such systems. If we observe a source with mean intensity I over a time interval ΔT, then we expect to see $I \times \Delta T$ photons on average. Poisson calculated the probability of observing N photons over this interval, $P(N, I\,\Delta T)$, to be given by

$$P(N, I\,\Delta T) = \frac{(I\,\Delta T)^N \times \mathrm{e}^{-I\,\Delta T}}{N!}. \tag{6.23}$$

Equation 6.23 is known as the Poisson distribution, and has a mean value of $I\,\Delta T$ with a standard deviation of $(I\,\Delta T)^{0.5}$. The corresponding variation in the number of photons observed from a light source over a given interval is known as the Poisson noise.

The PDS of an LMXB can be approximated by a series of power laws that apply for a particular frequency range: $P(f) \propto f^{-\alpha_f}$. For low accretion rates, $\alpha_f \sim 0$ for frequencies below a certain critical value, where $\alpha_f \sim 1$ up until a second, higher critical frequency, where $\alpha_f \sim 2$. Also, the random scatter in intensity is about 10–50%; we say that the *root mean square (rms) variability* is 10–50%. For high accretion rates, the $\alpha_f \sim 0$ regime does not exist; $\alpha_f = 1$ up until a certain frequency, and $\alpha_f \sim 2$ for higher frequencies. Furthermore, the random scatter is reduced to around 5–10% above Poisson noise. Figure 6.15 above shows an example PDS from Cygnus X-1 at low and high accretion rates.

Combining emission and variability

As well as showing the characteristic PDS described above, disc-accreting X-ray binaries have particular emission spectra associated with the different accretion rates. Therefore we can classify an X-ray binary if we can determine its PDS and emission spectrum.

At low accretion rates, LMXBs exhibit the same kind of variability and emission regardless of whether the accretor is a neutron star or a black hole; astronomers call this behaviour the low–hard state. Low-state emission is dominated by inverse Comptonization as described in Subsection 6.3.2. We see a power-law energy spectrum, with photon index Γ_p in the range ~ 1.4–2 up to the electron energy ($kT_e \sim 100$–$300\,\mathrm{keV}$), and see a Wien tail for energies above the electron energy.

Neutron star emission spectra at higher accretion rates are still dominated by inverse Comptonized emission, but a second, thermal component appears and becomes increasingly more prominent as the accretion rate gets higher. The photon index of the non-thermal component tends to get steeper (i.e. Γ_p increases), and the electron temperature decreases as the accretion rate increases, due to the higher influx of cooling photons onto the hot electrons.

Black hole LMXBs can exhibit similar high accretion rate spectra; however, they do also have a party trick that neutron star LMXBs can't match. They can enter a so-called high–soft state, where the emission is almost totally thermal, from the accretion disc. The disc black body spectra are characterized by inner disc temperatures of ~ 0.7–$2\,\mathrm{keV}$.

- What is the corresponding temperature in Kelvin?
- According to Equation 1.26, 1 eV is about $10^4\,\mathrm{K}$, and so here $T \simeq 10^7\,\mathrm{K}$.

The temperature is lower for more massive black holes, because the last stable orbit is further out (see Equation 1.11). One graphic demonstration of this was given in 2003 by Chris Done and Marek Gierliński. They generated colour–colour plots from hundreds of observations of Galactic LMXBs with known black hole or neutron star primaries. Figure 6.26 is a schematic representation of their results. In particular, there is a region that is inhabited exclusively by black hole systems.

A survey of transitions from low state to high state in Galactic neutron star LMXBs reveals a highly significant fact. At the transition, the 0.01 keV–1000 keV luminosity of the system is a fixed fraction of the Eddington luminosity, either $\sim 2\%$ or $\sim 10\%$, depending on the colour properties of the transition.

Thanks to the signature properties of the different accretion states and these regulated transitions from low to high states in neutron star LMXBs, we can identify black hole LMXBs from their X-ray properties alone. This is because the Eddington luminosity is proportional to the mass of the primary (see Equation 6.9). A black hole will have a higher Eddington luminosity than a neutron star, hence it can show low-state behaviour at higher luminosity than a neutron star.

Exercise 6.9   Known neutron star masses tend to be very close to $1.4\,\mathrm{M_\odot}$, although a couple have been found with masses as high as $2.1\,\mathrm{M_\odot}$. What is the highest luminosity low to high state transition that you could expect from a neutron star? ∎

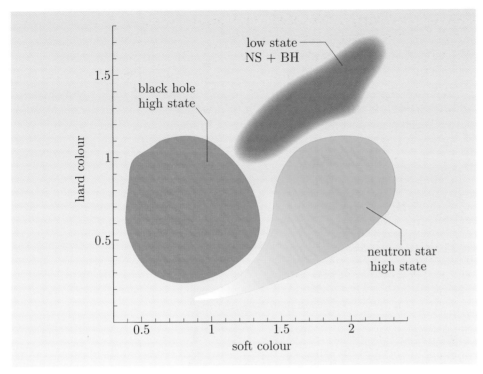

Figure 6.26 The hard colour versus soft colour plane with regions populated by different types of LMXBs, indicated schematically. The red region is occupied only by black hole LMXBs.

6.6 Ultraluminous X-ray sources

So far, we have discussed accretion onto neutron stars and stellar mass black holes, i.e. black holes that are formed when a star goes supernova.

There is a limit to how much mass a stellar mass black hole can have, since higher-mass stars are more difficult to form and their luminosity approaches the star's Eddington limit (see the box below).

Accreting systems that exceed the Eddington limit for stellar mass objects, and are not related to the galaxy nucleus, are called **ultraluminous X-ray sources** (ULXs). There is no generally accepted strict definition of what constitutes a ULX, but most would agree that a ULX requires a luminosity exceeding $\sim 2 \times 10^{32}$ W, the Eddington limit for a $15\,M_\odot$ black hole. However, a record-breaking black hole has recently been found in an X-ray binary called IC10 X-1 (the brightest X-ray source in a galaxy called IC10); its mass is in the range ~ 24–$33\,M_\odot$!

ULXs are found in regions of intense star formation, such as the arms of spiral galaxies or where two galaxies collide. There are no ULXs in the Milky Way, or in the Andromeda galaxy (the nearest spiral galaxy to our own). Therefore we must study objects that are 10^6 pc away or more, and therefore it is no surprise that the nature of ULXs is far from certain.

The Pistol Star: giant amongst giants

The Pistol Star (see Figure 6.27) is thought to have started on the main sequence with a mass of $200\,\mathrm{M_\odot}$, which is near the upper mass limit for a stable star according to current theory. Its luminosity is $> 10^6\,\mathrm{L_\odot}$, which is likely to be around the Eddington limit, hence the Pistol Star can barely hold itself together against the radiation pressure. Indeed, it may have lost as much as $\sim 100\,\mathrm{M_\odot}$ in the last 2×10^6 years!

Figure 6.27   The Pistol Nebula, containing the Pistol Star that created it, one of the most massive stars known, or indeed possible. This is a false-colour infrared image taken by the Hubble Space Telescope, and shows a field that is $\sim 1.5\,\mathrm{pc}$ across at the distance of the Pistol Star.

Some astronomers believe that ULXs may harbour intermediate-mass black holes that bridge the gap between black holes formed from stars and the supermassive black holes at the centres of galaxies. Others believe that ULXs may contain stellar mass black holes, but emit their radiation in a tight beam, like pulsars, rather than isotropically. Yet others believe that most of the emission from ULXs may come from an extended region that is locally sub-Eddington, and that ULXs may represent a short-lived period of exceptionally high mass transfer in X-ray binaries with stellar mass black hole primaries. However, since the ULXs are defined only by their luminosities, the term could describe a rag-bag mixture of sources encompassing all three types of ULX and others besides.

6.6.1   Intermediate-mass black holes

Perhaps the most straightforward interpretation of ULXs is that they contain black holes with masses ~ 50–$1000\,\mathrm{M_\odot}$. However, theoretical models describing the formation of black holes from stars have difficulty producing black holes with masses $\gtrsim 20\,\mathrm{M_\odot}$. This is because their progenitors are very high mass stars, and these are more difficult to form and their luminosity approaches the star's Eddington limit (see the box on the Pistol Star).

Furthermore, it is difficult to reconcile the expected emission from such systems with the observed ULX spectra. The closest approach of the accretion disc to the black hole is the last stable orbit, around $3\,R_S$ for a Schwarzschild black hole or $0.5\,R_S$ for a Kerr black hole with maximum spin, and R_S depends on the mass of the black hole (see Section 1.2). Using Equation 3.28 or Equation 1.20, we can estimate the temperature of the inner edge of the accretion disc. The resulting values are much lower than those observed in most ULXs (~ 1–2×10^7 K); this is another reason why most ULXs are unlikely to contain intermediate-mass black holes.

Exercise 6.10 (a) The Eddington rate $\dot{M}_{\rm Edd}$ is the mass accretion rate for which the accretion luminosity equals the Eddington luminosity. The Eddington rate is an upper limit for the mass accretion rate. Determine an expression for $\dot{M}_{\rm Edd}$ for Schwarzschild black holes.

(b) Use the expression for $\dot{M}_{\rm Edd}$ and Equation 1.17 (Example 1.10 in Subsection 1.4.1) to determine the peak temperature in an accretion disc around a black hole that accretes at the Eddington rate.

(c) What is the corresponding temperature for a ULX containing a $100\,{\rm M}_\odot$ black hole and accreting at the Eddington rate? How does this compare to observed temperatures of 1–2×10^7 K? ∎

6.6.2 Are ULXs beamed like pulsars?

Conventionally, we estimate the luminosity of a source at distance d from the received flux via

$$F = \frac{L}{4\pi d^2}, \quad \text{so} \quad L = F4\pi d^2. \tag{6.24}$$

This assumes that the power is emitted equally in all directions. If the emission from a particular system is confined to certain directions, or a given **solid angle** $\Delta\Omega$, our luminosity estimate for that system will be wrong, possibly by an order of magnitude or more. We already know that pulsars emit their radiation in beams, and some ULXs may be beamed also.

● What is the solid angle Ω subtended by the full sphere?

○ The surface area of a sphere with radius r is given by $4\pi r^2$. Hence $\Omega = 4\pi r^2/r^2 = 4\pi$.

We define a beaming factor b as

$$b = \frac{4\pi}{\Delta\Omega} \tag{6.25}$$

so that $b = 1$ when there is no beaming, and $b \gg 1$ indicates strong beaming. For any ULX, we can derive the minimum beaming factor, if it is beamed, by simply dividing the hypothetical luminosity deduced from Equation 6.24, by the Eddington limit of the accretor.

Exercise 6.11   A bright X-ray source is found in the spiral arm of a galaxy that is known to be $\sim 3.26 \times 10^6$ pc away. It has a 0.3–10 keV flux of 3.5×10^{-15} W m^{-2}.

(a) What is its 0.3–10 keV luminosity?

(b) Assuming that it is a beamed X-ray binary containing a 10 M$_\odot$ black hole, what is the minimum beaming factor?

(c) What solid angle does this correspond to? Is this a maximum or minimum? ■

6.6.3   Extended emission

The Eddington limit that we derived in Section 6.2 makes the idealized assumption that all the X-rays come from a point source, and the accretion is spherically symmetric. However, we have already learned from the dipping sources that the non-thermal emission from LMXBs comes from an extended region, with measured corona sizes ranging over $\sim 10^7$–10^9 m. Therefore the LMXBs can achieve total luminosities that significantly exceed L_{Edd} without breaking L_{Edd} in any one place. For example, the neutron star LMXB Sco X-1 regularly puts out $\sim 5 \times 10^{31}$ W of X-rays, and sometimes has short, violent outbursts that exceed 10^{32} W. In other words, Sco X-1 routinely operates at 3–7 times L_{Edd}! However, most of the X-ray emission is non-thermal, and hence comes from an extended region, and nowhere exceeds the Eddington limit locally.

Most ULXs could be explained by X-ray binary systems containing stellar mass black holes, if most of the emission came from an extended region. However, they still must be supplied with enough material to fuel these high luminosities, and one possibility is thermal timescale mass transfer. This is mass transfer that is forced to proceed on the thermal time of the donor star by the mass transfer induced expansion of the Roche lobe relative to the donor star. (Although the mass transfer is dynamically stable, it cannot proceed on a slower nuclear expansion timescale because $\zeta_{equilibrium} - \zeta_L < 0$; see Equation 2.28 in Section 2.6.) Thermal-timescale mass transfer occurs only when the mass ratio is large, and will correspondingly last for only a very short time, near the birth of the X-ray binary. Therefore many astronomers believe that ULXs are high-mass X-ray binaries that are caught in some short-lived phase (like unruly teenagers!). This is consistent with the observation that ULXs are found in regions of active star formation.

6.7   From X-ray binaries to AGN

As we have already discovered, each galaxy is thought to contain a supermassive black hole at its centre, with a mass of $\sim 10^6$–10^{11} M$_\odot$. In galaxies like the Milky Way, this beast is fairly tame and quiet. However, in many cases there are enough stars nearby to fuel a giant accretion disc, creating a powerful AGN that can in some cases consume 1 M$_\odot$ yr^{-1}!

6.7.1 Spectral and timing properties

In fact, most X-ray emission from AGN is non-thermal, and originates in inverse Compton scattering of cool photons on hot electrons in a corona; this is very like the emission spectra of LMXBs in the low–hard state . Indeed, typical AGN X-ray spectra resemble power laws with photon index ~ 1.4. The timing properties of AGN are also similar to those of LMXBs, albeit shifted down to lower frequencies. We can see this in Figure 6.28, where the power density spectra of two AGN are compared with high-state and low-state power density spectra of the black hole X-ray binary Cygnus X-1. Note that the y-axis shows frequency $\times$ power, rather than power. The frequency breaks in the AGN NGC4051 and MCG-6-30-15 occur at frequencies $\sim 10^{-4}$ Hz, while the corresponding break in the high-state PDS of Cygnus X-1 occurs at 10 Hz.

● Why might we expect variability in AGN to be over much longer timescales than for X-ray binaries?

○ The supermassive black holes at the centres of AGN are many orders of magnitude more massive than stellar mass black holes, so their last stable orbits are that much bigger. In-falling material is consequently slower, leading to slower variations in intensity.

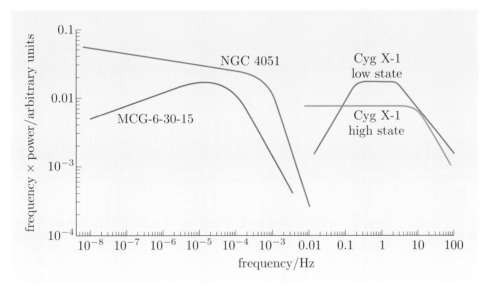

Figure 6.28 Here we compare the power density spectra of two AGN with the Galactic black hole binary Cyg X-1 in its low state and high state. Note that the y-axis is frequency $\times$ power rather than power. The PDS of NGC4051 is remarkably similar to Cyg X-1 in the high state, while MCG-6-30-15 has a PDS that is like Cyg X-1 in the low state, although the frequencies are several decades lower in both cases.

It is thought that the frequency of the second break in the low-state PDS, going from $P(f) \propto f^{-1}$ to $P(f) \propto f^{-2}$, is proportional to the mass of the black hole. We shall call this frequency f_{12}. The relationship between the mass of the supermassive black hole, M_{SMBH}, and f_{12} is not quite one-to-one, as different AGN with the same M_{SMBH} can have a wide range in f_{12}. However, this scatter appears to be due to differences in the accretion rate, which we can see as differences in luminosity. In 2006, Ian McHardy and co-workers found an empirical relation for AGN with luminosities ~ 0.001–$1\,L_{\mathrm{Edd}}$:

$$-\log_{10}\left(\frac{f_{12}}{\mathrm{Hz}}\right) \simeq 2.2 \pm 0.3 \times \log_{10}\left(\frac{M_{\mathrm{SMBH}}}{\mathrm{M}_\odot}\right) - 0.9^{+0.3}_{-0.2} \times \log_{10}\left(\frac{L}{\mathrm{L}_\odot}\right) - 2.4^{+0.2}_{-0.3}.$$

Remarkably, this result also seems to work for black hole X-ray binaries, demonstrating that disc accretion in AGN is closely related to disc accretion in

X-ray binaries, albeit on a much larger scale.

6.7.2   Fluorescence: measuring the black hole spin

As we discovered in Exercise 1.10, AGN discs have black body spectra that peak in the ultraviolet. Many of these photons are up-scattered to X-ray energies via inverse Compton scattering on hot electrons (Subsection 6.3.2). When X-rays hit the cold, optically thick accretion disc, they are reflected with a modified emission spectrum. The most conspicuous feature of the reflected light is an X-ray emission line at $6.4\,\text{keV}$. This is caused by the **fluorescence** of cold iron atoms.

> **Fluorescence**
>
> When an atom absorbs photons at one frequency, but re-emits them at a lower frequency, we call this fluorescence. It happens when the atom is excited to some high energy level, then decays to an intermediate energy level before returning to its original state. The energy of the emitted photon is independent of the energy of the incoming photon; the remaining energy is dealt with by other processes.

The $6.4\,\text{keV}$ iron K line is the most prominent X-ray emission line for two reasons. First, iron is the end-point of the stellar fusion cycle and therefore the most abundant element in the cosmos with an atomic number higher than 20. Second, the K fluorescence yield increases with atomic number.

These iron lines were first discovered in several AGN by Ginga. Once they were attributed to the disc, they were expected to be Doppler broadened, but the energy resolution of Ginga was too poor to confirm this. Later observations by ASCA, another Japanese X-ray satellite, demonstrated broadened iron lines for the first time, indicating speeds of $0.2c$. However, the broadening was found to be asymmetric, due to **gravitational redshift**. Hence fluorescent iron lines from material near the black hole would appear at lower energies than fluorescent lines from material that is further away, making the line profile asymmetric. Figure 6.29 shows an example $6.4\,\text{keV}$ line profile in an AGN spectrum.

The $6.4\,\text{keV}$ iron line profiles allow us to measure the black hole spin, because the last stable orbit for a spinning black hole is smaller than for a non-rotating one (see Section 1.2); hence the asymmetric gravitational line broadening will extend to lower energies for a rotating black hole. For example, Brenneman and Reynolds (2006) modelled the line profile seen in Figure 6.29 and found that the black hole in MCG-6-30-15 must be spinning, probably near its maximum rate.

Summary of Chapter 6

1. In X-ray astronomy, it is more common to describe the emission in terms of photon energy rather than wavelength or frequency. A photon with energy $1\,\text{keV}$ has frequency $\sim 2.4 \times 10^{17}\,\text{Hz}$ and wavelength $\sim 1.2\,\text{nm}$.

2. X-ray telescopes are quite unlike optical telescopes, because X-rays are scattered rather than focused when reflected from a mirror at incidence angles $\lesssim 88°$.

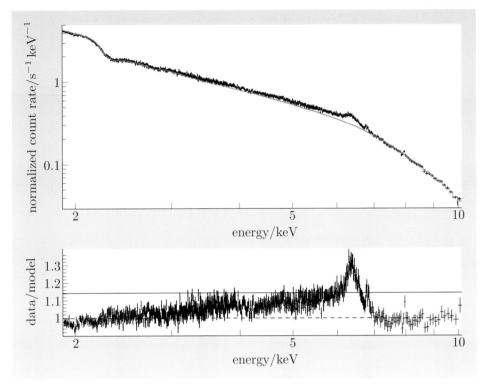

Figure 6.29 The top panel shows the 2–10 keV XMM-Newton spectrum of the AGN MCG-6-30-15 in the 2–10 keV band fitted with a power law. The bottom panel shows profile of the 6.4 keV line, expressed as the difference between the observed spectrum and the power-law model. This figure is taken from Brenneman and Reynolds (2006).

3. The Eddington luminosity L_{Edd} is the maximum theoretical luminosity for systems powered by spherical accretion. At this luminosity, the outward radiation pressure balances gravity. For hydrogen accretion,

$$L_{Edd} = \frac{4\pi G M m_{\mathrm{p}} c}{\sigma_{\mathrm{T}}} \simeq 1.26 \times 10^{31} \left(\frac{M}{M_\odot}\right) \mathrm{W}. \qquad \text{(Eqn 6.9)}$$

4. Cosmic X-ray emission suffers photo-electric absorption, that preferentially removes low-energy photons. This absorption can be caused by interstellar material along the line of sight, a nebula surrounding the X-ray source, or periodic occultation by a bulge in the accretion disc for high-inclination X-ray binaries. Hydrogen is the dominant absorber, and we express the absorption as N_{H}, the equivalent column density of neutral hydrogen along the line of sight. Heavier elements also contribute to the absorption, particularly for photon energies corresponding to an ionization state. The optical depth of the absorber to photons at energy E, $\tau(E)$, is given by

$$\tau(E) = N_{\mathrm{H}}\, \sigma(E), \qquad \text{(Eqn 6.14)}$$

where $\sigma(E)$ is an energy-dependent cross-section.

5. Most X-ray binaries have a strong non-thermal X-ray emission component. This arises from unsaturated inverse Compton scattering of cool photons on hot electrons in an accretion disc corona. Each photon takes its own path through the corona, but the mean number of scatterings is not sufficient to bring the photon energy up to the thermal energy of the electrons, kT_{e}. The resulting processed emission spectrum has a power-law energy distribution for energies lower than kT_{e}; the spectrum for energies higher than kT_{e} follows the Wien tail.

6. Analyzing the light curves of X-ray binaries at different energy bands can reveal a great deal of information. We can then manipulate these light curves to give X-ray colours that give a rough approximation of the emission spectrum. There is a region in colour space that is occupied only by black hole X-ray binaries in their 'high–soft' state. If an X-ray binary is observed with these colours, the accretor is very likely to be a black hole.

7. X-ray binaries can vary over timescales of milliseconds to years, hence variability analysis is a vitally important diagnostic tool for working out the physical processes that govern the observed behaviour.

8. Any light curve can be decomposed into sine and cosine waves with the appropriate amplitudes and frequencies. If we have a light curve with N evenly spaced data points with interval ΔT, then it is composed of waves with evenly spaced frequencies $k/N \Delta T$, where $k = 1, 2, \ldots, N/2$. The highest frequency, $1/2 \Delta T$, is called the Nyquist frequency. The power density spectrum (PDS) of a light curve is a recipe that gives the relative weight of each frequency in the sum; the power of each frequency, $P(f_k)$, is given by

$$P(f_k) \propto \left[\sum_{j=1}^{N} I_j \cos(2\pi f_k t_j) \right]^2 + \left[\sum_{j=1}^{N} I_j \sin(2\pi f_k t_j) \right]^2. \quad \text{(Eqn 6.19)}$$

However, the PDS cannot tell us the phase of each frequency component, so we cannot simply build the light curve from the PDS.

9. Some neutron star X-ray binaries exhibit X-ray bursts. These occur when accreted material builds up on the surface until a thermonuclear runaway is triggered, then all that fuel is burned in one go. This burst appears to emit a pure black body spectrum. Radius-expanding bursts are particularly interesting, as the luminosity reaches the Eddington limit and the photosphere of the neutron star is pushed out by the force of the blast. Since the luminosity must remain constant, the temperature drops as the photosphere radius increases; hence radius-expanding bursts are identifiable by a double peak, or a precursor burst. The X-ray peak of a radius-expanding burst occurs when the photosphere contracts to its original size; we can use it to measure the neutron star radius.

10. High-inclination X-ray binaries can exhibit periodic intensity dips caused by additional X-ray absorption by material in the outer accretion disc. Often, the absorption comes from the hot, inflated region where the accretion stream impacts on the accretion disc. Studies of the evolving emission spectra during dipping have shown that the accretion disc corona responsible for the non-thermal emission is extended. Furthermore, the diameter of the corona may be estimated from the ingress time of the dips.

11. All X-ray binaries exhibit remarkably similar variability and emission characteristics at low accretion rates (0.01–1000 keV luminosity $< 0.1 L_{\text{Edd}}$); however, L_{Edd} is proportional to the mass of the accretor. The most massive neutron stars known are $\sim 2.1 \, \text{M}_\odot$, hence if we observe low accretion rate behaviour at a 0.01–1000 keV luminosity $\geq 3 \times 10^{30}$ W, then the accretor is a candidate black hole.

12. Ultraluminous X-ray sources (ULXs) are found in star-forming regions of some galaxies, and are characterized by luminosities exceeding $\sim 2 \times 10^{32}$ W, the Eddington limit for a $15\,M_\odot$ black hole. Some ULXs could contain intermediate-mass black holes (IMBHs) that sit between the stellar mass black holes and supermassive black holes in the centres of galaxies. Other ULXs may beam their radiation (like pulsars) rather than emit isotropically. Yet others may have extended emission regions, so that the Eddington limit need not apply; these would likely be fed by thermal timescale mass transfer.

13. AGN discs are too cool for thermal X-ray emission; instead, AGN X-ray emission spectra are well described by a power law with photon index ~ 1.4. This non-thermal component comes from unsaturated inverse Compton scattering, just like the non-thermal emission from X-ray binaries.

14. The remarkable similarities in the power density spectra of AGN and X-ray binaries show that the discs around supermassive black holes are just like scaled-up versions of the discs found in X-ray binaries.

15. The $6.4\,\mathrm{keV}$ iron line profiles allow us to measure the black hole spin, because the last stable orbit for a spinning black hole is smaller than for a non-rotating one; hence the asymmetric gravitational line broadening will extend to lower energies for a rotating black hole.

Chapter 7 Relativistic outflows

Introduction

Outflows are a common feature of astrophysical objects. They take the form of quasi-steady stellar winds (e.g. the solar wind), unstable, outbursting activity (e.g. in X-ray binaries), or catastrophic explosions (e.g. supernovae). The amount of mass expelled, the rate at which this takes place, and the speed of the outflow combine to create a whole range of behaviours. A further division of outflows is based on their geometry: *winds* refers to nearly spherical outflows, while *jets* refers to elongated, collimated streams. Jets feature in many environments (extreme and less so) and they will be the focus of this chapter. Rather than giving a comprehensive overview of outflows, this chapter will focus on relativistic jet outflows that are common in high-energy astrophysical sources and more specifically in active galactic nuclei (AGN), in many X-ray binaries (also known as **microquasars**) and in gamma-ray bursts (GRBs), the subject of Chapter 8.

We start with a brief discussion of winds and explosions to place outflows in context. We then consider examples of outflows that attain speeds very close to the speed of light. This gives rise to a host of particular, counter-intuitive, phenomena that we discuss in detail. We move on to jets and touch on how they form and how they become and remain collimated, before examining in more detail how they dissipate the energy that they carry to eventually radiate it away with light curves and spectra that match observations. Shocks are the main mechanism of energy dissipation. We discuss their physics and proceed with examining processes that take place in them: the acceleration of particles to relativistic energies and the radiative mechanisms responsible for the emission.

7.1   Outflow conditions and types of outflow

Launching an outflow requires the presence of outward pressure that overcomes the gravitational pull of matter. In astrophysical systems, this pressure is provided by radiation, heat or magnetic fields. If the luminosity of an object exceeds the Eddington limit, the steady-state existence of this object is jeopardized. Radiation pressure will, in this case, drive a powerful wind or even a catastrophic explosion. The nature of the energy generation mechanism that ultimately leads to radiation is immaterial to the creation of the outflow. In main sequence or giant stars, it may be the nuclear fusion that sustains the luminosity. In proto-stars it is the gravitational potential energy released as the star contracts. Similarly, the collapse of the core of a massive star powers a supernova explosion. In high-energy systems, the gravitational potential energy that is generated by mass accretion feeds jets. Magnetic pressure is responsible for winds raised by the magnetospheres of pulsars and is also proposed for GRBs. The properties of the outflow do not depend on the type of pressure that is driving it. The most conspicuous case of an outflow is provided by the expansion of the early Universe following the Big Bang. The physics of relativistic gas dynamics that is necessary to describe the expansion and transition from radiation dominance to gas dominance is identical to that needed to describe outflows from stellar accreting systems, AGN or GRBs!

7.1.1 Winds

Stellar winds from very massive main sequence stars are clear evidence of radiation-driven outflows. They do not destroy the star but may cause the star to shed a large amount of mass in its lifetime. **Wolf–Rayet stars** are very massive ($M \gtrsim 20\,\mathrm{M_\odot}$) main sequence stars that expel mass copiously in strong winds. Their mass loss rate is $\sim 10^{-5}\,\mathrm{M_\odot\,yr^{-1}}$, a billion or so times higher than that of the Sun.

Exercise 7.1   (a) A star with an initial mass $20\,\mathrm{M_\odot}$ has a main sequence lifetime of about $5 \times 10^6\,\mathrm{yr}$. It is thought that such a star will undergo Wolf–Rayet type of winds for about a tenth of its lifetime. Estimate the mass that the star will lose over its Wolf–Rayet stage.

(b) Estimate the mass that the Sun will lose over its own lifetime ($10^{10}\,\mathrm{yr}$), given its mass loss rate of $10^{-14}\,\mathrm{M_\odot\,yr^{-1}}$. ■

The spectral types of Wolf–Rayet stars suggest surface temperatures in the range $25\,000$–$50\,000\,\mathrm{K}$. Their optical spectra also show very broad emission lines. The line broadening is due to the Doppler effect: radiation from material that is emitting at a certain wavelength λ_{em} (e.g. a spectral line) will be received at a wavelength λ_{rec} that is blueshifted if the material is moving towards the observer and redshifted if it is receding (see the discussion in Subsection 7.2.5). For a spherically symmetric wind, the radiation is blueshifted and redshifted by the same amount, thus resulting in a spectral line that is broader than its intrinsic width. The line width is given by Equation 5.1

$$\frac{\Delta\lambda}{\lambda_{em}} = \frac{v_{\parallel}}{c},$$

where $\Delta\lambda = |\lambda_{rec} - \lambda_{em}|$, and $v_{\parallel}$ is the component of the source velocity along the line of sight.

Exercise 7.2   Estimate the wind speeds implied by the observed equivalent widths of Wolf–Rayet stars, assuming a line width of $40\,\text{Å}$. Use a typical optical wavelength. ■

7.1.2 Explosive outflows

The quintessential example of an explosive outflow is a supernova explosion triggered by either the collapse of the core of a massive star to a neutron star or black hole, or the explosive burning of degenerate material in a white dwarf. In the former case, the fast collapse of the core sets off a shock front (see also Subsection 7.3.3 below) that propagates through the in-falling material, thus expelling the outer layers of the star. The details of how the collapse is reversed to an outflow are still poorly understood. It is clear, however, that a large amount of material (around $1\,\mathrm{M_\odot}$) is expelled, with speeds between $5000\,\mathrm{km\,s^{-1}}$ and $30\,000\,\mathrm{km\,s^{-1}}$. The outflow carries a huge amount of energy that is radiated away over the course of several weeks to months, finally leaving behind a remnant of shocked, filamentary interstellar matter. Type II supernovae are powered by radioactive decay during the initial weeks of their expansion. As this does not pertain to other flows, we do not discuss it further. Later stages of supernovae are powered by shocks that propagate into the surrounding medium and dissipate energy that can be radiated away. This process is common to various objects and will be discussed in detail in Subsection 7.3.2.

7.1.3 Relativistic outflows

An outflow is called relativistic if it attains speeds that are comparable to the speed of light. The most direct evidence that we have for relativistic outflows comes from radio observations of jets displaying apparent superluminal motion.

Jets

An astrophysical jet is an outflow that forms a stream. Jets are invoked in the lives of a variety of astrophysical systems (e.g. protostars, stellar binaries, galaxies). The jets that we consider here are visible across a broad part of the electromagnetic spectrum. Observations in the radio band are capable of high resolution and have thus been used to make detailed images of jets, such as those shown in Figure 7.1. Jets tend to form in pairs. When describing images of jets it is customary to use the terms *core*, to indicate the centre of symmetry of the system, *lobe* for the elongated extended emission part, and *hot spot* for the part where the formation broadens and ends in what appears to be a collision with the surrounding medium. Not all of these components are always visible. Depending on the orientation, as well as the jet's speed (see Subsection 7.2.3), we may be viewing both parts or just a single lobe.

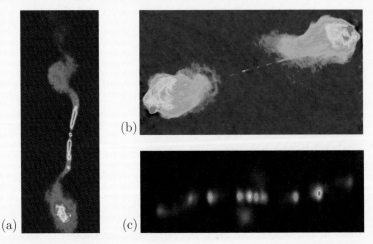

Figure 7.1   Radio images of jets. False colour is employed to render different intensities: (a) radio jets of 3C 449, an AGN system; (b) Cygnus A, another AGN system; (c) the jets of the Galactic binary source SS433.

In describing jets one has to account for the collimation, their manifestation across the spectral range and their association with systems that harbour a very compact object. More features arise when one considers the jets' behaviour in time, and their variability and spectral characteristics. In deciphering these, we shall resort to the pieces of physics that have been highlighted in the Introduction and are developed in the rest of this chapter.

Apparent superluminal motion is displayed by a number of jets of AGN as well as of X-ray binaries.

Exercise 7.3 From the image of GRS 1915+105 in Figure 7.2, estimate the speed at which each bright spot appears to be moving from the core (indicated by a cross). (At the source's distance of 40 000 ly, the angular size of 1 arcsec corresponds to a distance of 0.2 ly.) ∎

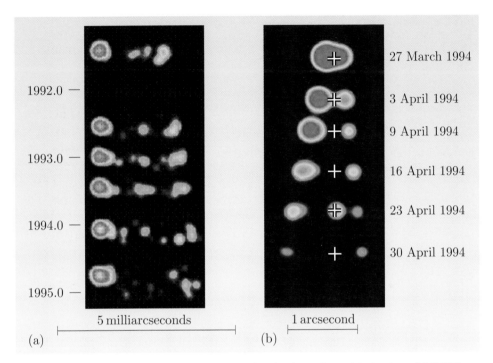

(a)   (b)

Figure 7.2 Examples of apparent superluminal motion: (a) the jet of 3C 279, an AGN source; (b) the jet of GRS 1915+105, a Galactic X-ray binary source.

Other evidence of relativistic outflows is more subtle and is based on inference (see Section 8.3).

7.2 Relativistic effects

Relativistic motion can deceive the observer. One such example is the illusion of superluminal motion that we have just discussed. Quite often, it is seemingly unphysical behaviour that betrays a relativistic outflow. We shall look into this after we set out the physics of relativistic motion that will equip us with the interpretational tools.

7.2.1 Lorentz transformations

In special relativity, things often look deceptively different from what they intrinsically are, so it is important to distinguish between physical quantities in the rest frame of the source and that of the observer. To describe the relationship between intrinsic properties (in the source frame) and observed properties (in the observer frame), one must resort to the relativistic **Lorentz transformations** linking the spacetime coordinates in the two frames.

To avoid clutter, from now on we shall omit the explicit dependence on V when writing γ.

For highly relativistic sources, it is convenient to express speed in terms of the Lorentz factor γ, which is related to the speed V by

$$\gamma(V) = \frac{1}{\sqrt{1 - (V/c)^2}}. \tag{7.1}$$

Thus the speed $0.999\,95c$ is indicated by $\gamma = 100$, since for $V \longrightarrow c$,

$$\frac{V}{c} = \sqrt{1 - \frac{1}{\gamma^2}} \simeq 1 - \frac{1}{2\gamma^2}. \tag{7.2}$$

Exercise 7.4   Obtain the approximate expression for V as a function of γ (Equation 7.2), for speeds approaching the speed of light, $V \longrightarrow c$. ■

The Lorentz transformations for the two frames of Figure 7.3 are

$$x' = \gamma\,(x - Vt), \tag{7.3}$$
$$y' = y, \tag{7.4}$$
$$z' = z, \tag{7.5}$$
$$t' = \gamma\left(t - \frac{Vx}{c^2}\right). \tag{7.6}$$

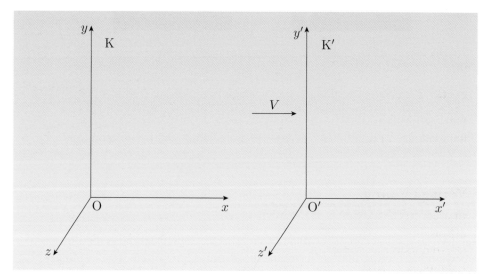

Figure 7.3   Two inertial frames K and K$'$ with a relative velocity V along the x-axis.

The inverse transformations (obtaining the coordinates in K from those in K$'$) can be obtained easily from Equations 7.3–7.6 by replacing V with $-V$:

$$x = \gamma_{_s}\left(x' + Vt'\right), \tag{7.7}$$
$$y = y', \tag{7.8}$$
$$z = z', \tag{7.9}$$
$$t = \gamma\left(t' + \frac{Vx'}{c^2}\right). \tag{7.10}$$

7.2.2 Special relativistic effects on lengths and durations

Immediate and experimentally confirmed implications of the Lorentz transformations are **length contraction** and **time dilation**: as seen by an observer in relative relativistic motion, lengths are foreshortened and clocks run more slowly. In the next two examples, we demonstrate how these effects come about. We use the frame set-up of Figure 7.3.

Worked Example 7.1

To demonstrate length contraction, i.e. that rods are measured shorter by a moving observer, consider a rod at rest in frame K'. The rod is oriented along the x'-axis, and its ends are at positions x'_1 and x'_2. In its own rest frame, the rod has length $\Delta l' = x'_2 - x'_1$ — called the **proper length**.

The proper length is also denoted as Δl_0.

Calculate the rod's length Δl observed in frame K moving at speed V with respect to frame K'.

Solution

In order to determine the length in frame K, the observer makes a simultaneous measurement of the positions of the rod ends, x_1 and x_2, at time t. Applying the Lorentz transformation Equation 7.3 to the coordinates of the ends of the rod, we get

$$\Delta l' = x'_2 - x'_1 = \gamma \left[x_2 - Vt - (x_1 - Vt) \right]$$

or

$$\Delta l = \Delta l'/\gamma, \tag{7.11}$$

which shows that the observer in frame K measures the length of the rod to be γ times smaller than the rod's proper length.

Worked Example 7.2

In order to demonstrate time dilation, consider an event that takes place at the origin O' of frame K' and lasts for $\Delta t' = t'_2 - t'_1$ in this frame. ($\Delta t'$ is referred to as the **proper time**.) Following the same method as in Worked Example 7.1, determine the duration Δt of this event as measured by an observer at rest in frame K.

The proper time is also denoted as $\Delta \tau$.

Solution

In order to make full use of the information given, we need the time transformation into frame K (Equation 7.10). We know that the positions of the event in frame K' are $x'_1 = x'_2 = 0$. Applying the time transformation to the start and end of the event in K, we obtain $\Delta t = t_2 - t_1 = \gamma(t'_2 - t'_1)$, so

$$\Delta t = \gamma \Delta t'. \tag{7.12}$$

The observer in motion will thus perceive the event as longer.

Length contraction and time dilation are not the only non-intuitive consequences of the Lorentz transformations that become appreciable at relativistic speeds. The following are observed phenomena due to a source that is approaching at a relativistic speed that can mislead the interpretations of observations:

- the radiation is concentrated into a narrow beam (**radiation beaming**);

- the source moves at a speed faster than that of light (**superluminal motion**);

- photons are received with frequencies higher than those emitted (**photon blueshift**);

- the source appears brighter (**luminosity boosting**);

- waiting time between events is briefer(!) or else photons arrive at a higher rate.

The understanding of these effects lies at the heart of deciphering sources that involve relativistic outflows such as AGN, GRBs and X-ray binaries with jets. We now discuss these effects in detail, and for motion in any direction (not just along the line of sight), starting with non-relativistic analogues where available. We will identify K′ as the source frame and express quantities measured in it as primed ′. Quantities in the observer frame K are unprimed.

7.2.3   Relativistic beaming

Aberration of light: the non-relativistic limit

Light aberration arises because motion of the observer changes the apparent direction towards a luminous object. The finite speed of light is the cause of the phenomenon, which takes place at any source speed. We have first-hand experience of aberration at non-relativistic speeds, for example in rain coming straight down from the sky (no side winds). Compared to when one is standing still, rain feels heavier as one starts walking, or running: rain that is falling straight down acquires a velocity component that corresponds to the walking speed, so it looks as if the rain is coming towards us from a forward position. Faces and any spectacles get wet faster, and umbrellas get tilted forward so as to intercept more raindrops. The effect is more pronounced as the pace quickens. A car's screen wipers have to work faster when one is driving at a higher speed in rain.

In strict analogy, travelling in a spaceship enjoying the vista of distant stars, we would see a skewed image if we were to travel at high speeds. Assume that stars are distributed isotropically about the observer. The view will remain isotropic if we travel slowly (as does the Earth). As we fire up the spaceship's engine, our starry sky will brighten right ahead. For this to be noticeable, our travel speed would have to be relativistic.

We shall show below how light aberration comes about for non-relativistic speeds, by considering the case of observing a star through an Earth-bound telescope. Figure 7.4 illustrates a telescope, pointed in the exact direction of a particular star, moving with speed v at angle θ to the direction to that star. The light from the star is moving at speed c, so it will take a time $t = l/c$ to traverse the length l of the telescope. During this time interval, the telescope will have travelled a distance vt. The component of this motion perpendicular to the light beam is $vt \sin \theta$, hence the light beam will strike the primary mirror of the telescope at a distance $vt \sin \theta$ from the axis of the telescope. This means that the observed direction to the star is shifted by an angle α (called the *aberration angle*), where

$$\tan \alpha \simeq \alpha = \frac{vt \sin \theta}{l} = \frac{vt \sin \theta}{ct} = \frac{v}{c} \sin \theta. \qquad (7.13)$$

● What units is α measured in?

○ α is measured in radians.

We have assumed that α is small, an assumption that is well-justified in the non-relativistic ($v \ll c$) limit.

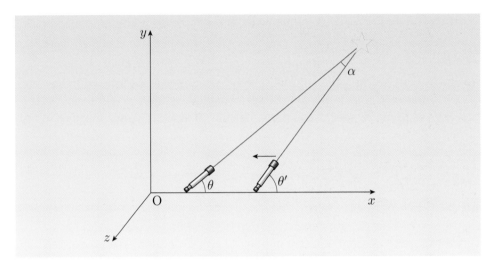

Figure 7.4 A telescope on the surface of Earth pointed at a star. The Earth's motion causes the star to appear in a direction slightly off its actual one.

Aberration is caused by the relative motion between source and observer. Our analysis would lead us to the same result if we took the telescope as fixed and the source moving.

Exercise 7.5 (a) The maximum possible value of aberration due to the Earth's orbital motion around the Sun is $21''$. Assuming that the orbit of the Earth is a perfect circle, calculate the Earth's orbital speed.

(b) Predict the maximum aberration that a telescope placed at the Earth's equator will measure due to the Earth's rotation. ∎

Aberration of light: the relativistic limit

The analysis above is strictly limited to the non-relativistic case. For a proper analysis that can be applied to all speeds, we have to turn to the Lorentz transformations. Measuring light aberration involves velocities. We therefore need to obtain the Lorentz transformations of velocities. Assuming that we know the velocity v' of an object in frame K$'$, where the source is at rest, we may obtain its velocity v in frame K, where the observer is at rest, by differentiating the appropriate Lorentz transformation equations (Equations 7.7–7.10).

Through the velocity transformation, we aim to determine how motion patterns that are known in the source frame appear in the observer frame. We usually have a good idea of the properties in the source frame despite the fact that we have no direct way of probing it (e.g. the simplest and often adopted assumption is that a source is expanding or emitting isotropically).

The velocity transformations are given by

$$v_x = \frac{v'_x + V}{1 + V v'_x/c^2},$$

(7.14)

$$v_y = \frac{v'_y}{\gamma(1 + V v'_x/c^2)},$$

(7.15)

$$v_z = \frac{v'_z}{\gamma(1 + V v'_x/c^2)}.$$

(7.16)

Exercise 7.6   The velocity components in frame K are given by $v_x = \mathrm{d}x/\mathrm{d}t$, etc. Differentiate the appropriate Lorentz transformation to obtain Equation 7.14. ■

We now need to see how the angle θ' that the velocity vector forms with the x'-axis in frame K$'$ transforms to the angle θ that the velocity vector forms with the x-axis in frame K. The x- and y-components of velocity define the tangent of the angle θ:

$$\tan\theta = \frac{v_y}{v_x} = \frac{v'_y}{\gamma(v'_x + V)} = \frac{1}{\gamma}\frac{v'\sin\theta'}{v'\cos\theta' + V},$$

(7.17)

where we have used $v'_y = v'\sin\theta'$ and $v'_x = v'\cos\theta'$.

The velocity and angle transformations that we have obtained are valid for any velocity $\boldsymbol{v}'$. Here we are interested in light — hence we set $v' = v = c$ as the speed of light in vacuum is the same in any frame. In this case Equation 7.17 becomes

$$\tan\theta = \frac{1}{\gamma}\frac{\sin\theta'}{\cos\theta' + V/c}.$$

(7.18)

Equation 7.18 expresses light aberration for any speed V.

The most interesting consequence of the velocity transformation is the effect of relativistic beaming: radiation emitted by a relativistically moving source is beamed in the direction of the motion. Let us quantify this beaming. Consider a photon that is emitted in frame K$'$ at right angles with respect to the line of sight, i.e. in the direction $\theta' = \pi/2$. Applying the angle transformation (Equation 7.18), this photon will be seen by the observer in frame K to travel in the direction given by

$$\tan\theta = \frac{\sin(\pi/2)}{\gamma[(\cos(\pi/2) + V/c]} = \frac{c}{\gamma V}.$$

For very large speeds ($V \to c$), the photon is directed towards the observer at a very small angle:

$$\tan\theta \simeq \theta \simeq \frac{1}{\gamma}.$$

(7.19)

So for a source that emits isotropically in its own rest frame, half the photons (those with $-\pi/2 < \theta' < \pi/2$) are seen to be narrowly concentrated in a cone of opening angle $2/\gamma$ (see Figure 7.5). Thus, if the source is viewed within an angle of $1/\gamma$ from the direction of the source's motion, it will appear much brighter than it intrinsically is. (The full power will seem to be emitted into a tiny solid angle.) Outside this cone, the source will be practically invisible.

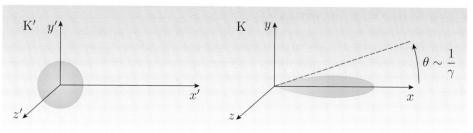

Figure 7.5 Radiation emitted isotropically in frame K′, the rest frame of the source, which is moving relativistically towards the observer, appears to be highly concentrated around the direction of the motion as seen by the observer in frame K.

Beaming is inferred in the observations of blobs of radiating material comprising the jets of AGN, in the fireball or jet-like outflow suggested in GRBs (Section 7.3 and Chapter 8), and at the microscopic level, occurs in synchrotron radiation (Subsection 7.3.4).

Worked Example 7.3

Estimate how much the observed flux is increased due to relativistic beaming for a source approaching the observer with velocity V.

(a) Apply this to a source of $V = 0.96c$. (This is the speed inferred for the jet of GRS 1915+105 — see Exercise 7.3 — as we shall see in Subsection 7.2.4.)

(b) Compare this result with the case of a source that is headed towards the observer at $\gamma = 300$. (This is the typical expansion factor of GRB jets, as we shall see in Chapter 8.)

Solution

The light from a source with luminosity L that emits isotropically is equally distributed over a spherical surface, so at distance d the flux (power received per unit surface area) is $L/4\pi d^2$. If the same power is, however, concentrated in a solid angle $\Delta\Omega$, the flux at the same distance will be $L/\Delta\Omega \, d^2$. Thus the brightness of the source gets boosted by a factor of $b = 4\pi/\Delta\Omega$ (see also Subsection 6.6.2).

Using spherical coordinates, the solid angle subtended by a cone with opening angle Θ is given by

$$\Delta\Omega = \int_0^{2\pi} \mathrm{d}\phi \int_0^{\Theta/2} \sin\theta \, \mathrm{d}\theta = 2\pi \left[-\cos\theta \right]_0^{\Theta/2},$$

so

$$\Delta\Omega = 2\pi(1 - \cos(\Theta/2)). \tag{7.20}$$

For small x we have $\cos x \simeq 1 - x^2/2$, so this becomes $\Delta\Omega \simeq \pi\Theta^2/4$. Here, $\Theta = 2/\gamma$ is small for large γ. So the brightness of the source is boosted by a factor $b = 4\pi/\Delta\Omega$, i.e.

$$b = 4\gamma^2. \tag{7.21}$$

(a) Applying this to a source approaching the observer at $V = 0.96c$, we get $\gamma = 1/\sqrt{1 - 0.96^2} \approx 3.6$ and $b \approx 51$.

(b) For a source moving at $\gamma = 300$, the brightness is boosted by 3.6×10^5.

7.2.4   Superluminal motion

A pure illusion created by relativistic motion is that of motion at speeds larger than the speed of light (Exercise 7.3). This is a different manifestation of the fact that photons emitted from a source that is moving towards the observer arrive with a time difference that is shorter than the one that they were emitted with. Consider a source moving relativistically at speed V making an angle θ with the line of sight (see Figure 7.6).

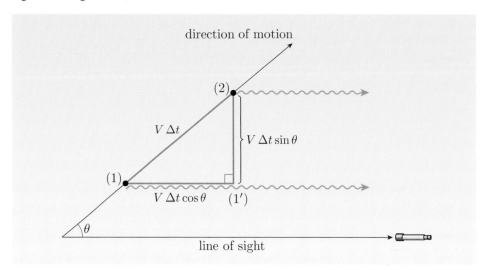

Figure 7.6   Two photons emitted at instances 1 and 2 from a source moving at a relativistic speed V that forms an angle θ with the line of sight.

As seen in the observer frame, the source emits a photon at position 1 and another one at position 2, a time Δt later. The later photon has been emitted at a distance $d = V \Delta t \cos \theta$ closer to the observer, so despite being emitted later, it can make up some of the time difference with the first photon when it reaches the observer. The two photons are received by the observer with a time difference of

$$\Delta t_{\text{rec}} = \Delta t - d/c = \Delta t \left(1 - \frac{V}{c} \cos \theta \right). \tag{7.22}$$

This applies equally to the time elapsed between distinct events, and to the duration of pulses. So events taking place in a source that is moving relativistically towards the observer are squeezed in time.

For a source approaching the observer along the line of sight ($\theta = 0$) with a highly relativistic speed, we can use the approximation $V/c \simeq 1 - 1/2\gamma^2$ (Equation 7.2) to rewrite the above equation as

$$\Delta t_{\text{rec}} = \frac{\Delta t}{2\gamma^2}. \tag{7.23}$$

Let us now revisit Figure 7.2 and Exercise 7.3. Using the images, an observer determines positions from the projection of the sources on the sky, through measurements of angular separations. Thus, in the time that the source has moved from position 1 to position 2 (an actual distance of $V \Delta t$), the observer sees the source having relocated from position 1' to position 2 (that are separated by $V \Delta t \sin \theta$). Hence the apparent velocity that the observer assigns to the source is

$$v_{\text{ap}} = \frac{V \Delta t \sin \theta}{\Delta t_{\text{rec}}} = \frac{V \sin \theta}{1 - (V/c) \cos \theta}. \tag{7.24}$$

For very small angles θ, we have $\sin\theta \simeq \theta$ and $\cos\theta \simeq 1$, so Equation 7.24 becomes

$$v_{ap} \simeq \frac{V\theta}{1 - V/c}. \qquad (7.25)$$

We have seen that the apparent speed can exceed the speed of light, $v_{ap} > c$. Using the last equation this translates into a lower limit for θ:

$$\theta > \frac{1 - V/c}{V/c}. \qquad (7.26)$$

Exercise 7.7 (a) What angle θ gives the maximum value of v_{ap}?

(b) Calculate the maximum value of v_{ap} from Equation 7.24. ∎

Exercise 7.7 shows that a source moving at relativistic speeds at a small angle with respect to the line of sight will appear to be superluminal, with an inferred speed reaching γV.

7.2.5 Doppler effect

We are familiar with the Doppler shift of sound waves in the non-relativistic domain. The changing pitch of a passing ambulance is the all too frequently quoted example. The sound wave frequency of the approaching source is higher than the frequency when the source is at rest. The emitted waves have to fit in a shrinking distance (see Figure 7.7), which increases their frequency. In complete analogy, a receding source is perceived as emitting lower frequencies than it intrinsically is. This is the *classical Doppler effect*, purely a result of relative motion along the line of sight.

For light, the effect of increased frequency is called blueshift, while the effect of decreased frequency is called redshift. The terminology draws from the phenomenon in the optical waveband, but it is applied to all types of emission, irrespective of the actual wavelength the source is observed at. Thus it is customary to refer to, for example, redshifted radio sources or blueshifted gamma-rays.

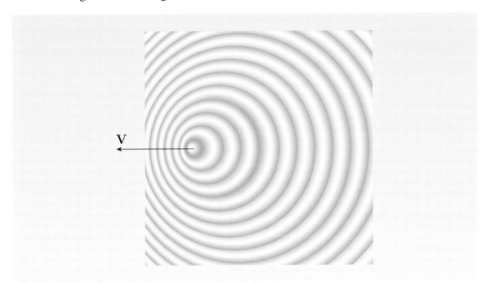

Figure 7.7 Demonstration of the classical Doppler effect. As the source is catching up with the emitted waves, the distance between wave crests shrinks, resulting in an increase in the received frequency.

For electromagnetic radiation, we can quantify the classical Doppler effect by recasting the time intervals in Equation 7.22 in terms of the frequency $\nu = 1/\Delta t$ of the wave.

In the special relativistic regime, the observed frequency of photons is affected by both the relative motion of source and receiver *and* the time dilation applied to the duration of any event.

Thus, expressing the time interval Δt separating the emission of two photons as seen in the observer frame in terms of the time interval $\Delta t'_{em}$ as seen in the source frame, i.e. $\Delta t = \gamma \, \Delta t'_{em}$ (Equation 7.12), we can write Equation 7.22 as

$$\Delta t_{rec} = \gamma \, \Delta t'_{em}[1 - (V/c)\cos\theta]. \tag{7.27}$$

The *Doppler factor* $\mathcal{D}$ defined as

$$\mathcal{D} = \gamma^{-1}[1 - (V/c)\cos\theta]^{-1} \tag{7.28}$$

provides a compact form in which to express relativistic transformations in the case of motion at any angle with the line of sight. With this definition, Equation 7.27 becomes

$$\Delta t_{rec} = \frac{\Delta t'_{em}}{\mathcal{D}}. \tag{7.29}$$

Recasting this in terms of frequencies, the *relativistic Doppler effect* can then be written as

$$\nu_{rec} = \mathcal{D}\,\nu'_{em}. \tag{7.30}$$

● Show that Equation 7.30 is equivalent to Equation 5.1 for non-relativistic source motion.

○ For non-relativistic motion $\gamma = 1$ and $\mathcal{D} = [1 - (v/c)\cos\theta]^{-1}$. The velocity component along the line of sight is $v_\| = v\cos\theta$. Equation 7.30 thus becomes $\nu_{rec} = \nu'_{em}/(1 - v/c)$. Noting that λ_{em} is the wavelength in the source rest frame, we have

$$\frac{\nu'_{em}}{\nu_{rec}} = \frac{\lambda_{rec}}{\lambda_{em}},$$

which leads to Equation 5.1.

Exercise 7.8 (a) Calculate the Doppler factor $\mathcal{D}$ for a source moving at right angles to the observer (i.e. $\theta = \pi/2$).

(b) Calculate the Doppler factor $\mathcal{D}$ if the source is moving head-on towards the observer.

(c) If the source is emitting monochromatic radiation of frequency ν_{em}, qualify the change in the observed frequency as blueshift or redshift for the angles considered in parts (a) and (b). ∎

As Exercise 7.8 illustrates, the magnitude and sign of the Doppler effect depend on the angle between the direction of the motion of the source and the line of sight. For a source approaching the observer head-on with a large Lorentz factor, we have $\mathcal{D} \simeq 2\gamma$, so Equation 7.29 — describing time intervals or the duration of pulses — becomes

$$\Delta t_{rec} \simeq \frac{\Delta t'_{em}}{2\gamma} = \frac{\Delta t}{2\gamma^2}, \tag{7.31}$$

while the relativistic Doppler effect (Equation 7.30) becomes

$$\nu_{\text{rec}} \simeq 2\gamma\,\nu'_{\text{em}}. \tag{7.32}$$

For a spherically, relativistically expanding source, it can be shown that photons emitted with a frequency ν'_{em} arrive at the observer with a frequency that is on average $\nu_{\text{rec}} \simeq \gamma\,\nu'_{\text{em}}$. They thus appear significantly more energetic.

7.2.6 Luminosity boosting

We saw earlier that relativistic beaming concentrates the light towards the direction of motion, making the source appear brighter. When it comes to inferring the luminosity of the source, further relativistic effects come into play. (In what follows, we neglect factors of a few and focus on the dependence on γ, which is the dominant factor.) The luminosity — the amount of energy carried by photons per unit time — is a property intrinsic to the source. We measure flux — the amount of energy received per unit time per unit frequency per unit area of the detector (see also the box in Subsection 7.3.4 below). In the case of relativistic motion towards the observer, the flux that we measure is boosted by a factor of γ^2 due to beaming (Equation 7.21) and a further Lorentz factor γ because of the Doppler shift of the rate of photon arrival (Equation 7.31). Moreover, photons emitted at a certain frequency get registered at a frequency that is γ times higher (Equation 7.32). If the emission is characterized by a power-law spectral distribution $F_{\nu'} \sim (\nu')^{-\alpha}$, the flux received at ν is $F_{\nu} \sim (\nu/\gamma)^{-\alpha}$ and hence further boosted by a factor of γ^{α}. Overall, the flux at a specific frequency will therefore be boosted by a factor of $\gamma^{3+\alpha}$. If we attribute this to a source that is emitting isotropically, as we have assumed here, the luminosity L_{ν} at this frequency that we infer for the source is higher than the intrinsic luminosity $L'_{\nu'}$ by the same factor:

$$L_{\nu} \simeq \gamma^{3+\alpha} L'_{\nu'}. \tag{7.33}$$

For relativistic motion that forms an angle θ with the line of sight, this relationship becomes

$$L_{\nu} \simeq \mathcal{D}^{3+\alpha} L'_{\nu'}. \tag{7.34}$$

Relativistic effects summary

Here, we bring together all the relevant relativistic effects. The following are — in a nutshell — the relativistic effects on observable quantities, in the case of highly relativistic motion along the line of sight. The unprimed quantities refer to the observer frame, and the primed quantities refer to the source frame.

$\Delta l = \Delta l'/\gamma$	length contraction	(Eqn 7.11)
$\Delta t = \gamma\,\Delta t'$	time dilation	(Eqn 7.12)
$\Delta t_{\text{rec}} = \Delta t'_{\text{em}}/2\gamma$		(Eqn 7.31)
$\quad \simeq \Delta t/2\gamma^2$	pulse duration	
$\nu_{\text{rec}} = 2\gamma\,\nu'_{\text{em}}$	Doppler effect	(Eqn 7.32)
$L_{\nu} \simeq \gamma^{3+\alpha} L'_{\nu'}$	luminosity boosting	(Eqn 7.33)

7.3   Energy extraction and emission

Jets were thought for a long time to be rather exceptional, but observations of ever-increasing elaboration have shown them to be fairly common. They are even an indispensable component of star formation. They seem to spring out of situations where gas with high angular momentum swirls down a steep gravitational well. Accretion discs provide such an environment. Relativistic jets are inferred in AGN, many X-ray binaries (Figure 7.1) and GRBs (see Chapter 8). In all three cases the underlying astrophysical objects are believed to harbour black holes, though some of the X-ray binaries with jets may involve a neutron star accretor.

Observational evidence suggests that these jets differ in their propagation speed, with AGN showing speeds of γ of the order of a few, X-ray binaries γ of up to about 10, and GRBs γ between 100 and 1000.

The jets provide a vehicle by which energy tapped from the accretion disc eventually emerges as radiation. For this conversion, energy must be extracted from the disc, transported outward, converted to thermal particle energy and radiated away. In what follows, we examine the stages of this energy conversion.

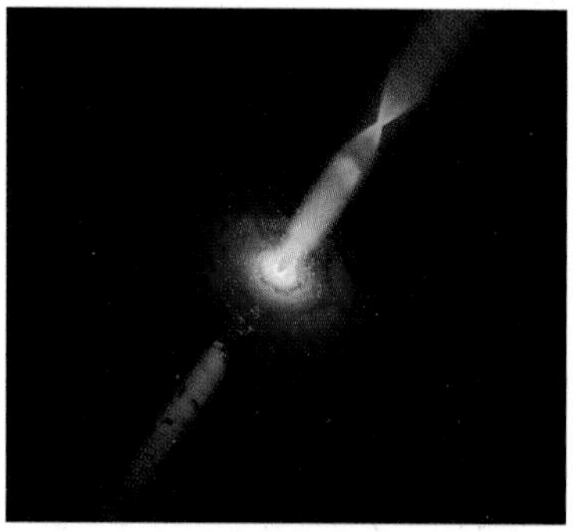

7.3.1   Jet formation and collimation

The formation of a jet demands a mechanism of acceleration and one of collimation: matter has to be accelerated to a high speed in order to escape the central gravitational potential and also attain relativistic speeds thereafter. Moreover, it has to be channelled in narrow streams and not diffuse thinly soon after its formation. These are areas of current research, and many questions remain open on both the nature and the efficiency of the mechanisms. In general, jets are associated with rotation and are thought to be launched along the rotation axis of the system in question (see Figure 7.8). They are often seen in systems that possess an accretion disc and are now used to infer the presence of such a disc.

Figure 7.8   Artist's conception of a system harbouring a jet.

The initial outflow may, in fact, be driven by radiation, thermal pressure or magnetic pressure. The efficiency of the first two mechanisms is probably too low to lead to jets that accelerate to relativistic speeds. Magnetic pressure may be key to setting off the flow initially. The flow can further accelerate through radiation pressure (see also Section 8.3 below).

Collimation is thought to be supplied by magnetic fields that essentially form a cavity in which the outflow propagates. The specifics of the field's geometry and how collimation is achieved are hotly disputed. However, the role of a magnetic field is thought of as indispensable as earlier considerations limited to gas dynamics have proven to be inadequate.

7.3.2   Dissipation in the flow

More likely than not, the flow in the jet is not smooth. It is conceivable that the energy is supplied to the jet in a variable manner. Differences in speed and energy will be ironed out in collisions that will dissipate any excess energy. Approximating such energy-injection episodes by blobs of certain mass and speed allows a schematic description of this dissipation process. Faster blobs (which may actually attain the form of a shell) ejected later will catch up with slower, earlier ones. The simplified model in the example below provides the basis for modelling energy dissipation in AGN and GRBs.

Worked Example 7.4

Consider two blobs with masses m_1 and m_2 ejected from a source in the same direction, with time difference Δt and speeds v_1 and v_2 (where $v_2 > v_1$ so that the second blob eventually catches up with the first one). All quantities are measured in the frame of the origin of ejection.

(a) Calculate the distance r_{dis} that the two blobs will have travelled when they collide.

(b) Assume that the two speeds, while being highly relativistic, differ by a very small factor only. (This expresses well the idea of fluctuations.) To make the calculation specific, express v_1 in terms of its Lorentz factor $\gamma \equiv \gamma(v_1)$ and set $v_1/v_2 \simeq v_1/c$. Obtain r_{dis} in the limit $\gamma \gg 1$.

Solution

(a) The distance that each blob will have travelled after time t when the blob ejected later catches up with the earlier one is given by

$$r_{dis} = v_1 t,$$
$$r_{dis} = v_2(t - \Delta t).$$

Combining these two equations, we can solve for t, to obtain $t = v_2 \Delta t/(v_1 - v_2)$. Substituting this result into the first equation, we get

$$r_{dis} = \Delta t \, \frac{v_1 v_2}{v_2 - v_1}. \tag{7.35}$$

(b) We have

$$r_{dis} = \Delta t \, \frac{v_1}{1 - v_1/v_2} \simeq \Delta t \, \frac{v_1}{1 - v_1/c}.$$

Using the approximation for large Lorentz factors, Equation 7.2, gives $1 - v_1/c \simeq 1/(2\gamma^2)$, so

$$r_{dis} \simeq \Delta t \, \frac{c(1 - 1/(2\gamma^2))}{1/(2\gamma^2)} = c \, \Delta t(2\gamma^2 - 1) \simeq 2\gamma^2 c \, \Delta t. \tag{7.36}$$

This analysis can be applied to any combination of masses and initial speeds. Such inelastic collisions convert the bulk kinetic energy of the flow to random particle energy. The efficiency of this conversion is a consideration that is central to astrophysical models involving shocks. Radiation mechanisms convert the random particle energy to radiation with some efficiency too. If the efficiency of energy conversion by the collisions is low, little energy will be available to be radiated in the first place. To find out the energy content and speed of motion of the post-collision blob, one has to consider conservation of the special relativistic energy and momentum. For an object of rest mass m and speed

$v = c(1 - 1/\gamma^2)^{1/2}$, its special relativistic energy is given by

$$E = \gamma m c^2,$$

while its momentum is

$$p = \gamma m v = \gamma \left(1 - \frac{1}{\gamma^2}\right)^{1/2} mc = (\gamma^2 - 1)^{1/2} mc.$$

Setting up the conservation of these two quantities prior to and after the collision, in the general case of two blobs with different masses m_1 and m_2, and Lorentz factors $\gamma_1 \equiv \gamma(v_1)$ and $\gamma_2 \equiv \gamma(v_2)$, that are cold (no thermal energy content), one can obtain the Lorentz factor γ_b of the post-collision blob and the efficiency ε of the conversion of bulk kinetic energy to thermal energy E_{th}. Expressing all quantities in the observer frame (which is the same as the frame in which the blob injection centre is at rest), the energy and momentum conservation equations are

$$\gamma_1 m_1 c^2 + \gamma_2 m_2 c^2 = \gamma_b (m_1 + m_2) c^2 + E_{th}, \tag{7.37}$$

$$(\gamma_1^2 - 1)^{1/2} m_1 + (\gamma_2^2 - 1)^{1/2} m_2 = (m_1 + m_2 + E_{th}/\gamma_b c^2)(\gamma_b^2 - 1)^{1/2}, \tag{7.38}$$

where $E_{th}/\gamma_b c^2$ is the equivalent mass of the thermal energy. The efficiency ε is defined as the ratio of thermal energy produced in the collision to the total energy available to the system initially:

$$\varepsilon = \frac{E_{th}}{(\gamma_1 m_1 + \gamma_2 m_2) c^2}. \tag{7.39}$$

One interesting limit of this model is the case where the two blobs are ejected with relativistic speeds (i.e. $\gamma_2 > \gamma_1 \gg 1$). This is applicable to collisions within the flow referred to as *internal*. In this case, the approximation used in Equation 7.2 gives

$$(\gamma^2 - 1)^{1/2} \simeq \gamma - \frac{1}{2\gamma} \tag{7.40}$$

and this holds for γ_1, γ_2 and γ_b. Combining Equations 7.37 and 7.38 to eliminate E_{th} in this approximation, gives

$$\gamma_{b,int} \simeq \sqrt{\frac{\gamma_1 m_1 + \gamma_2 m_2}{m_1/\gamma_1 + m_2/\gamma_2}}, \tag{7.41}$$

while using Equation 7.37 in Equation 7.39 gives

$$\varepsilon_{int} = 1 - \frac{(m_1 + m_2)\gamma_b}{\gamma_1 m_1 + \gamma_2 m_2}. \tag{7.42}$$

For a wide range of masses and speed differences, the efficiency is of the order of a few to 10%. Such efficiencies are appropriate for AGN and X-ray binary jets, as suggested by observations. However, if the mechanism is applicable to GRBs, then the required efficiencies are higher, implying comparable blob masses and speeds that correspond to Lorentz factor ratios of a few. Fine tuning of the ejection mechanism in terms of masses and energies may thus be required for GRBs.

The other limit of interest is when the slow-moving blob (blob 1) is assumed to be at rest. Then the model describes the situation whereby a fast-moving flow hits stationary cold material. This set-up actually describes what happens when the material of a supernova explosion hits the interstellar medium. The supernova explosion shell is moving at non-relativistic speeds. The equivalent situation involving relativistic motion describes the braking of the relativistic outflow generated by a GRB. The outflow then decelerates from γ_2 to γ_b. We now determine what amount of stationary mass m_1 is needed to cause a significant deceleration, say to $\gamma_b \lesssim \gamma_2/2$. Applying Equations 7.37 and 7.38 for $\gamma_1 = 1$, and making use of the approximation in Equation 7.2 for γ_2 and γ_b, shows that a mass $m_1 \gtrsim m_2/\gamma_2$ is required to yield a post-collision Lorentz factor of $\gamma_b \lesssim \gamma_2/2$, generating a thermal energy of $E_{th} \gtrsim \gamma_2 m_2 c^2/4$. These collisions are often called *external*.

Exercise 7.9 Calculate the efficiency of bulk to thermal energy conversion for the external collision described in the last paragraph. ∎

7.3.3 Relativistic shocks

The treatment of energy dissipation in terms of inelastic collisions is sufficient so long as the whole shell (blob) of material can be approximated as a point mass. In realistic situations, however, the flow has to be described as a fluid and the aforementioned collisions as shocks.

Shock conditions

In the following discussion of shocks we shall adopt a qualitative approach that will lead us to the results applicable to relativistic shocks, bypassing the derivation of equations.

Shocks

Pressure perturbations in a fluid propagate as sound waves that consist of successive compressions and rarefactions. The speed of propagation of such compressions/rarefactions is

$$c_s = \sqrt{\left(\frac{\partial P}{\partial \rho}\right)_s},\qquad(7.43)$$

where P is the pressure, ρ is the mass density, and the derivative is taken at constant entropy. This generalizes the definition for the isothermal sound speed that we presented in Chapter 3 (Equation 3.31). For an ideal monoatomic gas with a **polytropic equation of state** (i.e. $P \sim \rho^{\widehat{\gamma}_{ad}}$, where $\widehat{\gamma}_{ad}$ is the adiabatic index), this becomes

$$c_s = \sqrt{\frac{\widehat{\gamma}_{ad}\, P}{\rho}}.\qquad(7.44)$$

Pressure is an increasing function of density ($\widehat{\gamma}_{ad} \geq 1$). As a result, the speed of sound is higher in the higher-density regions of a fluid. In the course of propagation of a perturbation, compressions will tend to travel faster than rarefactions, eventually catching up with leading parts of a fluid. This creates a situation whereby the same location in a fluid would have multiple values of physical parameters (see Figure 7.9). As this is an impossible situation, a slim region develops across which the physical properties (e.g. density, velocity, temperature) change abruptly. This region is called a **shock front**, or simply shock. The shock front propagates supersonically. The fluid that is traversed by a shock front gets compressed, heats up and accelerates. The yet unshocked region of the fluid is usually referred to as *upstream*, and the region that has been shocked as *downstream*.

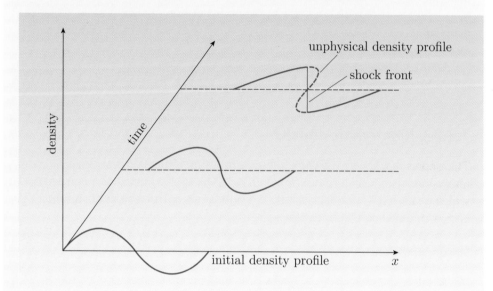

Figure 7.9   A sketch of how a wave propagating in a fluid develops to a shock.

Similarly, a shock front develops when a supersonic disturbance (such as the one caused by a supersonic aeroplane in air) propagates in a fluid.

The properties of an astrophysical fluid are determined by the densities of three physical quantities that relate to mass, energy and momentum. Across a shock front there is an effectively discontinuous change of the characteristics of the fluid. The relationships between the values of the physical quantities on either side of the shock front are derived from conservation laws and are called **jump conditions**. Mass (particle number) is conserved as there is nothing in a shock front that can create new mass. The same applies to the momentum flux, as there is no additional push generated by the shock. If the shock is not radiating (a reasonable first approximation), then energy is also conserved.

Before presenting the jump conditions, we apply additional restrictions to the types of shock that we consider here, focusing on shocks that develop in relativistic flows. Shocks are called relativistic if their speed, as measured from the yet unshocked fluid, corresponds to a high Lorentz factor. We examine the limit of strong shocks, those for which the square of the ratio of the fluid speed to the speed of sound in the fluid is substantially higher than 1. Moreover, the pre-shocked material is assumed to be cold (non-relativistic). Finally, the shocks considered in the sources of interest are *collisionless*: dissipation is due to electromagnetic fields present in the flow rather than collisions between particles. So the pressure in the fluid is provided by the magnetic field.

To express the jump conditions that link the particle number density n, speed V and energy density U across the shock front, we label the physical quantities by u for upstream (yet unshocked) and d for downstream (shocked) quantities and measure them in the rest frame of the respective fluid. Let $\gamma (\equiv \gamma(V))$ denote the flow Lorentz factor of the pre-shocked fluid, as seen in the observer frame. Assume that the flow is cold and that the particles of the post-shock material are heated to relativistic speeds (so the equation of state is relativistic, $\widehat{\gamma}_{ad} = 4/3$). Then, viewed from a frame that is *at rest* with the pre-shocked fluid, the post-shocked fluid acquires a speed corresponding to a very high Lorentz factor, of order γ:

$$\gamma_d \simeq \gamma/\sqrt{2}. \tag{7.45}$$

The jump condition for the particle number density is:

$$n_d' \simeq 4\gamma_d n_u'. \tag{7.46}$$

This shows that the flow highly compresses the fluid. The compression can be understood as a result of length contraction (Equation 7.11). A cubic unit volume will appear compressed by a factor $1/\gamma_d$ in the direction of motion. Similarly, the energy density increases by a factor of order γ^2:

$$U_d' \simeq 4\gamma_d^2 n_u' m_p c^2. \tag{7.47}$$

● Identify the two relativistic effects that give rise to the factor γ_d^2 in Equation 7.47.

○ U is an energy density. Just as for the particle number density n, the length contraction contributes a factor γ_d, while the energy of one particle (proton) is $\gamma_d m_p c^2$.

The expressions in Equations 7.45–7.47 are customarily used in the calculation of the dynamic and radiative properties of the relativistic shocks believed to be involved in GRBs. We shall return to them in Section 8.4 in the next chapter. In situations that lead to internal shocks as described earlier or shocks from the flow running against a stationary medium, shocks are produced in pairs. One shock moves forward into the stationary (or slow) fluid, and the other moves backwards within the as yet unshocked flow. They are separated by a contact discontinuity that, viewed by the observer, is moving in the direction of the flow with a Lorentz factor γ. These shocks are called *forward* and *reverse*.

Particle acceleration

We saw that the shock equations are derived from the conservation of a number of physical quantities across the shock front. What does change through a shock front is the entropy of the fluid. It increases downstream, signalling the conversion of streaming (bulk) kinetic energy to random (thermal) energy of the particles. This provides the energy reservoir that is tapped into by the radiation mechanisms. In astrophysical settings, the heat generated in shocks is radiated away with characteristic power-law spectra. The conversion of heat to radiation is mediated by the increase in the speeds of particles in a process known as **particle acceleration**. During this, electrons attain a power-law energy distribution:

$$N(\gamma_e) \propto \gamma_e^{-p} \quad \text{for } \gamma_e > \gamma_{min}, \tag{7.48}$$

where γ_e is the Lorentz factor of an electron of energy $\gamma_e m_e c^2$, $N(\gamma_e)$ is the number of electrons with energies in the γ_e to $\gamma_e + d\gamma_e$ range, p is the power-law index, and γ_{min} is the minimum energy that electrons will have reached by the end of the acceleration phase.

We use γ_e to indicate the Lorentz factor of electrons. The relativistic speed will have been acquired via particle acceleration mechanisms and describes motion that is random. This is to be distinguished from γ that we use to describe the Lorentz factor of the bulk motion of the flow. This motion is directed, like in an outflow.

Electrons are thought to be accelerated in shocks through the repeated crossing of the shock front. Electrons in the shocked fluid will have an initially thermal (Maxwellian) velocity distribution. Their random velocities will make some of them cross the shock front repeatedly, in a process that accelerates them to ever-increasing energy. It can be shown (though this is beyond the scope of this book) that each crossing of the shock front makes an electron gain a fraction v_{ud}/c of their initial energy, where v_{ud} is the relative velocity of the upstream and downstream fluids. Application of this to non-relativistic shocks shows that after multiple crossings, electrons assume a power-law distribution with $p = 2$, while simulations of relativistic shocks suggest that p is in the range of 2.2–2.3. Observational evidence from supernovae, AGN and GRBs corroborates these findings. Nevertheless, the details of this process are poorly understood.

7.3.4 Non-thermal radiative processes

Radiation from relativistic sources, as well as many highly energetic settings in astrophysical objects, is often characterized by power-law spectra spanning a very broad range of frequencies. The radiation mechanisms behind these are non-thermal; the most prevalent are synchrotron radiation and inverse Compton scattering, which we have already met in Subsection 6.3.2 in the context of ADC sources. In this subsection we discuss the physical mechanisms and their observed signatures.

Measures of light

A number of quantities are used to describe the amount of light emitted by or received from a source. We review here the definitions of those quantities that are used in this book.

The familiar luminosity L of a source measures the *emitted* power, i.e. the amount of energy E divided by the time over which this energy is emitted. A useful concept is the luminosity in specific bands, for example L_γ in gamma-rays, or $L_{0.1-10\,keV}$ for the X-ray band between 0.1 and 10 keV.

Quantities that describe the amount of radiation *received* include the flux F_ν, which measures the power received per unit area per unit frequency. Corresponding to the luminosity in a specific band is the power received per unit area, νF_ν. The significance of this quantity becomes apparent when one considers the spectral distribution of the received flux. The power $P_{[\nu_1,\nu_2]}$ received per unit area in a *finite* frequency interval $[\nu_1, \nu_2]$ is given by

$$P_{[\nu_1,\nu_2]} = \int_{\nu_1}^{\nu_2} F_\nu \, d\nu.$$

This is just the area under the curve F_ν for ν between frequencies ν_1 and ν_2. However, spectral energy distributions that extend over many decades in frequency are commonly displayed on a logarithmic frequency scale, i.e. as a function of $\log_{10} \nu$. Noting that $\mathrm{d}\log_{10}\nu \propto \mathrm{d}\nu/\nu$, the received power can also be written as

$$P_{[\nu_1,\nu_2]} \propto \int_{\log_{10}\nu_1}^{\log_{10}\nu_2} \nu F_\nu \, \mathrm{d}(\log_{10}\nu).$$

So the area under the curve νF_ν for $\log_{10}\nu$ between $\log_{10}\nu_1$ and $\log_{10}\nu_2$ signifies the power received in this frequency interval.

Time-integrated versions of F_ν and νF_ν define fluences S_ν and νS_ν, respectively. The total energy emitted, E, and the bolometric (i.e. integrated across the full spectrum) fluence S are related via

$$S = \frac{E}{4\pi d^2}, \tag{7.49}$$

where d is the distance to the source. Luminosity and flux are measures of power, while fluence is a measure of energy.

For sources that are moving relativistically, the time over which the radiation is observed may be very different from the time over which it was emitted (Equation 7.23). The spectral distribution of the energy is accordingly disparate. (The total energy received, however, is the same as the total energy emitted, as relativistic transformations do not violate physical laws.) Thus it is practical to have quantities that describe both the source's intrinsic properties (such as luminosity L and energy E) and its appearance (such as flux F_ν, or νF_ν, and fluence S_ν, or νS_ν).

In high-energy astronomy, photons are tagged by their energy content $h\nu$ (where h is the Planck constant), rather than frequency. Hence photon frequencies are often referred to by their energy equivalent in eV. Photons are the energy quanta in which radiation is produced and propagates. The Doppler and aberration effects are examples of how relative motion alters the flux and photon frequency received. The total number of photons, however, is a count not subjected to such effects. The total number of photons received from a source is the total number of photons produced. Thus the photon number is a meaningful *invariant* quantity.

Densities, rather than total values, determine the radiative properties of a source. Densities are not invariant, but are linked to total numbers in a straightforward manner. The relationship of the photon number density n_ν with the source luminosity is especially useful. For a spherically symmetric source of radius R and luminosity L_ν, the energy flux through its surface, $L_\nu/4\pi R^2$, is equal to the photon energy streaming at the speed of light c through a unit surface area, $n_\nu h\nu c$, from which we obtain

$$n_\nu = \frac{L_\nu}{ch\nu 4\pi R^2}. \tag{7.50}$$

Synchrotron radiation

An electron that enters a magnetic field B with velocity v experiences a Lorentz force $ev \times B$ that accelerates it in a direction perpendicular to the plane defined by the magnetic field and the velocity. As a result, the electron is set in a helical orbit about the magnetic field direction. Due to the acceleration, the electron radiates with the frequency of its revolving motion. This is known as **cyclotron radiation** and it is *monochromatic* (i.e. of a single frequency). The cyclotron frequency is obtained by requiring the Lorentz force to act as the centripetal force of the electron's helical motion:

$$\nu_{\mathrm{cy}} = \frac{1}{2\pi}\frac{eB_\perp}{m_{\mathrm{e}}}, \tag{7.51}$$

where $B_\perp$ is the magnetic field strength perpendicular to the velocity of the electron.

Exercise 7.10   Derive Equation 7.51 by equating the centripetal force (see, for example, Subsection 1.3.1) and the Lorentz force quoted above.   ■

If the electron speed is relativistic (of Lorentz factor γ_{e}), its radiation becomes strongly beamed in the constantly changing direction (Subsection 7.2.3). The pulse seen by the observer is thus substantially shrunk, resulting in a broader spectrum. The spectrum (Figure 7.10) is extended and assumes a power-law shape of slope $\frac{1}{3}$ in F_ν vs ν, peaking at a characteristic frequency

$$\nu_{\mathrm{sy}} \simeq \frac{1}{3}\gamma_{\mathrm{e}}^2\frac{1}{2\pi}\frac{eB_\perp}{m_{\mathrm{e}}c}. \tag{7.52}$$

As the energy of a relativistic electron is proportional to $\gamma_{\mathrm{e}}m_{\mathrm{e}}c^2$, the frequency of this **synchrotron radiation** increases with the square of the energy. Thus higher-energy electrons emit at substantially higher frequencies.

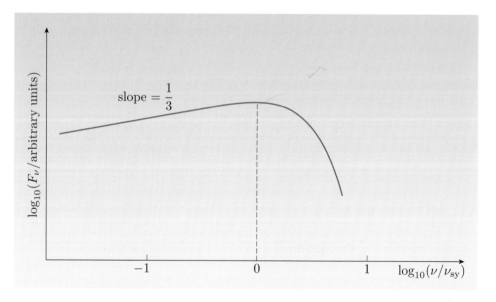

Figure 7.10   A sketch of the synchrotron spectrum of a single electron.

The power P_{sy} emitted in synchrotron radiation from a single electron depends on the energy of the electron and the energy density U_B of the magnetic field:

$$P_{\mathrm{sy}} = \tfrac{4}{3}\sigma_{\mathrm{T}}\, c(\gamma_{\mathrm{e}}^2 - 1)U_B, \tag{7.53}$$

where σ_{T} is the Thomson cross-section and for a magnetic field B we have $U_B \propto B^2$. Equation 7.53 shows that more energetic electrons are contributing more to the power emitted, as well as radiating at higher frequencies.

We may now consider the spectrum from an ensemble of electrons. As we've seen in Subsection 7.3.3, electrons in shocks are accelerated to a power-law distribution in energy. In the presence of a magnetic field, they will emit synchrotron radiation. The observed synchrotron emission is the sum of the emission from individual electrons, with different energies and moving at different angles to the magnetic field. Hence the contribution of each individual electron peaks at the corresponding value of ν_{sy}. At each frequency ν, the emission comes predominantly from the electrons that have $\nu_{sy} \simeq \nu$. Consequently, if the distribution of electron energies follows a power law, then the observed synchrotron spectrum will also be a power law as we demonstrate below.

Worked Example 7.5

Considering a power-law distribution of electrons (Equation 7.48), calculate the power-law index of the synchrotron spectrum.

Solution

The flux emitted at a particular frequency ν is determined by the number of electrons emitting at that frequency, $N(\nu)$, and the power P_{sy} emitted by each of these. This can be written as

$$F_\nu \propto P_{sy}\, N(\nu), \qquad (7.54)$$

where the power P_{sy} is given by Equation 7.53, and the number of electrons is determined by Equation 7.48. We must, however, transform this distribution from the number of electrons with a particular energy γ_e to the number of electrons emitting at a particular frequency ν. For this, it is sufficient to note that the number of electrons within a range of energies can be expressed equivalently using different labels, so that

$$N(\gamma_e)\, d\gamma_e = N(\nu)\, d\nu, \quad \text{or} \quad N(\nu) = N(\gamma_e)\frac{d\gamma_e}{d\nu}.$$

In order to evaluate $d\gamma_e/d\nu$, we need an expression that links γ_e with ν. This is given by Equation 7.52, since we have argued that emission at ν comes predominantly from electrons whose emission peaks at this frequency. Differentiating Equation 7.52, we get $d\nu/d\gamma_e \propto 2\gamma_e$, which can be inserted into the last equation to give

$$N(\nu) \propto N(\gamma_e)/\gamma_e.$$

For highly relativistic electrons, Equation 7.53 suggests that $P_{sy} \sim \gamma_e^2$. Collecting these expressions for P_{sy} and N in terms of γ_e, and substituting into Equation 7.54, we obtain

$$F_\nu \propto \gamma_e^2\, N(\gamma_e)\frac{1}{\gamma_e} \propto \gamma_e \times \gamma_e^{-p} = \gamma_e^{-(p-1)}. \qquad (7.55)$$

The last step in our derivation is to recast the last expression in terms of frequency. For that we need to use Equation 7.52 again to express γ_e as a function of ν: $\gamma_e \propto \nu^{1/2}$. The spectrum of the synchrotron radiation from an ensemble of electrons with a power-law distribution thus becomes

$$F_\nu \propto (\nu^{1/2})^{-(p-1)} = \nu^{-(p-1)/2}. \qquad (7.56)$$

The spectrum obtained in this worked example is sketched in Figure 7.11.

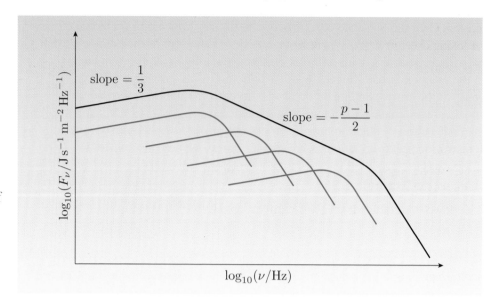

Figure 7.11   A sketch of the synchrotron spectrum from a power-law energy distribution of electrons, plotted as log flux versus log frequency. The low-frequency slope is $\frac{1}{3}$. The high-frequency spectrum has a slope of $(p-1)/2$.

Observed spectra are often described as power laws with an index α (see Chapter 8), as $F_\nu \propto \nu^{-\alpha}$. In the example above we linked the observed spectral flux F_ν with the distribution of the underlying electron population. The relationship

$$\alpha = (p-1)/2 \qquad (7.57)$$

is between an observationally determined quantity, α, and a model-derived quantity, p. This is an example of the interplay between observations and theoretical understanding. Spectral studies of sources where the emission is identified as synchrotron radiation and the spectral index can be confidently identified with that of Equation 7.57 provide the motivation and check for particle acceleration models.

● In Subsection 7.3.3 we saw that non-relativistic shock simulations give electron power-law distributions with $p \approx 2$, while relativistic shocks give $p \approx 2.3$. What are the high-frequency spectral indices of synchrotron radiation from the respective sources?

○ With the use of Equation 7.57, non-relativistic shocks should produce synchrotron spectra with $\alpha \approx 0.5$, while relativistic shocks will show *steeper* spectral indices, i.e. $\alpha \approx 0.65$.

Inverse Compton scattering

The other prominent radiation mechanism present in high-energy astrophysical settings, and always found in conjunction with synchrotron emission, is inverse Compton scattering.

Compton scattering describes the radiation resulting from the scattering (collision) of a photon and an electron. In low-energy settings (i.e. electrons moving slowly and photons carrying little energy), this simply results in the reshuffling (scattering) of the directions of photons. This is the familiar Thomson scattering. Energetic photons tend to lose energy in their interactions with less energetic electrons, in a process called Compton scattering. If, however, the electron is relativistic, it imparts energy to the photon. The process is called *inverse* Compton scattering to highlight the fact that the photon is gaining instead of losing energy (it gets *up-scattered*).

Application of the relativistic energy–momentum conservation (which is beyond the scope of the present text) supplies the frequency of the scattered photon ν_{ic} following the collision with an electron of Lorentz factor γ_e:

$$\nu_{ic} = \tfrac{4}{3}(\gamma_e^2 - 1)\nu_0, \tag{7.58}$$

where ν_0 is the frequency of the photon prior to scattering. In close analogy with synchrotron emission, more energetic electrons produce substantially higher frequency emission.

The power emitted by a single electron of energy proportional to γ_e is

$$P_{ic} = \tfrac{4}{3}\sigma_T c(\gamma_e^2 - 1)U_{rad}, \tag{7.59}$$

where U_{rad} is the energy density of the radiation (photons) that is available for scattering.

- ● In Equation 7.59, identify the factors that make the power of the inverse Compton scattering higher.

- ○ The emitted power increases with increasing energy of the electrons γ_e, and with increasing density of the radiation field.

There is very close resemblance between the relationships describing synchrotron radiation and inverse Compton scattering. This is not accidental. It derives from the facts that, in the proper (quantum electrodynamical) description, the two processes are equivalent (both are scattering), and there is no distinction between the magnetic and radiation fields. The analogy does not stop there. It can be shown that the energy spectrum of inverse Compton scattering by an ensemble of electrons with a power-law distribution obeys a power law with the index given by Equation 7.57, irrespective of the shape of the spectrum of the radiation field (see Figure 7.12).

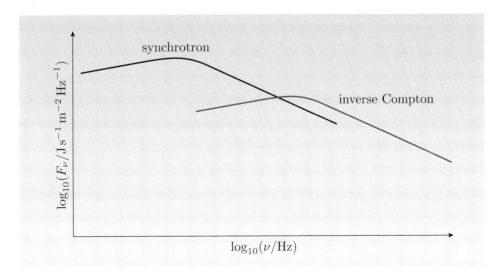

Figure 7.12 A sketch of the spectrum of synchrotron radiation and the inverse Compton radiation up-scattered from a power-law distribution of electrons.

It is evident that if the appropriate conditions for synchrotron radiation are satisfied (i.e. a source contains a magnetic field and has a supply of relativistic electrons), then inverse Compton emission will be inevitable. The relativistic electrons producing synchrotron radiation will also up-scatter these photons. This is often referred to as self-inverse Compton scattering and is a substantial contributor to high-frequency emission in all AGN, X-ray binaries and GRBs.

Exercise 7.11   Determine the ratio of the power emitted by a single electron of synchrotron radiation and inverse Compton scattering.   ∎

Summary of Chapter 7

In this chapter we have glimpsed some of the features of relativistic outflows and examined some of the tools necessary for their investigation. We have looked at observable effects of relativistic motion. We have further laid out the basics of the physical processes that are involved in the extraction, propagation, conversion and radiation of the energy from a central source that is most likely powered by an accretion disc. These constitute the building blocks of models for AGN, microquasars and GRBs. Specific points that we have covered include:

1. Outflows are common in astrophysical settings. They describe particle and energy flows that escape the gravitational bound of the host object and require some source of outward pressure. This pressure may be supplied by radiation, heat or magnetic fields.

2. Sources with luminosity above the Eddington limit are prime locations of outflows.

3. On the basis of their geometry, two broad classes of outflows can be distinguished: winds (predominantly spherically symmetric) and jets (elongated streams). Winds are found in many types of massive stars. Jets are present in systems that have an axial symmetry, and have been empirically found to be associated with accretion discs.

4. Outflows can result from an explosion (as in supernovae). Outflows that attain speeds close to the speed of light are called relativistic. Such outflows are suspected in AGN, many X-ray binaries and GRBs.

5. The Lorentz transformations are the coordinate transformations in special relativity that establish a physically consistent description of phenomena by observers in different inertial frames of reference. By applying these transformations, we can disentangle observations to probe the intrinsic properties of the sources.

6. Highly relativistic motion is best described by the Lorentz factor γ, which is a dimensionless expression of speed and appears in the Lorentz transformations in a simple form. For speed V, the Lorentz factor is defined as $\gamma = 1/\sqrt{1 - (V/c)^2}$, where c is the speed of light.

7. Special relativistic effects, some of which lead to apparently unphysical deductions, include the following:

    • *Length contraction*: a moving observer perceives lengths as shorter than they intrinsically are:

$$\Delta l = \Delta l'/\gamma. \qquad \text{(Eqn 7.11)}$$

- *Time dilation*: a moving observer perceives the durations of events as longer than they intrinsically are:

$$\Delta t = \gamma \, \Delta t'. \qquad \text{(Eqn 7.12)}$$

- *Pulse duration shortening*: photons that are emitted at specific intervals from a source moving relativistically towards the observer are received in much shorter time intervals:

$$\Delta t_{\text{rec}} \simeq \Delta t'_{\text{em}}/2\gamma = \Delta t/2\gamma^2. \qquad \text{(Eqn 7.31)}$$

- *Light beaming*: the light from a relativistically moving source is beamed in a narrow cone along the line of motion.

- *Superluminal motion*: a relativistically moving source may appear to move at a speed faster than the speed of light.

- *Relativistic Doppler shift*: the frequency of emission from a source moving relativistically towards the observer is strongly blueshifted:

$$\nu_{\text{rec}} = 2\gamma \, \nu'_{\text{em}}. \qquad \text{(Eqn 7.32)}$$

- *Luminosity boosting*: due to the combined effects of *beaming*, *Doppler shift* and *pulse duration shortening*, the apparent luminosity of a source moving relativistically towards the observer is strongly enhanced:

$$L_\nu \simeq \gamma^{3+\alpha} L'_{\nu'}. \qquad \text{(Eqn 7.33)}$$

8. The mechanisms by which jets are formed and maintain collimation are poorly understood. However, the jets' presence is associated with rotation and seems to require high radiation or magnetic pressure for setting off and magnetic fields to remain collimated.

9. Small perturbations in an outflow result in shocks that dissipate bulk kinetic energy, converting it to random kinetic energy of the particles. Shocks are also formed as the flow is impeded by the ambient medium into which it moves. Shocks within the flow are called internal, while those resulting from the interaction with the surrounding medium are called external.

10. Some basic physical properties of shocks can be described by approximating shocks with inelastic collisions of blobs of material.

11. Shocks develop in fluids when varying fluid conditions lead to parts of the fluid moving at a speed greater than its speed of sound. The shock front marks a sharp transition in the values of the fluid's physical properties.

12. The relationships between the physical properties of the fluid across the shock front are determined by the jump conditions: expressions that link the values of the physical properties in the pre- and post-shocked fluid, derived from the conservation laws of mass, momentum and energy.

13. Strong, relativistic shocks develop in flows that are moving with high Lorentz factors γ. The speed as well as mass and energy densities of the shocked fluid are enhanced by factors that are functions of γ. This greatly increases the radiative potential of these flows.

14. The high density of heat generated in a shock is channelled to particles that, through successive accelerations, achieve highly relativistic speeds. Particles

undergoing this shock acceleration process acquire a power-law distribution in energies. Such power laws are thought to be characteristic of shocks:

$$N(\gamma_e) \propto \gamma_e^{-p} \quad \text{for } \gamma_e > \gamma_{\min}. \tag{Eqn 7.48}$$

15. The energy stored in relativistic particles is radiated away via non-thermal radiation processes. Such processes result in power-law frequency spectra $F_\nu \propto \nu^{-\alpha}$.

16. The most prevalent non-thermal radiation processes anticipated in relativistic outflows are synchrotron emission and inverse Compton scattering.

17. Synchrotron radiation results from the acceleration of relativistic electrons in magnetic fields. The frequency spectrum of a single electron is a broad power law with spectral index $\frac{1}{3}$ peaking at a characteristic frequency that is proportional to the magnetic field and the square of the electron energy:

$$\nu_{\text{sy}} = \frac{1}{3}\gamma_e^2 \frac{1}{2\pi} \frac{eB_\perp}{m_e c}. \tag{Eqn 7.52}$$

18. Inverse Compton scattering radiation results from the scattering of photons by relativistic electrons. The typical frequency is proportional to the square of the electron energy and the frequency of the scattered radiation:

$$\nu_{\text{ic}} = \tfrac{4}{3}(\gamma_e^2 - 1)\nu_0. \tag{Eqn 7.58}$$

19. Both the synchrotron and inverse Compton scattering spectra from a power-law distribution of electrons are power laws, with a high-energy spectral index related to the energy index of the electron distribution by

$$\alpha = (p - 1)/2. \tag{Eqn 7.57}$$

The relative intensity depends on the relative strength of the radiation and magnetic field energy densities:

$$P_{\text{sy}}/P_{\text{ic}} = U_B/U_{\text{rad}}.$$

Chapter 8   Gamma-ray bursts

Introduction

One has to resort to superlatives when describing gamma-ray bursts (GRBs): the brightest, most violent and most energetic cosmic explosions, the most distant objects ever seen, the hardest astrophysical mystery to crack, to list but a few. This chapter will introduce GRBs and sketch the current understanding of the sources, with the proviso that this is far from settled or exhausted. Though most of the things discussed here are well-founded, this is a field evolving in real time. Chances are that some of the current theoretical understanding will be superseded in the years to come, as more elaborate observations complete the picture. Yet it is unlikely that it will be totally overthrown.

After a brief historical context, we lay out the observed properties of GRBs, deduce some features of the sources along with the implied constraints on the models, describe the basic model, and use it to interpret the observations and present the astrophysical objects that are the favoured candidates for giving rise to GRBs. We briefly mention predictions of other signatures of GRBs and the prospect of detecting them. We finish with a look at GRBs as cosmological tools.

8.1   Discovery and breakthroughs

GRBs have been a greatly perplexing puzzle for some 30 years since their serendipitous discovery in the midst of the Cold War. These intense gamma-ray flashes, coming unanticipated from any direction in the sky, blind detectors for a brief few seconds, to leave no trace behind. They were first detected in 1967 by the US Vela satellites put in high orbit as part of the enforcement of the Nuclear Test Ban Treaty. Gamma-rays are notoriously hard to focus by detectors, thus giving precious little information on the location of the sources in the sky. Shortly after the discovery announcement — in 1973 — dozens of models were constructed on the basis of sparse data. The proposed sources lay anywhere from our Solar System neighbourhood to the edge of the visible Universe. They ranged from dust grains smashing onto our solar windscreen, to black holes being born out of a catastrophic collapse, or transmuting exotic particles.

A key clue to the nature of the sources lies in their isotropic sky distribution. During the 1990s, the BATSE detector (see the box below) collected a sufficiently large sample to reject some of the proposals and fuel the debate on the origin of GRBs. Still it was not possible to differentiate between sources at the outskirts of the visible Universe and those that populated a presumed **Galactic halo** of the Milky Way. The issue was settled in 1997 with the detection of an **afterglow**: a long-lasting, fading emission that follows the GRB, in X-rays. The improved accuracy of the source position supplied by X-ray detectors, along with the lingering afterglow emission, permitted optical telescopes to pinpoint the source. Remarkably, this led to the measurement of the redshift of the galaxy hosting the burst. Such redshift measurements demonstrated that the sources are indeed at cosmological distances. The study of GRB afterglows has flourished since, shedding ample light on the potential underlying astrophysical objects. More recently, GRBs have broken yet another record, that of the most distant object ever observed.

A public debate was held in 1995 to commemorate the 1920 Great Debate on the scale of the Universe. Astrophysicists Bodhan Paczyński and Don Lamb argued in favour of a cosmological and a Galactic origin, respectively, in an animated debate moderated by Martin Rees.

Key GRB missions and discovery highlights

As the Earth is sheltered from gamma-rays, the study of GRBs has relied mostly on space missions. The military Vela satellites of 1967–71 discovered GRBs (as well as X-ray bursts), identifying them as short, highly energetic events, with no preference to any particular sky location. The stagnation of the following 20 years was overcome in 1991, with the launch of the *Compton Gamma-Ray Observatory* (CGRO), one of NASA's great observatories. The BATSE experiment on board CGRO (operating in the 10 keV to 1 MeV energy range) detected roughly one burst per day in its 9 years of operation, establishing beyond doubt the isotropy of the bursts' sky distribution and characterizing their light curves and spectra.

The long sought holy grail of a 'counterpart' (detection at wavelengths other than gamma-rays) was achieved in 1997 by the Italian/Dutch *BeppoSAX* satellite. BeppoSAX (covering energies between 0.1 keV and 200 keV) could detect a fast-fading X-ray source within hours of the GRB, with a positional accuracy of arcminutes. Relaying the position to ground-based observatories that chased the optical counterparts, allowed the determination of the host's redshift and established the cosmological origin of GRBs. A serendipitous discovery enabled by BeppoSAX was the association of some GRBs with a simultaneous supernova. Numerous optical ground-based observatories (most notably robotic wide-field collections of photolenses like *ROTSE*) have been instrumental in the discoveries since. Fast communications, automated response and broad availability of telescopes have made successful monitoring possible. ROTSE captured the first optical flash concurrent with a GRB.

Swift (covering energies from the ultraviolet to 10 keV) has been the latest wonder-mission that owes its success to its ability to rapidly slew its X-ray and ultraviolet telescopes to a GRB first glimpsed by its wide-field gamma-ray sensor. It has returned multi-wavelength afterglows for several bursts, in particular those lasting a fraction of a second. It has also shown a taste for record setting. Among the most spectacular discoveries that it has enabled are the detection of the brightest object ever seen (a GRB outshining the brightest supernova by a factor of millions) and most distant objects in the Universe: as of mid-2009, the record is held by a GRB that erupted when the Universe was a mere 800 million years old (some 6% of its current age!). *Fermi*, launched in 2008, is probing the high-energy (above 100 MeV) emission from GRBs. Ground-based Çerenkov-type detectors, like *HESS*, poke into even higher energies (above tens of GeV).

8.2   Observed properties

GRBs are named after the date of their detection by the convention GRByymmdd. For example, the first burst accurately localized via its afterglow was the now famous GRB970228, detected on 28 February 1997.

The *prompt* emission of GRBs is short-lived and seen as gamma-rays with energies of hundreds of keV. It is followed by an afterglow, emission that persists over weeks to months in progressively lower photon energies (X-rays, optical and radio band). The prompt and afterglow stages have distinct temporal and spectral characteristics that we study in turn.

8.2.1   Prompt emission

Timescales and light curves

GRBs last anywhere between a few milliseconds to tens of minutes. This alone is a remarkable property not displayed by any other astrophysical object. Moreover, their light curve morphology is exceptionally diverse (see Figure 8.1), showing single or multiple spikes, in tight sequence or with long intervening pauses, that may rise fast and decay slowly or be rather symmetric. In general, they exhibit millisecond variability.

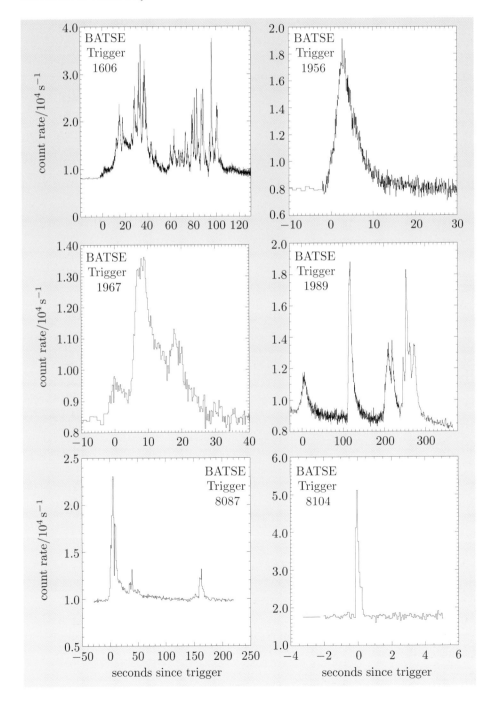

Figure 8.1   A selection of GRB light curves collected by BATSE, displaying a vast diversity of shapes and wide range of durations. Note the fast variability shown by most bursts.

Typical values of fluence (see the box entitled 'Measures of light' in Subsection 7.3.4) are 10^{-10}–$10^{-7}\,\mathrm{J\,m^{-2}}$ for the bulk of the GRB sources, with the brightest reaching values as high as $10^{-5}\,\mathrm{J\,m^{-2}}$. The lower limit is set by current detectors' sensitivity; the true minimum fluence could be smaller still.

Spectra

GRBs are detected in gamma-rays ranging from tens of keV to tens of GeV, with most of their energy around hundreds of keV to a few MeV. Their spectra, extending over a broad frequency range, have the shape of a *broken-power law* (see Figure 8.2), described by

$$F_\nu \propto \begin{cases} (\nu/\nu_{\mathrm{p}})^{\alpha_{\mathrm{l}}} & \text{for } \nu \leq \nu_{\mathrm{p}}, \\ (\nu/\nu_{\mathrm{p}})^{-\alpha_{\mathrm{h}}} & \text{for } \nu > \nu_{\mathrm{p}}, \end{cases} \qquad (8.1)$$

where ν_{p} is the break frequency, corresponding to the photon energy where the power law changes from a power law of index α_{l} to one of index $-\alpha_{\mathrm{h}}$ (with $\alpha_{\mathrm{l}}, \alpha_{\mathrm{h}} > 0$). The break frequency is (usually) also the peak of the νF_ν distribution. Equation 8.1 merely defines a simple function that approximates the observed spectrum and carries no physical meaning. The parameters $\nu_{\mathrm{p}}, \alpha_{\mathrm{l}}, \alpha_{\mathrm{h}}$, however, are expected to be linked to physical processes and properties (see Subsections 7.3.4, 8.4.2 and 8.4.3).

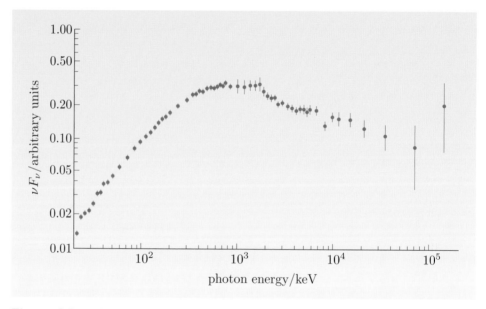

Figure 8.2 The energy spectrum of GRB910503, which is typical of GRB spectra extending from 20 keV to 200 MeV. It is a power law peaking at about 2 MeV.

The break frequency is far from unique. In BATSE data, it spans the range corresponding to energies between 100 keV and 500 keV, with the spectra of most bursts breaking around 300 keV. However, there is at least one subclass (called *X-ray flashes*) with peak photon energies as low as 30 keV. Typical values of the energy spectral slopes are $\alpha_{\mathrm{l}} \approx 0$ and $\alpha_{\mathrm{h}} \approx \frac{4}{3}$, though both cover a substantial range.

A burst often consists of a sequence of pulses. The peak energy of individual pulses shifts to lower values with time, a trend that we call **spectral softening**.

Pulses are also seen to last longer in lower energies. Such trends are interesting as the pulse characteristics reflect the physical conditions of the emitting source.

Classification

The accumulation of a large sample of observed GRBs has revealed two subclasses of bursts in terms of their duration and spectra (see Figure 8.3). Roughly, a quarter of bursts last less than 2 s, and these are known as *short bursts*, while the rest are called *long bursts*. The typical duration of a short burst is 0.5 s, while that of a long burst is 30 s.

A detector-based measure of spectral shape is the *hardness ratio*, defined as the fraction of the fluence measured in a high-energy channel versus a low-energy channel (see also Chapter 6, Subsection 6.4.1). These channels may be different for different instruments. The significance of a high ratio, though, is that more energy is emitted in high photon energies than in lower ones, implying more energetic processes. Such bursts are called *hard*, to be distinguished from *soft* bursts, which have low hardness ratios. In practice, hard spectra usually have a high peak photon energy ν_p, or a shallow high photon energy power law (i.e. a small value of α_h). Short bursts are generally harder than the long ones as shown in Figure 8.3(b).

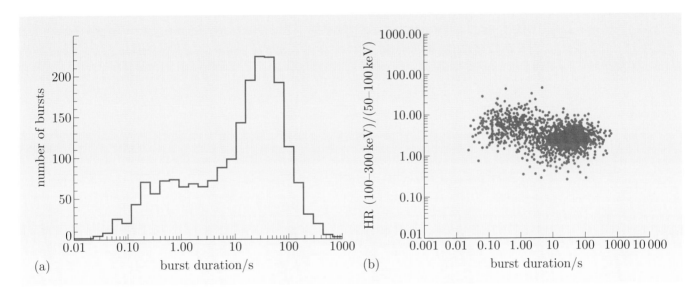

Figure 8.3   (a) The burst duration distribution and (b) hardness ratio versus duration of BATSE bursts. The trends hold for different measures of duration and hardness.

Distribution in space

The *isotropy* of the sky distribution of GRBs (Figure 8.4) is striking. It is evident that GRBs are distributed differently than Galactic sources (e.g. pulsars are concentrated along the Galactic plane, which in this graph coincides with the horizontal axis).

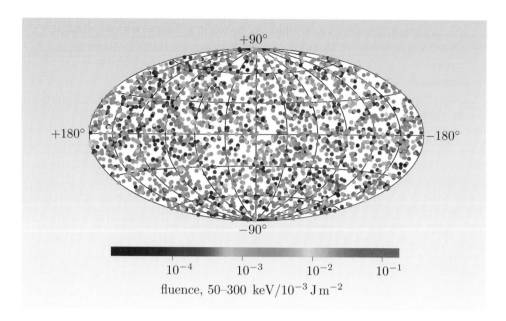

Figure 8.4 The sky distribution of GRBs, as recorded by BATSE over its 9 years of operation. The distribution is in Galactic coordinates, like Figure 1.16. Note the perfect isotropy.

Information about the distance distribution of the sources, albeit indirect, comes from the so-called **log N– log S graph** where the number of sources with fluence in excess of a certain value is plotted versus this fluence (see Figure 8.5). Such graphs, common in the studies of extragalactic sources, can be produced for any instrument. The relationship is therefore only relative and carries no scale information.

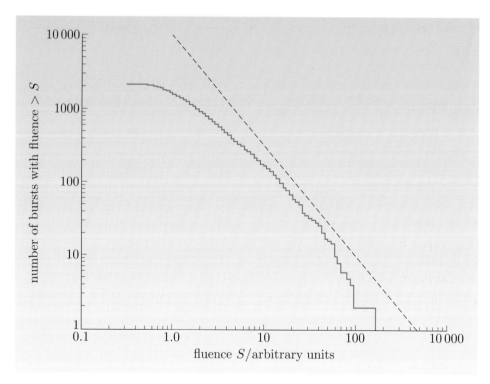

Figure 8.5 The number of BATSE bursts with fluence above a certain value S, as a function of fluence, in logarithmic representation. The lower fluence limit of the curve is the detection limit of BATSE. Note the deviation from a uniform distribution with slope $-\frac{3}{2}$, indicated by the dashed straight line.

● If **standard candle** sources (i.e. sources with a fixed luminosity) are uniformly distributed within a spherical volume of radius d, what is the shape of their log N–log S distribution?

○ We have to determine how the number of sources N with fluence values above S scales with S. For a uniform distribution, the number of sources is proportional to the volume of the enclosed sphere, $N \propto d^3$. Because the luminosity of the sources is constant, the fluence depends on the distance only. The fluence, like the flux, obeys the familiar $S \propto d^{-2}$ law. Eliminating the d-dependence, we find that $N \propto S^{-3/2}$.

A deviation from the $-\frac{3}{2}$ slope, such as the one shown in Figure 8.5, indicates a scarcity of fainter bursts, which could come about in various ways. The assumptions underlying the answer above are suggestive: the distribution may not be homogeneous (e.g. we sample the edge of it, or the sources are not as frequent further away); the sources may not be standard candles (e.g. sources were dimmer in the past); or the space is not Euclidian (i.e. the volume is not proportional to d^3). The measured distribution in itself cannot distinguish between these possibilities. Combined, the sky and the $\log N$–$\log S$ distributions are consistent with any arrangement of sources about the Earth that is *isotropic* (i.e. the same in all directions) and *inhomogeneous* (i.e. varying with distance d). Such locations are the **Oort cloud** of comets (at a fraction of a pc), a presumed extended Galactic halo (no closer than ~50 kpc, otherwise the Sun's offset from the Galactic centre should lend the sky distribution a skew), or the far reaches of the Universe.

Exercise 8.1   Given a typical burst fluence of $S = 10^{-9}\,\mathrm{J\,m^{-2}}$, calculate the energy emitted if the source lies (a) at the presumed halo of our Galaxy ($d \approx 50\,\mathrm{kpc}$), (b) at a cosmological distance ($d \approx 3\,\mathrm{Gpc}$).

Consider the typical energy of a supernova ($\approx 10^{44}\,\mathrm{J}$), that of an X-ray burst, $E_{\mathrm{XRB}} \simeq 10^{33}\,\mathrm{J}$ (Subsection 6.5.1), and the mass energy of a solar mass object, $M_\odot c^2$. How does the energy of the proposed sources calculated in (a) and (b) compare to these values? ■

Counterparts

Unlike other sources of gamma-rays that emit significantly also at lower energies, GRBs are primarily a gamma-ray phenomenon. This is not purely due to observational bias, though the difficulty in pointing any kind of telescope to a vaguely defined position in the sky within seconds plays a dominant role. Detections at other wavelengths make notable exceptions. We note them here not only for standing out but because they challenge models.

The prospect of an optical burst concurrent with the gamma-ray signal has always been tantalizing, as it would tie down the GRB location on the sky. In order to estimate the magnitude of an optical flash, one may assume that the observed spectrum (see Equation 8.1) simply extends to lower frequencies. For typical parameters this gives an optical magnitude dimmer than 18. By the standards of appropriate Earth-bound optical telescopes, this is rather faint. ROTSE, the most suitable dedicated instrument in the late 1990s, could achieve magnitude 15.5 provided that the source persisted for half an hour. Nevertheless, the large positional error of the GRB as well as the time taken for an optical telescope to be pointed to a GRB were more severe limitations. Since 1999, with the surprise detection of a 9th-magnitude optical flare from GRB990123, several optical flashes have been observed. They all started within a few minutes from the gamma-ray trigger, and while the gamma-ray burst was still underway. More recently, the optical flash of GRB080319B reached a peak magnitude of 5.3, earning the nickname 'the naked-eye GRB'. In fact, GRB080319B was the visually most luminous event witnessed on Earth to date!

● GRB080319B was at redshift 1 ($d \approx 3 \, \text{Gpc}$) and had a gamma-ray fluence of $10^{-5} \, \text{J m}^{-2}$. Revisit Exercise 8.1 to estimate the total energy emitted by the source.

○ GRB080319B is at the same distance as the source in Exercise 8.1(b), but its fluence is 4 orders of magnitude larger. Hence the implied emitted energy is $E \approx 10^4 \times 10^{44} \, \text{J} = 10^{48} \, \text{J} \approx 5 \, \text{M}_\odot c^2$!

Despite these very impressive cases, optical flashes are seen in only a handful of bursts and are moderate in magnitude.

8.2.2   Afterglows

The significance of afterglow detections lies in the enormous improvement in the accuracy with which the sky position of the source can be determined. The positional accuracy of degrees for gamma-rays improves to arcmin for X-ray sources, thus enabling the identification of optical counterparts that can be pinpointed to within arcsec. The latter, in turn, allow the redshift determination of the host galaxies, providing measures of the luminosity and total energy output, as well as clues on the geometry of the source. Finally, the position of radio sources can be pinned down to a fraction of a milliarcsec! Accounting for observing conditions, nearly all GRBs are followed by an X-ray afterglow. However, only half of them show an optical afterglow, and a third a radio afterglow. It is not trivial to disentangle the effects that contribute to this. They may include intrinsic properties of the source, but also extinction by **circumstellar dust** or the **intergalactic medium** that can easily obscure optical sources.

Unlike the erratically variable prompt emission, the afterglow does not fluctuate. The flux follows power laws in both frequency and time:

$$F_\nu(t) \propto \nu^{-\alpha} t^{-\kappa}, \tag{8.2}$$

where α and κ are positive, indicating flux that decreases with time at a given frequency (dimming), maintaining a power-law spectrum at every time. The values of α and κ are different in different spectral bands and also change with time resulting in broken-power laws. Moreover, emission first appears at high frequencies and progresses to lower frequencies with time, i.e. the afterglow displays spectral softening. There are no qualitative differences between the afterglow properties of short and long GRBs.

X-ray, optical and radio light curves

A distinct drop in intensity from the prompt emission signals the onset of the X-ray afterglow. There is substantial variation in the early behaviour of the X-ray light curve: there may be a hump with a sharp decline, followed by a plateau, or even a modest or substantial flare. Eventually the light curve displays a power-law decay (with κ in the range 1–1.5) that may become even steeper at a later stage (e.g. κ in the range 2–3), to disappear within a month. Figure 8.6 shows some examples of X-ray afterglow light curves.

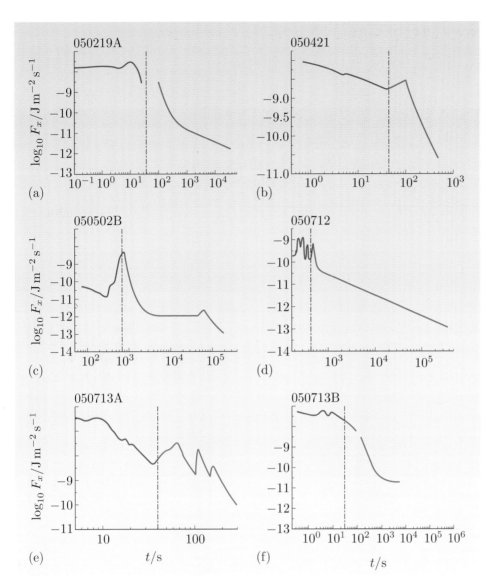

Figure 8.6 A schematic representation of a selection of typical light curves of X-ray afterglows. Light curves of the burst (prompt emission) are shown in blue while the red portions are part of the X-ray afterglow. Cases (b), (c), (d) and (e) are displaying some flaring.

The X-ray behaviour is mirrored in the optical wavelengths and then in the radio wavelengths, but the light curves in the different bands are not identical. The optical afterglow may persist for months, though it is often obscured by dust, a feature that can diagnose the burst's birth environment. It also shows **polarization** that can potentially identify the emission mechanism.

Radio afterglows rise within days to weeks after the GRB and then decay slowly, eventually blending in with the background within a year. An intriguing and unique feature is the wide fluctuations seen in the radio light curve early on that eventually die out. This has provided model-independent confirmation of the source's relativistic expansion (see Section 8.3).

Some afterglows display a simultaneous steepening in their light curves in all frequency bands. These are known as *achromatic breaks* (see Figure 8.7). Due to their frequency-independence they are attributed to geometric effects (see Subsection 8.4.4).

Figure 8.7   An example of a burst showing an achromatic break. Light curves in different spectral ranges break simultaneously.

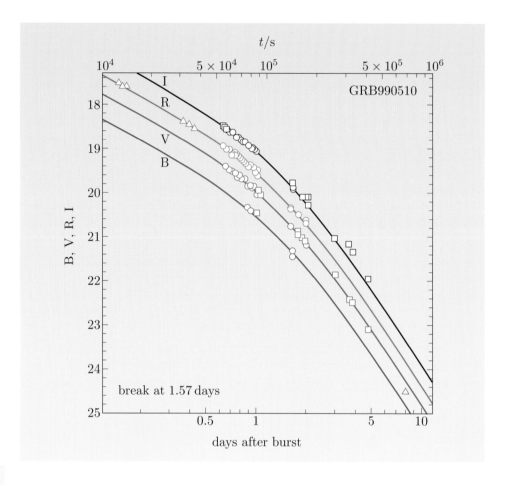

Supernovae (SNe) come in two varieties. Type II SNe show H lines in their spectra, while type I SNe do not. Type I SNe are further subdivided into types Ia, Ib and Ic. Type Ia show Si lines that the others do not. Type Ib show He lines, while type Ic do not. Both type Ic and type II SNe are found in the spiral arms of galaxies that are made up of very young stars. Thus type Ic SNe are interpreted as the result of the core collapse (like type SN II) of a massive (young) star that has shed its hydrogen and helium layers.

Hosts and nurseries

Optical afterglows have made it possible to identify the host galaxies of GRBs. There is a quite clear distinction between the types of hosts of long and short bursts. Long bursts are found in the brightest (i.e. star-forming) regions of small, low-luminosity and low-metallicity irregular galaxies. Short GRBs, on the other hand, are found in a variety of extragalactic environments. Furthermore, long GRBs lie at very high redshifts. In contrast, the less common short GRBs seem to be intrinsically dimmer by a factor of 10–100. They are thus not detectable as far away as long GRBs. Their $\log N$–$\log S$ distribution, in fact, shows no deviation from a $-\frac{3}{2}$ slope, hinting at a nearby population.

Association with supernovae

Yet another surprise came with the detection of a type Ic supernova concurrent with some of the long GRBs (see Figure 8.8 for an example). Such coincident detections are only occasional. Nothing in the prompt emission betrays the presence of a supernova, but it can be identified in the light curve of the optical afterglow. The optical light curve shows a monotonic decay at first and then acquires a 'bump' about 1–3 weeks after the burst. This is the timescale over which the optical light output from a supernova peaks. Such a bump can be distinguished from similar deviations from a smooth decay in the afterglow light curve on the basis of its spectrum: it is unmistakably that of a type Ic supernova.

The handful of GRB/type Ic supernova associations confirmed so far all come from sources that are nearby. The bursts are quite faint by GRB standards and would have remained undetected had they occurred at larger distances. Apart from their low gamma-ray luminosity, the bursts are typical GRBs, undistinguishable from the rest. On the other hand, the majority of observed type Ic supernovae do not produce a GRB. Those that do are particularly bright compared to their peers. Most notably, they have been seen as bright radio sources a few days after the burst. Their radio brightness can be explained only if the shock wave associated with the supernova event is relativistic. In all GRB/type Ic supernova associations, the supernova is substantially dimmer than its GRB counterpart. However, a low-luminosity source that persists for a long time can deliver more energy than a brilliant flash. The long-lived supernova thus releases significantly more energy than the GRB. A trend of potent supernova/feeble GRB is emerging.

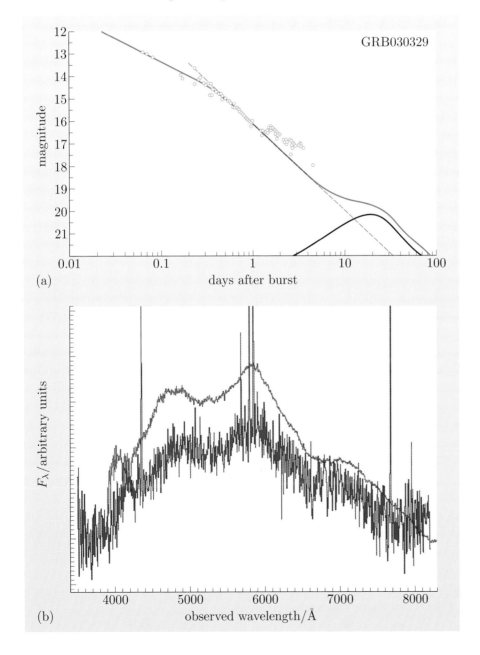

Figure 8.8 A GRB with a concurrent associated supernova. (a) The combined light curve is shown (in red), along with the predicted supernova light curve (in black), which rises slowly and peaks several days after the burst onset. (b) The optical spectra of the GRB (in red) and the supernova (in blue) at the time of the peak of the supernova light are virtually identical.

225

8.3 Constraints on GRB models

In this section we discuss the properties of GRB sources that can be deduced straight from the observations, in a model-independent way. These properties must be satisfied by any model for GRBs.

Cosmological distance

The isotropy and inhomogeneity displayed by the GRBs' spatial distributions do not point to a specific location; they are, however, typical of cosmological sources (e.g. quasars). Redshift determinations delivered the unambiguous proof that the sources lie at cosmological distances. The observed GRB host redshifts span the whole range and extend further than any other astronomical object observed to date.

Relativistic expansion

Seated at the far reaches of space, GRBs have enormous energies (see Exercise 8.1).

Exercise 8.2   GRB light curves show variability on timescales Δt_{var} in the range 1–10 ms. Applying the variability timescale argument of Subsection 1.3.2, estimate the size Δr of the emitting region. Based on the resulting size, suggest a possible source candidate. ∎

Exercise 8.2 demonstrates that the source of GRBs must be very compact. But a compact source with such a huge luminosity should not be emitting gamma-rays! High-energy photons are subject to the production of electron–positron (e^-–e^+) pairs on collision if their energies $h\nu_1$ and $h\nu_2$ satisfy the relation

$$h^2 \nu_1 \nu_2 (1 - \cos\theta_{12}) \geq 2(m_\text{e}c^2)^2,$$

where θ_{12} is the angle at which they are approaching each other. Taking the least demanding case of a head-on collision ($\theta_{12} = \pi$), target photons of energy $h\nu_1$ will produce pairs if they collide with photons of energy $h\nu_2 \geq (m_\text{e}c^2)^2/h\nu_1$. This allows us an estimate of the optical depth to pair production:

$$\tau_{\gamma\gamma} \simeq \sigma_\text{T} n_\gamma \Delta r, \tag{8.3}$$

where σ_T is the Thomson cross-section, n_γ is the number density of sufficiently energetic photons, and Δr is the size of the source. (See also Equation 6.5 in Chapter 6 for the definition of optical depth.)

Exercise 8.3   For a typical source with $L_\gamma \approx 10^{44}\,\text{J s}^{-1}$, $\Delta t_{\text{var}} = 10\,\text{ms}$, and typical photon energy $h\nu = 0.5\,\text{MeV}$, estimate the optical depth to pair production, $\tau_{\gamma\gamma}$. Consult Equation 7.50. ∎

The enormous optical depth highlighted by Exercise 8.3 suggests that all photons above 0.5 MeV should be consumed in a catastrophic so-called photon-pair cascade: in crossing the region of size Δr, a photon will interact on average 10^{13} times with other photons of appropriate energy for pair production. Each such interaction will result in an e^-–e^+ pair, converting all radiation to pairs. Yet we see photons of hundreds of MeV, even up to 10 GeV in GRBs! This is known as the *compactness problem* and seemed for a long time to contradict the cosmological origin of the sources.

● Compare the luminosity of a source at cosmological distances, from the energy obtained in Exercise 8.1, with the Eddington luminosity of the objects inferred by Exercise 8.2.

○ For a typical burst duration of 10 s, the implied GRB luminosity of $10^{43}\,\mathrm{J\,s^{-1}}$ is many orders of magnitude higher than the Eddington luminosity of an object of a few $M_\odot$ (see Equation 6.9 and Exercise 6.2).

As we saw in Chapter 7 (Section 7.1), such a highly super-Eddington luminosity sets off a fast expansion. In fact, the compactness problem is alleviated if the expansion is relativistic. Section 7.2 discussed the effects that relativistic motion has on observed quantities: photons are blueshifted, event durations are shortened and radiation is beamed. Consequently, relativistic expansion modifies the optical depth as the photons appear more energetic than they intrinsically are and the source size is larger than the observations imply. The higher the expansion speed, the more pronounced these effects are. We shall see shortly what the survival, and hence detection, of high-energy photons from a GRB implies for the magnitude of the expansion speed. For that, we need an estimate of the magnitude of the optical depth in the case of relativistic expansion.

There are two reference frames that we employ in our considerations: the *observer rest frame*, which is the frame in which both the observer and the origin of the expanding source (neglecting cosmological expansion) are at rest, and the *flow rest frame*, which follows the radial expansion of the source. For brevity, we shall refer to these as the *observer* and the *flow* frame, respectively.

Worked Example 8.1

Taking into account the aforementioned relativistic effects, recalculate the optical depth to pair production for a relativistically expanding source.

Solution

It is easier to calculate the optical depth to pair production in the flow frame. Equation 8.3 is written as $\tau_{\gamma\gamma} \simeq \sigma_T n'_\gamma \Delta r'$. The primed quantities are measured in the flow frame, and our task is to link them to the observed quantities. We consider each quantity in turn. The size of the emitting region (i.e. the expanding fireball), as seen in the observer frame, is Δr. In the flow frame, the corresponding distance is $\Delta r'$. We determine both Δr and $\Delta r'$ from the observed variability timescale Δt_{var}, as we did in Exercise 8.3 where we neglected the relativistic expansion. Noting the relation between photon emission and arrival times (Equation 7.32 with $\Delta t_{rec} = \Delta t_{var}$) we thus obtain in the flow frame

$$\Delta r' = 2\gamma c \, \Delta t_{var},$$

while in the observer frame we get

$$\Delta r = 2\gamma^2 c \, \Delta t_{var}.$$

The photon number density (Equation 7.50) in the flow frame is $n'_\gamma = L'_\gamma / 4\pi r_0^2 ch\nu'$ (where the area of the sphere is not affected by relativistic length contraction as it is perpendicular to the direction of the motion). By virtue of the relativistic Doppler shift (see the statement after Equation 7.32), the photon frequency is blueshifted, $\nu \simeq \gamma\nu'$. Finally, in Subsection 7.2.6, we saw that for a flux spectrum $F_\nu \propto \nu^{-\alpha_h}$, the apparent luminosity is boosted by a factor of $\gamma^{3+\alpha_h}$ (Equation 7.33), i.e. $L'_\gamma \simeq L_\gamma / \gamma^{3+\alpha_h}$. Bringing these together, we obtain

$$\tau_{\gamma\gamma} \simeq \sigma_T \frac{L_\gamma / \gamma^{3+\alpha_h}}{4\pi(\Delta r)^2 ch(\nu/\gamma)} 2\gamma c \, \Delta t_{var} = \frac{2\sigma_T L_\gamma \Delta t_{var}}{4\pi(\Delta r)^2 h\nu\gamma^{1+\alpha_h}}.$$

Being a pure number, the optical depth is the same in any frame.

and so

$$\tau_{\gamma\gamma} \simeq \left(\frac{\sigma_{\mathrm{T}} L_\gamma}{4\pi c^2 \Delta t_{\mathrm{rec}}\, h\nu} \right) \frac{1}{2\gamma^{5+\alpha_{\mathrm{h}}}}. \tag{8.4}$$

The term in parentheses is the optical depth for a source at rest (Exercise 8.3). Thus, for typical values of the observed quantities, the optical depth becomes

$$\tau_{\gamma\gamma} \simeq 3.5 \times 10^{12} \gamma^{-(5+\alpha_{\mathrm{h}})}. \tag{8.5}$$

We see that, for relativistic motion, the optical depth depends strongly on γ. If high-energy photons escape, the optical depth must be small. Requiring $\tau_{\gamma\gamma} < 1$ and taking $\alpha_{\mathrm{h}} = \frac{4}{3}$ (see Section 8.2.1), Equation 8.5 gives $3.5 \times 10^{12} \gamma^{-(5+4/3)} < 1$, i.e. $3.5 \times 10^{12} \gamma^{-19/3} < 1$, so $\gamma > (3.5 \times 10^{12})^{3/19}$, which gives $\gamma \gtrsim 100$! We have been rather casual with factors of a few — this is customary in astrophysics, where we are often just interested in the order of magnitude. This γ value is impressively high and nowhere close to the Lorentz factors in AGN or X-ray binary jets (see Section 7.3).

Notwithstanding this powerful argument, the relativistic expansion has been demonstrated through unambiguous evidence for superluminal motion. We saw earlier that the radio afterglow light curve shows significant variability. This is the radio equivalent of **scintillation**, the twinkling of stars in the night sky. Much like stars twinkle but planets do not, a point-like radio source is jittery while an extended one is not. The crucial observation is that this twinkling ceases at some point during the afterglow. This indicates that the moment when the apparent size of the source becomes larger than the scale R_{def} of light-ray deflections at a particular frequency, is set by the lumpiness of the **interstellar medium**. For observations at a radio frequency of $10\,\mathrm{GHz}$, the corresponding limit for a source at a redshift of 1 is $R_{\mathrm{def}} \approx 10^{15}\,\mathrm{m}$.

Exercise 8.4   The first source to show such a scintillation quenching, GRB970508, did so at 4 weeks after the prompt burst. Assume that the expansion speed V was constant throughout this period. Calculate V if the source is at a redshift of 1.   ∎

This is a clear demonstration of relativistic expansion. Though the implied γ is only moderate, the flow is expected to have slowed down by this point. This reinforces the conclusion of a relativistic expansion at earlier stages.

Progenitor rates

From the frequency at which we detect GRBs, we can deduce the rate at which the actual events leading to GRBs occur. The argument is widely applicable to all sorts of astrophysical sources.

Worked Example 8.2

We observe about one GRB per day. Due to observational limitations, we catch about 1 out of 3. Assuming that the bursts are uniformly distributed in space out to redshift 3 (corresponding to a distance of $d \approx 6 \, \mathrm{Gpc}$) and that the mean distance between galaxies is $5 \, \mathrm{Mpc}$, calculate the actual rate of GRBs expressed as events per galaxy per 10^6 years.

Solution

We can approach this in different ways. One way is to estimate the number of galaxies as the product of the number density of galaxies times the volume of a sphere of radius d. The number density can be approximated by the inverse of the volume of space between two neighboring galaxies. This gives $N_{\mathrm{gal}} \approx (6 \, \mathrm{Gpc}/5 \, \mathrm{Mpc})^3 \approx 2 \times 10^9$. Then, dividing the inferred daily rate (3 per day) by the number of galaxies, we obtain the galactic daily rate: 1.5×10^{-9}. There is little meaning in a daily rate (it is minuscule), so to convert it to a yearly rate, we need to multiply by the number of days in a year. This gives approximately 0.5×10^{-6} GRBs per galaxy per year. Given our rough approach, we cannot be confident about factors of a few. It is thus appropriate to quote the range 0.1–1 GRBs per galaxy per megayear. Clearly this is only a lower limit as there are bound to be more GRBs than we observe.

To summarize, observational evidence has tied GRBs to sources at cosmological distances that involve highly relativistic outflows ($\gamma \gtrsim 100$), with apparent luminosities L_γ of approximately 10^{42}–$10^{47} \, \mathrm{J \, s^{-1}}$. They occur at a rate of at least 10^{-7} per galaxy per year.

8.4 The fireball model

According to the understanding of the GRB sources that we have developed so far, some catastrophic stellar explosive event (leaving behind what is called the *central engine*) deposits a huge amount of energy into a very small volume of space. This is filled with an optically thick soup of photons, e^-–e^+ pairs and, possibly, baryons that expands relativistically — it is a *fireball*. The fireball gives rise to all the observable features of the prompt emission and the afterglow without betraying the central engine.

The term **baryon**, from the Greek word for heavy, is used to denote particles that have inertia. It refers to normal matter, the main constituents of which are protons and neutrons.

While our detailed discussion of the fireball below is based on a spherically symmetric geometry, it is equally applicable to a conical section. Such a section is a good approximation for a jet. The key stages of the fireball evolution are sketched in Figure 8.9 (overleaf). In what follows, we describe the physics of these stages and derive the corresponding typical radii.

8.4.1 Relativistic expansion and transparency

The fireball sets off when a large amount of energy E_0 mixed with a small amount of matter M (where $E_0 \gg M c^2$) is deposited in a small volume (i.e. a sphere of radius $r_0 \lesssim c \Delta t_{\mathrm{var}}$, where Δt_{var} is the observed variability timescale). The fireball expands in the form of a spherical shell of thickness Δr and radius r. The frequent collisions between particles and the abundant and highly energetic

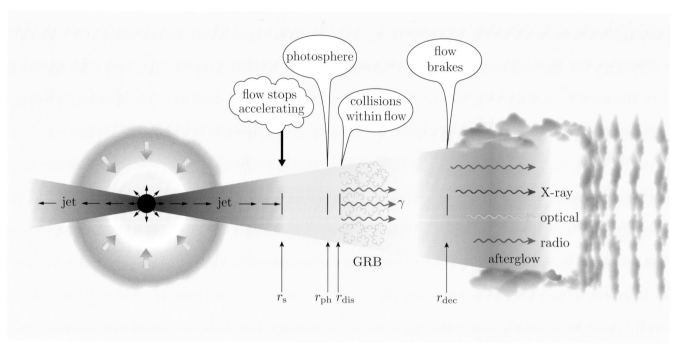

Figure 8.9 A sketch of the fireball, the prevailing paradigm for the generation of a GRB and its afterglow. There is growing evidence that the flow pattern has the shape of a cone rather than a sphere.

photons accelerate the particles. This process is very effective as long as the photons cannot escape, i.e. the fireball is *optically thick*. It can be shown that in this case the Lorentz factor γ of the expansion grows as

$$\gamma(r) \propto r, \tag{8.6}$$

while the temperature of the fireball drops.

● The fireball energy is equal to the energy in the initial explosion, $E_0 + Mc^2$. The expansion accelerates for as long as this energy is being converted to kinetic energy of the fireball's constituents. When all of it has been converted to kinetic energy of the baryons, γ saturates to its maximum value γ_s and the fireball will coast at the corresponding constant speed. Applying energy conservation, calculate γ_s.

○ At each stage, the total energy is the sum of photon and baryon energy. Applying energy conservation to the initial stage and the 'saturation' stages, we get

$$\gamma_s M c^2 = E_0 + Mc^2 \quad \Rightarrow \quad \gamma_s = \frac{E_0}{Mc^2} + 1.$$

Since $E_0/Mc^2 \gg 1$, γ attains the maximum value of

$$\gamma_s \simeq \frac{E_0}{Mc^2}. \tag{8.7}$$

We may assume that the expansion starts from rest ($\gamma_0 = 1$) and obtain the distance from the explosion centre at which the expansion saturates by applying Equation 8.6: $\gamma_s/\gamma_0 = r_s/r_0$, so the **saturation radius** (r_s) is

$$r_s = \gamma_s r_0.$$

Taking $r_0 \simeq c\,\Delta t_{\text{var}}$, we may express this in terms of typical values of the parameters:

$$r_s \simeq 300 \left(\frac{\gamma_s}{300}\right) (3 \times 10^8 \,\text{m s}^{-1}) \left(\frac{\Delta t_{\text{var}}}{10\,\text{ms}}\right) 10^{-2}\,\text{s}$$

$$\simeq 10^9 \left(\frac{\gamma_s}{300}\right) \left(\frac{\Delta t_{\text{var}}}{10\,\text{ms}}\right)\,\text{m}. \tag{8.8}$$

Implicit in our discussion so far has been that the fireball is opaque, i.e. optically thick. The electron–positron pairs will have annihilated at an earlier stage, and opacity is provided by the electrons associated with the baryons: the photons, trapped in the flow, scatter off electrons and thus accelerate them (Compton scattering; see Subsection 7.3.4). The protons are accelerated in turn via their electrostatic coupling to the electrons.

If the photons break free from the flow before the available energy is converted to kinetic energy of the protons, the fireball will not achieve a high speed and the energy will have been lost in a radiation burst with a nearly black body spectrum. As such spectra are not seen in GRBs, we are led to conclude that the fireball becomes transparent only after its expansion has saturated. Clearly, a fireball with very little mass will become transparent early on in its expansion. The two counterbalancing requirements of the fireball not being quenched by pair production or spent in a brief thermal burst place tight constraints on how much mass (or *baryon loading*) is acceptable. This turns out to be $10^{-7} \lesssim M/M_\odot \lesssim 0.5 \times 10^{-6}$, a pretty low and finely-tuned amount of mass for an explosion of this much energy that models for the energy extraction have to address. The amount of mass present determines the value of the Lorentz factor at saturation (via Equation 8.7), so these arguments restrict the Lorentz factor to values between 100 and a few thousand for typical energy values.

Having established that the **photosphere** has to lie beyond the saturation radius, we complete the discussion of this stage of the evolution of a fireball with the calculation of the location of the photosphere. We may assume equal numbers of electrons and protons, their respective number densities being $n'_e \approx n'_p \approx M/(m_p V')$, where $V' = 4\pi r^2 \Delta r'$ is the volume of a spherical shell of width $\Delta r'$ in the flow frame, and m_p is the proton mass. Casting the optical depth at a distance r (Equation 8.3) in terms of flow frame quantities, we get

The photosphere of a source is the layer or shell with optical depth $\tau \simeq 1$, i.e. photons emitted in the photosphere escape from the source.

$$\tau(r) = \sigma_T n'_e \Delta r' = \sigma_T \frac{M}{m_p V'} \Delta r' = \frac{\sigma_T M}{4\pi m_p} \frac{1}{r^2}. \tag{8.9}$$

We obtain the radius of the photosphere by setting $\tau(r_{\text{ph}}) \approx 1$ in Equation 8.9 and using Equation 8.7 to express M in terms of E_0 and γ_s. This gives

$$r_{\text{ph}} = \left(\frac{\sigma_T M}{4\pi m_p}\right)^{1/2} = \left(\frac{E_0 \sigma_T}{4\pi m_p c^2 \gamma_s}\right)^{1/2}.$$

Inserting typical values for the energy, $E_0 = 10^{44}\,\text{J}$, and the Lorentz factor, $\gamma = 300$, we get the radius of the photosphere:

$$r_{\text{ph}} \approx 10^{11} \left(\frac{E_0}{10^{44}\,\text{J}}\right)^{1/2} \left(\frac{300}{\gamma_s}\right)^{1/2}\,\text{m}. \tag{8.10}$$

8.4.2   Dissipation within the flow: prompt emission

So far we have discussed how the initial explosion energy is converted to kinetic energy of the flow. A mechanism is now required to convert this energy to thermal energy of the electrons, which will then radiate it away. In the fireball model, such a mechanism is provided by internal shocks (see Subsection 7.3.2).

Exercise 8.5   If energy is dissipated in internal shocks that develop due to small variations in the flow's speed, as described by Equation 7.36, calculate the distance r_{dis} from the centre of the explosion at which dissipation takes place. Assume that the energy injection is variable on a timescale $\Delta t_{var} = 10\,ms$, and that the flow has a Lorentz factor $\gamma = 300$.   ∎

This exercise shows that we can rewrite Equation 7.36 as the **dissipation radius** r_{dis}:

$$r_{dis} \simeq 5 \times 10^{11}\,\text{m} \left(\frac{\gamma_s}{300}\right)^2 \left(\frac{\Delta t_{var}}{10\,\text{ms}}\right). \qquad (8.11)$$

For these values of the parameters, internal shocks develop beyond the fireball photosphere (Equation 8.10). As the energy dissipation occurs in a region that is optically thin, the emitted radiation is non-thermal. The radiative mechanisms at work are synchrotron radiation and inverse Compton scattering (Subsection 7.3.4), whose spectra are broken-power laws, consistent with the observed GRB spectra. Detailed comparisons between the observed and model values of the break frequency and the slopes of the spectra provide information on the processes of particle acceleration (the values of p and γ_{min} of Equation 7.48 as hinted at in Subsection 7.3.3) and magnetic field generation. These considerations are beyond the scope of the present chapter.

Internal shocks are not the only proposed means of energy dissipation in the flow. Alternatives exist, some involving strong magnetic fields. The emission processes still involve scattering of photons by electrons. These alternative models are, however, less extensively studied at present.

8.4.3   Braking the fireball: afterglow

Following the prompt emission, enough energy remains in the flow to drive the expansion. The fireball sweeps up any material that it encounters: the interstellar medium (ISM) or any debris ejected by the progenitor before the burst. As a result, the fireball slows down. This drives a powerful relativistic shock (also called *blast wave*, or *forward* shock) through the ambient material. In response, a *reverse* shock propagates through the fireball (see Subsection 7.3.3). Both shocks heat up the material that they encounter, making the heat available for radiation. Although these shocks develop from the onset of the outflow, their effect on the fireball doesn't become manifest until enough material has piled up to brake the expansion of the outflow. As discussed in Subsection 7.3.2, the collision of a relativistic outflow with a stationary medium generates substantial thermal energy when the swept-up mass m_1 reaches a fraction $1/\gamma_s$ of the mass m_2 carried in the fireball.

Worked Example 8.3

For a spherical outflow with mass m_2 expanding into a medium of constant particle number density n, calculate the radius r_{dec} of the fireball at which deceleration becomes important.

Solution

Let m_1 denote the mass of the swept-up material. The mass of the ambient medium of particle number density n contained within a sphere of radius r and volume $V = (4\pi/3)r^3$ is $m_1 = m_p n V = (4\pi/3)nm_p r^3$. As argued above, at r_{dec} this becomes $m_1 = m_2/\gamma_s$. Equating the right-hand sides of the two expressions giving m, we obtain $(4\pi/3)nm_p r_{dec}^3 = m_2/\gamma_s = E_0/(\gamma_s^2 c^2)$, where in the last step we used Equation 8.7 to replace m_2. Solving for r_{dec} gives the **deceleration radius**:

$$r_{dec} = \left(\frac{3}{4\pi m_p c^2}\right)^{1/3}\left(\frac{E_0}{n\gamma_s^2}\right)^{1/3}. \tag{8.12}$$

Exercise 8.6 (a) The ISM consists of about 1 hydrogen atom in each cubic centimetre. Obtain the value of the number density in SI units.

(b) For a fireball of energy $E_0 = 10^{44}$ J and $\gamma_s = 300$, expanding into the ISM, obtain the value of the radius at which deceleration becomes important. ∎

This exercise shows that we can write Equation 8.12 as

$$r_{dec} \simeq 10^{14}\,\text{m}\left[\left(\frac{E_0}{10^{44}\,\text{J}}\right)\left(\frac{10^6\,\text{m}^{-3}}{n}\right)\left(\frac{300}{\gamma_s}\right)^2\right]^{1/3}. \tag{8.13}$$

It took the fireball the time $\Delta t \simeq r_{dec}/c$ to attain the size where the deceleration effects spark the start of the afterglow. Yet the observer places the onset of the afterglow at a time t_{dec} after the burst. According to Equation 7.23 this is given as

$$t_{dec} \simeq \frac{\Delta t}{2\gamma^2} = \frac{r_{dec}}{2\gamma_s^2 c}. \tag{8.14}$$

Exercise 8.7 Calculate the deceleration timescale t_{dec} for the values of the parameters of the previous exercise. ∎

Typical observed values of t_{dec} are of the order of seconds, suggesting that the afterglow may already begin during the prompt emission.

● For an afterglow of a flow with $\gamma_s = 300$ that starts $10\,\text{s}$ after the start of the burst, how long did the events leading up to it take?

○ Equation 8.14 links the observed time t_{dec} with the time that the fireball would take to attain a size r_{dec} expanding at nearly the speed of light, r_{dec}/c. The flow was growing for about $2\gamma_s^2 = 180\,000$ times longer, that is ~ 20 days!

The fireball's expansion at constant speed lasts until r_{dec} is reached. From then on, γ will be dropping fast due to the drag by the piled-up ambient material and

The term *adiabatic* is here somewhat loosely defined to mean that little energy is radiated away compared to the energy that is still trapped in the flow. This happens if the particles are losing energy due to radiation more slowly than they cool down due to the expanding volume that they occupy. In the reverse situation, the flow is described as radiative.

possible radiative losses, leading to a fading source. Light curves of afterglows show an inverse time-dependence, consistent with the view that at least part of the evolution of the afterglow is adiabatic. Application of energy and momentum conservation gives the evolution of γ in the limits of adiabatic and radiative regimes and for a general ambient medium density profile. (For example, while the particle number density n is constant if the explosion happened in the ISM, a pre-ejected wind from the progenitor will have created an $n \propto r^{-2}$ profile.) To assist in the interpretation of the afterglow light curves and spectra, it is useful to express both the Lorentz factor γ and the size of the fireball r in terms of the observer time, t. The following scaling relationships hold:

$$r(t) \propto \begin{cases} (t/n)^{1/4}, \\ (t/n^6)^{1/7} \end{cases} \quad \text{and} \quad \gamma(t) \propto \begin{cases} (nt^3)^{-1/8}, & \text{Adiabatic} \\ (nt^3)^{-1/7}. & \text{Radiative} \end{cases} \quad (8.15)$$

These quantities enter in the modelling of light curves and spectra, and provide the stage for any quantitative testing of the fireball model. The power-law decay of the Lorentz factor of the flow ensures that all observable quantities will display a smooth, power-law decay.

Turning to the spectral features, the time-dependence of the characteristic frequencies of synchrotron radiation will also be determined by the type of the fireball evolution. In the following example, we shall demonstrate the spectral softening of the afterglow.

Worked Example 8.4

Assuming that the afterglow radiation is due to synchrotron emission, calculate the time-dependence of the synchrotron peak frequency, for adiabatic evolution in a constant-density environment. You may assume that the magnetic field in the flow rest frame is $B' \propto \gamma \sqrt{n}$ (where γ is the Lorentz factor of the flow). This is a consequence of the supposition that the magnetic field energy ($\propto B^2$) may be amplified to a fraction of the internal energy ($\propto \gamma^2 n$, see Equation 7.47) of the flow. Further, assume that the electrons are accelerated to a peak energy $\gamma_{\min} \simeq \gamma$ (see Equation 7.48), as suggested by the lack of low-energy emission.

Solution

According to Equation 7.52, the characteristic frequency of an electron of Lorentz factor $\gamma_{\min}$ in a magnetic field B' is proportional to $B'\gamma_{\min}^2$. This is the frequency in the rest frame of the flow. For a flow moving at a Lorentz factor γ, the observed synchrotron frequency will be Doppler boosted by a factor of γ (Equation 7.32), or

$$\nu_{\min,\text{sy}} \propto \gamma\gamma_{\min}^2 B'.$$

Since we are only interested in the time-dependence of this frequency, we may work with proportionalities and retain only any dependence on γ, which carries the time-dependence of all quantities (Equation 8.15). Hence $\nu_{\min,\text{sy}} \propto \gamma\gamma^2\gamma = \gamma^4$. For adiabatic evolution in a constant-density medium, Equation 8.15 gives $\gamma \propto t^{-3/8}$. This makes the evolution of the peak synchrotron frequency $\nu_{\min,\text{sy}} \propto t^{-3/2}$.

The spectral softening demonstrated in this example also explains the dimming of the burst in a given spectral range. The rate at which afterglow light curves fade and spectra soften (the indices in Equation 8.2) has thus been linked with the physical conditions and processes operating in the fireball. Figure 8.10 summarizes the evolution of the different characteristic frequencies of synchrotron radiation.

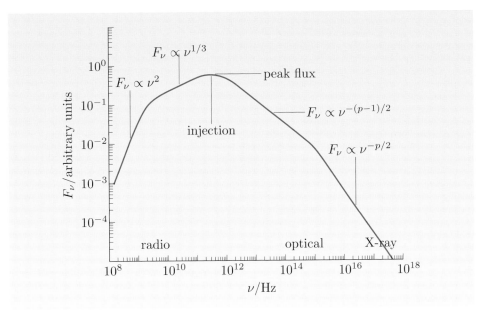

Figure 8.10 A sketch of the synchrotron spectrum at the onset of the afterglow, indicating the corresponding spectral range on the horizontal axis. The characteristic slopes and energy breaks are typical of synchrotron emission. The sketch includes modifications to the simple synchrotron spectrum presented in Subsection 7.3.4, which we do not discuss here.

While the afterglow emission is attributed to the forward shock that propagates in the surrounding medium, the reverse shock propagating through the fireball material offers a plausible interpretation of the optical flash. For a bright optical flash, the reverse shock has to be very efficient in amplifying magnetic fields and accelerating particles. Such special circumstances may explain the relative scarcity of optical flashes.

8.4.4 Jet effects and late evolution

As we hinted at earlier, nothing in our analysis changes if, instead of a sphere, the flow forms two opposite conical jets with opening angle $\Theta_J > 2/\gamma$. To adapt our analysis to the jet geometry, we have to account for the channelling of the burst energy into a solid angle $\Delta\Omega < 4\pi$. If we assume spherical symmetry and infer a burst energy E_{iso} from the observed fluence, then the actual burst energy E_0 is smaller by a factor $\Delta\Omega/4\pi$. The solid angle $\Delta\Omega$ subtended by an opening angle Θ_J is given by Equation 7.20. Accounting for both jets, this becomes $\Delta\Omega = 4\pi(1 - \cos(\Theta_J/2))$. Hence all expressions that we have quoted so far will hold if E_0 is taken to be

$$E_0 = \frac{\Delta\Omega}{4\pi} E_{iso} = [1 - \cos(\Theta_J/2)] E_{iso}.$$

For small Θ_J we can make use of the expansion $\cos x \simeq 1 - x^2/2$ for $x = \Theta/2 \ll 1$, to obtain

$$E_0 \simeq E_{\text{iso}} \frac{\Theta_J^2}{8}. \tag{8.16}$$

If the outflow forms a jet, it will be visible only to observers that intercept the physical cone of the jet (Figure 8.11). Moreover, because of relativistic light aberration (see Subsection 7.2.3), the outflow's emission is strongly beamed in the direction of the motion. So even those observers within the jet's opening angle will only see a fraction of it, with opening angle $2/\gamma$, beamed along the line of sight. As the fireball decelerates and γ decreases, an increasing fraction of it becomes visible. It is obvious that at a stage when $2/\gamma \approx \Theta_J$, the whole source will have been unveiled. As a result, the rate at which the visible surface of the fireball increases, in the course of the expansion, slows down.

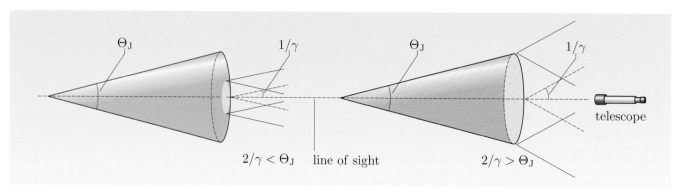

Figure 8.11   A sketch of the jet and the luminous area seen by the observer due to relativistic beaming, for the special case where the jet axis coincides with the line of sight: (a) when only part of the jet is seen ($2/\gamma < \Theta_J$), and (b) when the full jet is seen ($2/\gamma > \Theta_J$).

The intensity of the received radiation depends on the intrinsic properties of the emitting region and also on the surface area that is contributing to the emission. The physical conditions in the jet do not change discontinuously at the transition point where the whole jet becomes visible. Therefore the change of the light curve to a steeper slope is independent of frequency. This purely geometric effect will be seen as a simultaneous break in the light curves in all wavelengths: an *achromatic break*. From the timing of the break, the jet opening angle can be calculated. Most fireballs are collimated to jets of about $10°$ opening angle. Intriguingly, it seems that more luminous bursts are also more collimated.

Exercise 8.8   Consider a GRB where the emission is confined to a jet of opening angle $\Theta_J = 10°$.

(a) Revisit the result of Exercise 8.1. What is the source energy requirement?

(b) Revisit the result of Worked Example 8.2. What is the inferred source rate?   ∎

In the very late stages, the spent fireball becomes spherical and slows down to a non-relativistic expansion. This stage resembles the late evolution of supernovae. Unlike supernovae, the GRB fireball is not expected to leave behind any appreciable remnant. After all, it has been travelling light.

8.5   The central engine

The central engine of a GRB is enshrouded by the fireball. The evolution of the fireball is quite independent of how the central engine operates. The only imprints that the central engine makes on the observed phenomena are the time over which it operates (manifest as the duration of the prompt emission and subsequent flaring), the total energy it generates, the mass it ejects, and, possibly, the variability pattern. Thus clues to the nature of the progenitors and the mechanisms of energy extraction are scarce and concealed.

8.5.1   Progenitors

Applying the energy correction due to the collimation in a jet (Equation 8.16) to the observed afterglows displaying achromatic breaks has the remarkable result of reducing the actual energy of the sources to a very narrow range clustered around 3×10^{44} J. This is remarkable on two accounts. First, it brings the total energy requirements for the progenitor close to that of a supernova (see Exercise 8.1), thus alleviating the energy budget crisis at the heart of the extreme luminosities and distances observed. Second, it suggests that at least a subset of long bursts (to which the bulk of observations pertain) may be standard candles, which would make them powerful cosmological distance scale measuring tools and place stringent constraints on the nature of the progenitor. Apart from being plausible, the progenitor candidates must form at the appropriate rate and be able to supply a large amount of energy, with a dash of mass, in a catastrophic collapse. The energy and variability arguments presented earlier point to black holes or neutron stars as likely progenitors.

Clues from the hosts

As discussed already, long GRBs are found in regions where massive young stars lie. A few of them are clearly associated with type Ic supernovae. In contrast, short bursts are not selective in where they reside. Additionally, applying the external shock model predictions to the broadband light curves suggests that the bursts are expanding in a constant-density medium that is significantly denser for the long bursts.

Long GRBs are linked with young stellar populations, like type II supernovae, while short bursts are found everywhere, like type Ia supernovae. Type II supernovae are thought to be the products of the core collapse of massive stars into a neutron star or black hole, while type Ia supernovae are likely to occur in compact binaries with a white dwarf. Thus the current most popular GRB progenitor scenarios involve a hypernova (the supernova analogue of very massive stars) for long bursts and a compact binary merger for short bursts (see Figure 8.12 overleaf). There is the distinct possibility, however, that multiple types of progenitors are implicated in the production of either long or short bursts.

Long bursts: hypernova

The collapse of the core of a very massive star into a black hole has been called a hypernova as it is reminiscent of a supernova, in superlative. The key difference is

The descriptive term *collapsar* is used interchangeably with *hypernova*.

that while a supernova imparts the energy of the collapse to an approximately $1\,M_\odot$ outflow that will expand at tens of thousands of $km\,s^{-1}$, the hypernova

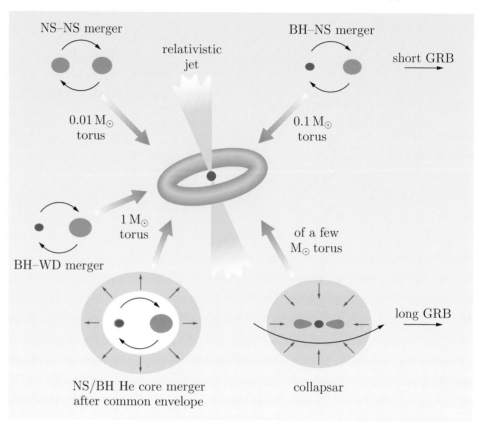

Figure 8.12 A sketch of the predominant paradigm for GRB progenitors. The accreting black hole at the core of the engine can result from the merger of two compact stars that initially form a binary system, or the core collapse of a very massive $(M > 20\,M_\odot)$ star. The top two systems are thought to give rise to the short bursts, the bottom two to long bursts.

does so to a very small amount of mass that ends up expanding relativistically. Wolf–Rayet stars that are very massive, rapidly rotating and lose mass in intense winds, contain massive cores that could make hypernovae.

We first consider the energy available from a hypernova, and how it can be harnessed to power a fireball. The burst is powered by the gravitational potential energy of the collapsing stellar core, or possibly the rotational energy of the newly formed black hole. In either case, the stellar debris forms an accreting torus around the black hole, and this torus supplies the energy on the accretion timescale. The formation of such a torus requires a large amount of angular momentum; this may pose a problem for stars that have undergone severe mass and angular momentum loss via winds prior to the hypernova event. The torus should be sustained for at least the duration of the prompt burst, possibly for longer, to feed later flaring activity in the afterglow.

One important remaining question is whether such a stellar collapse can create the clean collimated fireball that is required. More specifically, whether the jet formed in the core as the star collapses can break free through the stellar envelope and carry the energy out, building up to high Lorentz factors while staying collimated. This is a very complicated problem, barely tractable by state-of-the art computer simulations. Increasingly sophisticated simulations are just approaching realistic descriptions (Figure 8.13).

We now consider the number of hypernova progenitors. Stars with suitably high mass are in fact much more abundant than the GRB progenitors implied by the

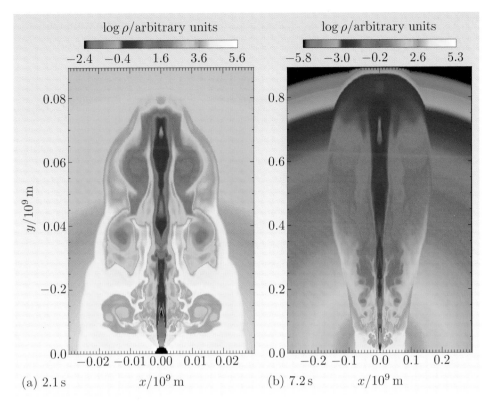

(a) 2.1 s $x/10^9$ m

(b) 7.2 s $x/10^9$ m

Figure 8.13 Simulation of the jet break-out at relativistic speeds from the collapsing core of a Wolf–Rayet star to a black hole, shown at two instances. The density ρ is lowest and the speed fastest along the jet axis. The jet emerges in (b), engulfed by a slower cocoon.

observed GRB rate. This is the case even when the effect of beaming of the GRB sources is taken into account. This demonstrates that the hypernova origin is a viable scenario, but also suggests that for such a collapse to give rise to a GRB, the circumstances have to be rather fortuitous.

Short bursts: neutron star–neutron star, neutron star–black hole binary coalescence

The orbital decay of double neutron star (NS) binaries due to gravitational radiation is well established (see Section 2.9). If the decay continues for long enough and the orbital separation becomes sufficiently small, the two stars merge, and a black hole (BH) is predicted to form. The same outcome is likely for NS–BH binary mergers. The coalescence of compact binaries thus seems a plausible mechanism for GRBs. In particular, these have been linked to short GRBs. The total energy released (equivalent to the rest mass energy of a few solar masses) is sufficient to power a GRB. As the environment may be cleaner than for the case of the collapse of a massive star (which is likely to have ejected some of its layers before the collapse), the outflow could be less mass loaded and thus reach a higher Lorentz factor (see Equation 8.7), which would make such sources spectrally harder, due to Doppler boosting. Broadly speaking this is consistent with the short bursts showing higher spectral hardness (see Figure 8.3). Again a jet has to break out of the newly formed compact object. Numerical simulations of binary mergers face challenges similar to those discussed above for the core collapse.

● Based on the observed number of compact binaries in our Galaxy, one can estimate the merger rate for NS–NS and NS–BH binaries to be a few times 10^{-5} per galaxy per year. Compare this with the observed GRB rate obtained in Worked Example 8.2. In view of that result, are these sources viable GRB candidates?

○ In Worked Example 8.2 we estimated a GRB rate of $(0.1–1) \times 10^{-6}$ GRBs per galaxy per year, so there seem to be between a few tens and hundreds of potential progenitors for each observed burst. This can accommodate beaming of the flow into jets (see Exercise 8.8).

The time lag between the formation of the progenitor main-sequence binary and the coalescence of a compact binary with an NS or BH component is of the order of 10^8–10^9 years. Thus these systems are expected to be merging only in recent cosmological times, in nearby galaxies. This is consistent with the derived distances of short bursts. Further, a NS–BH compact binary involves at least one supernova in its formation. Supernova explosions impart a kick velocity to the nascent NS of several hundreds of $km \, s^{-1}$ (Section 2.8) that is expected to send them off the galactic plane. This again is consistent with the numerous short GRBs found at the outskirts of galaxies.

8.5.2   Energy extraction

We saw in Section 8.4 that the very energetic fireball is slightly peppered with some baryons. Getting so much energy out without quenching the burst by overloading it with mass requires some special mechanism. The requirement for a clean (albeit not spick and span!) fireball has led to the consideration of a rapidly rotating object at the heart of the explosion. Both of the favoured progenitor scenarios involve a spinning black hole. This rotation creates a centrifugal barrier that matter will have a hard time penetrating. Conceivably, then, the deposited energy will escape in a collimated geometry, along the rotational axis, consistent with the formation of jets. The two main mechanisms invoked to extract the energy are **neutrino** annihilation and magnetic stressing. Neither mechanism has been proven efficient in this context, and their discussion is beyond the scope of this text. Their potential to cook up a clean fireball is qualitatively sketched here.

Neutrinos that may be produced in an accretion disc will have paths that cross more often close to the rotation axis of the disc and will thus interact with higher probability than elsewhere. The product of their interaction is an e^-–e^+ pair. The annihilation and recombination of such pairs power the fireball.

Magnetic fields can be important in a newly formed black hole that is surrounded by a torus of matter of the progenitor star. Magnetic field lines are anchored in the torus and get stretched as the black hole spins. The rotational energy of the black hole is thus transferred to the surrounding material via the twisting and breaking of magnetic field lines.

8.6   Other signatures and applications

We turn now to future developments by examining model predictions and the potential of using GRBs as cosmological tools.

The different progenitor scenarios along with the shock acceleration model predict other types of — hitherto unobserved — emission, as discussed below.

Very high energy photons, cosmic rays and neutrinos

If the synchrotron interpretation for the gamma-ray emission is valid, then the inverse Compton scattering component may reach energies well into the hundreds of GeV. Shocks accelerate protons along with electrons. Protons can reach energies of 10^{20} eV for the conditions implied by GRBs. Energetic protons are detected as **cosmic rays**. The contribution to the observed background cosmic ray distribution in energies 10^{18}–10^{20} eV is not properly accounted for, and GRBs may just provide a sufficient supply.

Neutrinos are implicated in the energy deposition at the central engine (Subsection 8.5.2). They are produced copiously during the core collapse as in supernovae and can readily escape, reaching us even before the gamma-rays. This is a potential diagnostic for the nature of GRB progenitors. Further neutrinos can be produced in the interactions of the accelerated protons.

Gravitational waves

Both hypernovae and compact binaries are potential gravitational wave sources (see Sections 2.9 and 2.5), with the latter expected to be stronger. A gravitational wave signal will spread spherically symmetrically and should thus reach us even from sources whose jets are beamed away from us and are therefore not detected as GRBs. If a signal coincident with a GRB were detected, this would singlehandedly prove the implication of a catastrophic collapse in the formation of a GRB, and the existence of gravitational radiation — something to write home about!

GRBs as yardsticks

The redshift distribution of GRBs reaches further than that of any other known class of objects in the Universe (see Figure 8.14). Hence GRBs might be used as the most comprehensive distance signposts. This would be straightforward if GRBs were standard candles (like type Ia supernovae) — yet a robust correlation between burst properties still evades us. The hope is that in a large enough sample of GRBs with known redshifts, some GRB properties will obey a definite relationship that would allow the independent determination of their luminosity and hence distance (as, for example, is the case for Cepheids). If such a relationship can be found, they may end up being the ultimate step in the distance ladder. GRBs could then be used to test cosmological models. Since they probe the earliest stages of the Universe, they could gauge the evolution of cosmological parameters too.

GRBs as beacons

The light from the most remote GRBs travels through most of the Universe on its way to us. It shines on obstacles, gets scattered and absorbed; in short, it can be used as a diagnostic of the Universe's state. It can probe the early star formation, the re-ionization epoch and the intervening medium, and identify the cosmological model, including the role of the cosmological constant (dark energy).

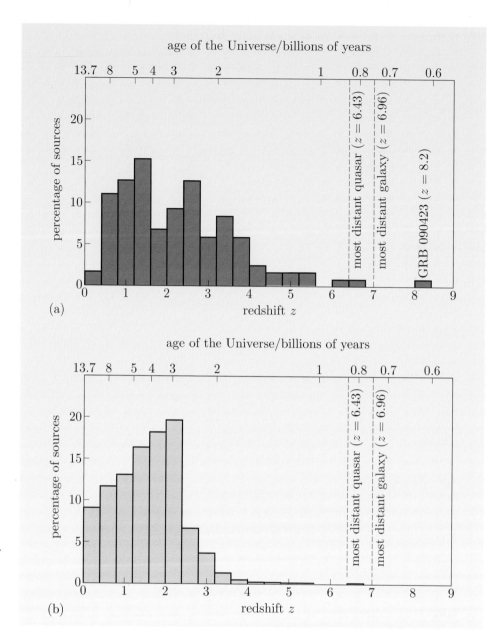

Figure 8.14 The percentage of (a) GRBs and (b) quasar sources as a function of redshift. The distribution of GRBs reaches further in the Universe than any other type of source observed to date.

Epilogue

As our observatories' capability for quick follow up of a gamma-ray trigger is improved, we shall increasingly be regarding GRBs as continuous, multi-wavelength transient events, much like supernovae. We may not abandon the term GRBs; after all, it is in gamma-rays that they exhibit their unique and extreme properties. But afterglows will be routinely used to characterize each burst. A conceivable future holds GRBs as the ultimate 'standard candles' to probe the deepest reaches of the Universe. Some of the key parameters determined from the prompt emission (e.g. the break frequency ν_p, the gamma-ray luminosity L_γ) and those determined from the afterglow stages (e.g. temporal evolution indices, host redshift, jet opening angle, total energy) will be used to calibrate types of sources. Increasing numbers of observed sources will also more clearly delineate subclasses and pick out salient features. GRBs may be seen as a stage in the life of a number of different source types, some of which are likely to be linked to the death throes of massive or paired stars and the birth cries of black holes. Or GRBs may be shown to be a generic feature of all things highly magnetized, much like X-rays are a tell-tale signature of matter sizzling when dropped in strong gravitational fields. This most puzzling and persistent of all astrophysical mysteries will eventually be tamed to a reliable and prolific work-horse in the disciplines of cosmology and astrophysics. Rather uniquely, GRBs may be put to the task of deciphering the warps and twists of spacetime, from stellar black holes to the cosmological horizon. This chapter is still to be written.

Summary of Chapter 8

1. Gamma-ray bursts (GRBs) are intense flashes of gamma-rays that last anywhere between milliseconds and tens of minutes. They are usually followed by an afterglow: prolonged, dwindling radiation in wavebands of progressively lower frequency. A GRB thus displays two distinct phases: the prompt emission and the afterglow.

2. The short duration of the bursts along with the inability of gamma-ray detectors to obtain precise directional information has precluded the identification of counterparts, thus stalling our understanding of the phenomenon for three decades. The detection of afterglows eventually allowed the determination of accurate positions of the sources, proof of their cosmological origin, and the identification of hosts.

3. The GRB distribution in the sky is isotropic and non-homogeneous. Redshift measurements show that GRBs make up the population that extends to the most distant reaches of the Universe.

4. The prompt emission is confined to gamma-rays (although a few sources have now been seen briefly in X-rays and most notably in the optical range). Afterglows are seen for up to days in X-rays, up to weeks in the optical range and up to several months in the radio band.

5. GRB energy spectra are extended broken-power laws that peak in photon energies in the range of hundreds of keV. Spectral softening is usually seen during a burst as well as within pulses.

6. There are two GRB classes: the short bursts and the long bursts. The group divide is at about 2 seconds. Short bursts are on average spectrally harder and fainter than long bursts. The afterglow properties of the two groups are very similar.

7. Long bursts are found in star-forming regions in the spiral arms of galaxies and have been seen at very high redshifts. Short bursts are found in any type of galaxy or environment. They form a relatively nearby population.

8. The super-Eddington luminosities and the observed very high photon energies imply relativistic expansion of the source — a fireball. The outflow may be confined in jets that can be approximated as conical sections of a spherical fireball.

9. The relativistically expanding fireball may consist of pure magnetic or radiation energy and may contain a small amount of particles. In a fireball that does carry matter, the expansion accelerates until all energy is passed on to matter and then carries on at a constant speed. The fireball becomes optically thin at a later stage.

10. The prompt emission requires energy dissipation in the outflow. Internal shocks may provide the required mechanism. Synchrotron radiation and inverse Compton scattering from relativistic electrons that are accelerated in the shocks can account for the observed features.

11. The afterglow is attributed to the shock formed when the flow is decelerated by the ambient material that it is sweeping up. This generates a forward shock, processing ambient material, and a reverse shock that propagates backwards in the flow.

12. The key stages in the evolution of the fireball are taking place at the following radii (evaluated for typical values of the parameters):

   - *saturation* of the expansion:

$$r_s \simeq 10^9 \, \frac{\gamma_s}{300} \frac{\Delta t_{\text{var}}}{10 \, \text{ms}} \, \text{m};$$

   (Eqn 8.8)

   - *photosphere*:

$$r_{\text{ph}} \simeq 10^{11} \left(\frac{E_0}{10^{44} \, \text{J}} \right)^{1/2} \left(\frac{300}{\gamma_s} \right)^{1/2} \, \text{m};$$

   (Eqn 8.10)

   - *dissipation* due to internal shocks:

$$r_{\text{dis}} \simeq 5 \times 10^{11} \, \text{m} \left(\frac{\gamma_s}{300} \right)^2 \left(\frac{\Delta t_{\text{var}}}{10 \, \text{ms}} \right);$$

   (Eqn 8.11)

   - *deceleration* leading to external shocks:

$$r_{\text{dec}} \simeq 10^{14} \, \text{m} \left[\left(\frac{E_0}{10^{44} \, \text{J}} \right) \left(\frac{10^{-6} \, \text{m}^{-3}}{n} \right) \left(\frac{300}{\gamma_s} \right)^2 \right]^{1/3}.$$

   (Eqn 8.13)

13. As it decelerates, the forward shock gives rise to the ever dimming and softening afterglow. This explains well the observed power-law segments of the GRB light curve. Appropriate conditions in the reverse shock may produce optical prompt emission.

14. Radio afterglow observations have provided a direct measurement of the size of the fireball and demonstrated its relativistic expansion.

15. The late steepening of the light curve that is simultaneously observed in all wavebands (the so-called achromatic break) is attributed to a jet. The inferred jet opening angles are clustered around $10°$.

16. Short and long GRBs are thought to have distinct progenitors. A hypernova (the core collapse of a very massive star) is the prevalent paradigm for long GRBs, while a compact binary merger is the favoured model for short GRBs. Both scenarios end up with a stellar mass black hole that accretes material over the timescale of the burst. However, each group may comprise several classes of progenitors beyond these systems.

17. Models of GRB progenitor systems that involve the violent formation of a black hole predict the creation of a burst of neutrinos and of gravitational waves immediately prior to the GRB.

18. Yet untested predictions of the fireball model include bursts at very high frequency and a persistent contribution to the cosmic ray background at energies 10^{18}–10^{20} eV.

19. GRBs are the most distant objects detected in the Universe. This makes them potentially powerful distance scale indicators, cosmological model checks and probes of the intergalactic medium.

References and further reading

This short compilation lists those books and journal articles that are either referred to explicitly in the text, or played a significant role in its preparation, or are a good stepping stone for further reading. Many of the journal papers and some of the book chapters are freely available on the internet as preprints — the final drafts before publication — from the `arXiv.org` preprint server.

No attempt is made to provide a comprehensive literature review of the subjects covered in this book.

Textbooks

1. Eggleton, P. P. (2006), *Evolutionary Processes in Binary and Multiple Stars*, CUP, ISBN 978-0521-85557-0.

2. Frank, J., King, A. R. and Raine, D. (2002), *Accretion Power in Astrophysics*, (3rd edition), CUP, ISBN 978-0521-62957-7.

3. Harwit, M. (1988), *Astrophysical Concepts*, Springer–Verlag, ISBN 978-0387-32943-7.

4. Haswell, C. A. (2010) *Transiting Exoplanets*, CUP, ISBN 978-0521-13938-0.

5. Lambourne, R. J. A. (2010) *Relativity, Gravitation and Cosmology*, CUP, ISBN 978-0521-13138-4.

6. Lewin, W. and van der Klis, M. (2006), *Compact Stellar X-ray Sources*, CUP, ISBN 978-0521-82659-4.

7. Peterson, B. M. (1997), *An Introduction to Active Galactic Nuclei*, CUP, ISBN 0521-47911-8.

8. Rosswog, S. and Brüggen, M. (2007), *Introduction to high-energy astrophysics*, CUP ISBN 978-0521-85769-7.

9. Ryan, S. G. and Norton, A. J. (2010) *Stellar Evolution and Nucleosynthesis*, CUP, ISBN 978-0521-13320-3.

10. Rybicki, G. B. and Lightman, A. P. (2004), *Radiative processes in Astrophysics* (2nd edition), Wiley-VCH Verlag ISBN 978-0471-82759-7.

11. Serjeant, S. (2010) *Observational Cosmology*, CUP, ISBN 978-0521-15715-5.

12. Wheeler, J. C. (2007), *Cosmic Catastrophes: Exploding Stars, Black Holes, and Mapping the Universe*, CUP, ISBN 978-0521-85714-7.

Chapters 1 and 2

1. Eggleton, P. P. (1983), 'Approximations to the radii of Roche lobes', *Astrophysical Journal*, **268**, 368–369.

2. King, A. R. (1988), 'The evolution of compact binaries', *Quarterly Journal of the Royal Astronomical Society*, **29**, 1–25.

3. Paczyński, B. (1971), 'Evolutionary Processes in Close Binary Systems', *Annual Review of Astronomy and Astrophysics*, **9**, 183.

4. Ritter, H. (2008), 'Formation and Evolution of Cataclysmic Variables', *Mem. Soc. Astron. Italiana*, Proceedings of the School of Astrophysics 'Francesco Lucchin', (arXiv:0809.1800v1 [astro-ph]).

5. Taam, R. E. and Sandquist, E. L. (2000), 'Common Envelope Evolution of Massive Binary Stars', *Annual Review of Astronomy and Astrophysics*, **38**, 113–141.

6. Webbink, R. F. (2008), 'Common Envelope Evolution Redux', in *Short-Period Binary Stars: Observations, Analyses and Results*, Edited by E. F. Milone, D. A. Leahy and D. W. Hobill. Springer–Verlag, Berlin, 233.

Chapters 3 and 4

1. Balbus, S. A. and Hawley, J. F. (1991), 'A powerful local shear instability in weakly magnetized disks. I - Linear analysis. II - Nonlinear evolution', *Astrophysical Journal*, **376**, 214–233.

2. Lasota, J-P. (2001), 'The disc instability model of dwarf novae and low-mass X-ray binary transients', *New Astronomy Reviews*, **45**, 449–508 (arXiv:astro-ph/0102072v1).

3. Ritter, H. and Kolb, U. (2003), 'Catalogue of cataclysmic binaries, low-mass X-ray binaries and related objects', (seventh edition), *Astronomy and Astrophysics*, **404**, 301–303 (updated online every 6 months).

4. Shakura, N. I. and Sunyaev, R. A. (1973) 'Black holes in binary systems - Observational appearance', *Astronomy and Astrophysics*, **24**, 337–355.

5. Spruit, H. C. (2000), 'Accretion Disks', *The neutron star black hole connection* (NATO ASI ELOUNDA 1999), eds. C. Kouveliotou et al., (arXiv:astro-ph/0003144v2).

Chapter 5

1. Baptista, R., Morales-Rueda, L., Harlaftis, E. T., Marsh, T. R. and Steeghs, D. (2005) 'Tracing the spiral arms in IP Pegasi', *Astronomy and Astrophysics*, **444**, 201–211.

2. Horne, K. (1985), 'Images of accretion discs. I - The eclipse mapping method', *Monthly Notices of the Royal Astronomical Society*, **213**, 129–141.

3. Marsh, T. R. (2005), 'Doppler Tomography', *Astrophysics and Space Science*, **296**, 403–415.

4. Steeghs, D., Harlaftis, E. T. and Horne, K. (1997) 'Spiral structure in the accretion disc of the binary IP Pegasi', *Monthly Notices of the Royal Astronomical Society*, **290**, L28–L32.

Chapter 6

1. Brenneman, L. W. and Reynolds, C. S. (2006), 'Constraining Black Hole Spin via X-Ray Spectroscopy', *The Astrophysical Journal*, **652**, 1028–1043.

2. Done, C. and Gierliński, M. (2003), 'Observing the effects of the event horizon in black holes', *Monthly Notices of the Royal Astronomical Society*, **342**, 1041–1055.

3. Lewin, W. and van der Klis, M. (2006), *Compact Stellar X-ray Sources*, CUP, ISBN 0521-82659-4. This textbook is a collection of review papers by

experts in the field, covering many different aspects of accreting X-ray sources. We quote the most relevant preprints below.

- Fabbiano, G. and White, N. E., 'Compact Stellar X-ray Sources in Normal Galaxies ' (arXiv:astro-ph/0307077v1).

- Tauris, T. M. and van den Heuvel, E., 'Formation and Evolution of Compact Stellar X-ray Sources ' (arXiv:astro-ph/0303456v1).

- Charles, P. A. and Coe, M. J., 'Optical, ultraviolet and infrared observations of X-ray binaries ' (arXiv:astro-ph/0308020v2).

- King, A. R., 'Accretion in Compact Binaries ' (arXiv:astro-ph/0301118v2).

- van der Klis, M., 'A review of rapid X-ray variability in X-ray binaries' arXiv:astro-ph/0410551v1.

- McClintock, J. E. and Remillard, R. A., 'Black Hole Binaries ' (arXiv:astro-ph/0306213v4).

- Strohmayer, T. and Bildsten, L., 'New Views of Thermonuclear Bursts' (arXiv:astro-ph/0301544v2).

4. McHardy, I. M., Koerding, E., Knigge, C., Uttley, P. and Fender, R. P. (2006), 'Active galactic nuclei as scaled-up Galactic black holes', *Nature*, **444**, 730–732.

5. Predehl, P. and Schmitt, J. H. M. M. (1995), 'X-raying the interstellar medium: ROSAT observations of dust scattering halos', *Astronomy and Astrophysics*, **293**, 889–905.

6. Schwarzschild, M. and Härm, R. (1965), 'Thermal Instability in Non-Degenerate Stars', *Astrophysical Journal*, **142**, 855.

Chapters 7 and 8

1. Harwit, M. (1988), Chapter 5: Photons and fast particles, *Astrophysical Concepts*, Springer–Verlag, ISBN 978-0387-32943-7.

2. Harwit, M. (1988), Sections 6.17–6.22 of Chapter 6: Electromagnetic processes in space, *Astrophysical Concepts*, Springer–Verlag, ISBN 978-0387-32943-7.

3. Katz, J. I. (2002), *The Biggest Bangs*, Oxford University Press, ISBN 0-19-514570-4.

4. Mészaros, P. (2008), Gamma ray burst theory, *Scholarpedia* 3(3): 4337, http://www.scholarpedia.org/

5. Mészaros, P. (2006), Gamma Ray Bursts, *Reports on Progress in Physics*, **69**, 2259–2321.

6. Piran, T. (1999), Gamma Ray Bursts and the fireball model, *Physics Reports*, **314**, 575–667.

7. Piran, T. (2004), The Physics of gamma-ray bursts, *Reviews of Modern Physics*, **76**, 1143.

8. Rybicki, G. B. and Lightman, A. P. (2004), Chapter 6: Synchrotron Radiation, *Radiative processes in Astrophysics*, Wiley VCH, ISBN 978-0471-82759-7.

9. Rybicki, G. B. and Lightman, A. P. (2004), Chapter 7: Compton Scattering, *Radiative processes in Astrophysics*, Wiley VCH, ISBN 978-0471-82759-7.

Appendix

Table A.1 Common SI unit conversions and derived units.

Quantity	Unit	Conversion
speed	m s^{-1}	
acceleration	m s^{-2}	
angular speed	rad s^{-1}	
angular acceleration	rad s^{-2}	
linear momentum	kg m s^{-1}	
angular momentum	$\text{kg m}^2\,\text{s}^{-1}$	
force	newton (N)	$1\,\text{N} = 1\,\text{kg m s}^{-2}$
energy	joule (J)	$1\,\text{J} = 1\,\text{N m} = 1\,\text{kg m}^2\,\text{s}^{-2}$
power	watt (W)	$1\,\text{W} = 1\,\text{J s}^{-1} = 1\,\text{kg m}^2\,\text{s}^{-3}$
pressure	pascal (Pa)	$1\,\text{Pa} = 1\,\text{N m}^{-2} = 1\,\text{kg m}^{-1}\,\text{s}^{-2}$
frequency	hertz (Hz)	$1\,\text{Hz} = 1\,\text{s}^{-1}$
charge	coulomb (C)	$1\,\text{C} = 1\,\text{A s}$
potential difference	volt (V)	$1\,\text{V} = 1\,\text{J C}^{-1} = 1\,\text{kg m}^2\,\text{s}^{-3}\,\text{A}^{-1}$
electric field	N C^{-1}	$1\,\text{N C}^{-1} = 1\,\text{V m}^{-1} = 1\,\text{kg m s}^{-3}\,\text{A}^{-1}$
magnetic field	tesla (T)	$1\,\text{T} = 1\,\text{N s m}^{-1}\,\text{C}^{-1} = 1\,\text{kg s}^{-2}\,\text{A}^{-1}$

Table A.2 Other unit conversions.

wavelength
1 nanometre (nm) $= 10\,\text{Å} = 10^{-9}\,\text{m}$
1 ångstrom $= 0.1\,\text{nm} = 10^{-10}\,\text{m}$

mass–energy equivalence
$1\,\text{kg} = 8.99 \times 10^{16}\,\text{J}/c^2$ (c in m s^{-1})
$1\,\text{kg} = 5.61 \times 10^{35}\,\text{eV}/c^2$ (c in m s^{-1})

angular measure
$1° = 60\,\text{arcmin} = 3600\,\text{arcsec}$
$1° = 0.017\,45\,\text{radian}$
$1\,\text{radian} = 57.30°$

distance
1 astronomical unit (AU) $= 1.496 \times 10^{11}\,\text{m}$
1 light-year (ly) $= 9.461 \times 10^{15}\,\text{m} = 0.307\,\text{pc}$
1 parsec (pc) $= 3.086 \times 10^{16}\,\text{m} = 3.26\,\text{ly}$

temperature
absolute zero: $0\,\text{K} = -273.15\,°\text{C}$
$0\,°\text{C} = 273.15\,\text{K}$

energy
$1\,\text{eV} = 1.602 \times 10^{-19}\,\text{J}$
$1\,\text{J} = 6.242 \times 10^{18}\,\text{eV}$

spectral flux density
1 jansky (Jy) $= 10^{-26}\,\text{W m}^{-2}\,\text{Hz}^{-1}$
$1\,\text{W m}^{-2}\,\text{Hz}^{-1} = 10^{26}\,\text{Jy}$

cross-sectional area
$1\,\text{barn} = 10^{-28}\,\text{m}^2$
$1\,\text{m}^2 = 10^{28}\,\text{barn}$

cgs units
$1\,\text{erg} = 10^{-7}\,\text{J}$
$1\,\text{dyne} = 10^{-5}\,\text{N}$
$1\,\text{gauss} = 10^{-4}\,\text{T}$
$1\,\text{emu} = 10\,\text{C}$

pressure
$1\,\text{bar} = 10^5\,\text{Pa}$
$1\,\text{Pa} = 10^{-5}\,\text{bar}$
1 atmosphere $= 1.013\,25\,\text{bar}$
1 atmosphere $= 1.013\,25 \times 10^5\,\text{Pa}$

Table A.3 Constants.

Name of constant	Symbol	SI value
Fundamental constants		
gravitational constant	G	$6.673 \times 10^{-11}\,\mathrm{N\,m^2\,kg^{-2}}$
Boltzmann's constant	k	$1.381 \times 10^{-23}\,\mathrm{J\,K^{-1}}$
speed of light in vacuum	c	$2.998 \times 10^{8}\,\mathrm{m\,s^{-1}}$
Planck's constant	h	$6.626 \times 10^{-34}\,\mathrm{J\,s}$
	$\hbar = h/2\pi$	$1.055 \times 10^{-34}\,\mathrm{J\,s}$
fine structure constant	$\alpha = e^2/4\pi\varepsilon_0\hbar c$	$1/137.0$
Stefan–Boltzmann constant	σ	$5.671 \times 10^{-8}\,\mathrm{J\,m^{-2}\,K^{-4}\,s^{-1}}$
Thomson cross-section	σ_{T}	$6.652 \times 10^{-29}\,\mathrm{m^2}$
permittivity of free space	ε_0	$8.854 \times 10^{-12}\,\mathrm{C^2\,N^{-1}\,m^{-2}}$
permeability of free space	μ_0	$4\pi \times 10^{-7}\,\mathrm{T\,m\,A^{-1}}$
Particle constants		
charge of proton	e	$1.602 \times 10^{-19}\,\mathrm{C}$
charge of electron	$-e$	$-1.602 \times 10^{-19}\,\mathrm{C}$
electron rest mass	m_{e}	$9.109 \times 10^{-31}\,\mathrm{kg}$
		$= 0.511\,\mathrm{MeV}/c^2$
proton rest mass	m_{p}	$1.673 \times 10^{-27}\,\mathrm{kg}$
		$= 938.3\,\mathrm{MeV}/c^2$
neutron rest mass	m_{n}	$1.675 \times 10^{-27}\,\mathrm{kg}$
		$= 939.6\,\mathrm{MeV}/c^2$
atomic mass unit	u	$1.661 \times 10^{-27}\,\mathrm{kg}$
Astronomical constants		
mass of the Sun	$\mathrm{M}_\odot$	$1.99 \times 10^{30}\,\mathrm{kg}$
radius of the Sun	$\mathrm{R}_\odot$	$6.96 \times 10^{8}\,\mathrm{m}$
luminosity of the sun	$\mathrm{L}_\odot$	$3.83 \times 10^{26}\,\mathrm{W}$
mass of the Earth	$\mathrm{M}_\oplus$	$5.97 \times 10^{24}\,\mathrm{kg}$
radius of the Earth	$\mathrm{R}_\oplus$	$6.37 \times 10^{6}\,\mathrm{m}$
mass of Jupiter	M_{J}	$1.90 \times 10^{27}\,\mathrm{kg}$
radius of Jupiter	R_{J}	$7.15 \times 10^{7}\,\mathrm{m}$
astronomical unit	AU	$1.496 \times 10^{11}\,\mathrm{m}$
light-year	ly	$9.461 \times 10^{15}\,\mathrm{m}$
parsec	pc	$3.086 \times 10^{16}\,\mathrm{m}$
Hubble parameter	H_0	$(70.4 \pm 1.5)\,\mathrm{km\,s^{-1}\,Mpc^{-1}}$
		$(2.28 \pm 0.05) \times 10^{-18}\,\mathrm{s^{-1}}$
age of Universe	t_0	$(13.73 \pm 0.15) \times 10^{9}\,\mathrm{years}$
critical density	$\rho_{\mathrm{crit},0}$	$(9.30 \pm 0.40) \times 10^{-27}\,\mathrm{kg\,m^{-3}}$
dark energy density parameter	Ω_Λ	$(73.2 \pm 1.8)\%$
matter density parameter	Ω_{m}	$(26.8 \pm 1.8)\%$
baryonic matter density parameter	Ω_{b}	$(4.4 \pm 0.2)\%$
non-baryonic matter density parameter	Ω_{c}	$(22.3 \pm 0.9)\%$
curvature density parameter	Ω_{k}	$(-1.4 \pm 1.7)\%$
deceleration parameter	q_0	-0.595 ± 0.025

Solutions to exercises

Exercise 1.1 (a) As the mass and mass accretion rate are the same in both cases, the ratio of the accretion luminosities is simply the inverse ratio of the radii:

$$\frac{L_{\text{acc, WD}}}{L_{\text{acc, Sun}}} = 1 \frac{R_\odot}{R_{\text{WD}}} = 100.$$

The luminosity $L_{\text{acc, WD}} = 100\, L_\odot$ much exceeds the white dwarf's intrinsic luminosity except for the very youngest, hottest white dwarfs.

(b) For the neutron star the accretion luminosity is also larger than the Sun's by a factor

$$\frac{M_{\text{NS}}/R_{\text{NS}}}{1\, M_\odot/R_\odot} = \frac{1.4}{1} \times \frac{6.96 \times 10^8\,\text{m}}{20 \times 10^3\,\text{m}} \approx 4.9 \times 10^4.$$

So the accretion luminosity is a few times $10^4\, L_\odot$; this is not much below the luminosity of the brightest, most massive stars.

Exercise 1.2 The accretion efficiency η_{acc} is defined by $L_{\text{acc}} = \eta_{\text{acc}} \dot{M} c^2$.

Equating this to Equation 1.3 and solving for η_{acc} gives

$$\eta_{\text{acc}} = \frac{GM}{Rc^2}.$$

For a neutron star with mass $1\, M_\odot = 1.99 \times 10^{30}\,\text{kg}$ and radius $10\,\text{km} = 10^4\,\text{m}$, this is

$$\eta_{\text{acc}} = \frac{6.673 \times 10^{-11}\,\text{N m}^2\,\text{kg}^{-2} \times 1.99 \times 10^{30}\,\text{kg}}{1 \times 10^4\,\text{m} \times (2.998 \times 10^8\,\text{m s}^{-1})^2} \approx 0.15.$$

Exercise 1.3 (a) The mass defect $\Delta m = 4.40 \times 10^{-29}\,\text{kg}$ involved in the fusion of four protons into one helium nucleus translates into an energy gain of $\Delta E = \Delta m c^2$ per four protons. The energy input is the mass energy of the four protons, $4 m_{\text{p}} c^2$, so the efficiency = gain/input is

$$\eta_{\text{H}} = \frac{\Delta m c^2}{4 m_{\text{p}} c^2} = \frac{4.40 \times 10^{-29}\,\text{kg}}{4 \times 1.673 \times 10^{-27}\,\text{kg}} \approx 0.0066.$$

(b) From part (a), the efficiency of hydrogen burning, the most common nuclear fusion reaction in the Universe, is only $\eta_{\text{H}} \approx 0.7\%$. In other words, if one kilogram of hydrogen accretes onto a neutron star, it liberates about 20 times more energy (in the form of heat and radiation, say) than if this kilogram of hydrogen undergoes nuclear fusion into helium.

There is no other process in the Universe that could persistently sustain the conversion of such a large fraction of mass energy ($\gtrsim 10\%$) into energy for a macroscopic amount of mass. There are processes with 100% efficiency such as the annihilation of electron–positron pairs (see Chapters 7 and 8), but these involve antimatter, which is not abundant in the known Universe.

Exercise 1.4 The accretion disc luminosity is

$$L_{\text{disc}} = \frac{1}{2} \frac{GM\dot{M}}{R}, \tag{Eqn 1.8}$$

where R is the inner disc radius. Equating this with $L_{acc} = \eta_{acc}\dot{M}c^2$ and solving for η_{acc} gives

$$\eta_{acc} = \frac{1}{2}\frac{GM}{Rc^2}.$$

We set $R = 3R_S$ and use Equation 1.11 to obtain

$$\eta_{acc} = \frac{GM}{6R_Sc^2} = \frac{GM}{12GMc^2/c^2} = \frac{1}{12} \approx 0.083 = 8.3\%.$$

Exercise 1.5 On the right-hand side, the first term is the gravitational potential of the primary star. The denominator is the magnitude of the vector pointing from the primary to the point of reference.

The second term is the corresponding gravitational potential of the secondary.

The third term describes the effect of the centrifugal force. The quantity $(\boldsymbol{\omega} \times (\boldsymbol{r} - \boldsymbol{r}_c))^2$ is the scalar product of the vector $\boldsymbol{\omega} \times (\boldsymbol{r} - \boldsymbol{r}_c)$ with itself. The vector $\boldsymbol{\omega} \times (\boldsymbol{r} - \boldsymbol{r}_c)$ has the magnitude $\omega r_\perp$, where $r_\perp$ is the distance of the point of reference from the rotational axis. The vector $\boldsymbol{\omega}$ is parallel to the rotational axis and has magnitude ω, the orbital angular speed.

Exercise 1.6 If the two stars with masses M_1 and M_2 are at $x = 0$ and $x = a$, respectively, and the centre of mass is at $x = x_c$, then $M_1 x_c = M_2(a - x_c)$, so that $(M_1 + M_2)x_c = M_2 a$, and hence

$$x_c = \frac{M_2}{M_1 + M_2}a = \frac{M_2}{M}a,$$

where $M = M_1 + M_2$ is the total binary mass. This is also the distance a_1 of the primary from the centre of mass. The distance of the secondary from the centre of mass is $a_2 = a - x_c = (M_1/M)a$.

Equation 1.16 describing the Roche potential contains the following vectors: $\boldsymbol{r} = (x, 0, 0)$, $\boldsymbol{r}_1 = (0, 0, 0)$, $\boldsymbol{r}_2 = (a, 0, 0)$, $\boldsymbol{r}_c = (x_c, 0, 0)$ and $\boldsymbol{\omega} = (0, 0, \omega)$. So we have $|\boldsymbol{r} - \boldsymbol{r}_1| = x$, $|\boldsymbol{r} - \boldsymbol{r}_2| = a - x$, and

$$(\boldsymbol{\omega} \times (\boldsymbol{r} - \boldsymbol{r}_c))^2 = \omega^2(x - x_c)^2 = \omega^2\left(x - \frac{M_2}{M}a\right)^2.$$

Therefore, for $0 < x < a$, the Roche potential as a function of the coordinate x is

$$\Phi_R(x) = -\frac{GM_1}{x} - \frac{GM_2}{a - x} - \frac{1}{2}\omega^2\left(x - \frac{M_2}{M}a\right)^2.$$

Now in the x-direction, $\boldsymbol{\nabla}\Phi_R = d\Phi_R(x)/dx$, so

$$\frac{d\Phi_R(x)}{dx} = +\frac{GM_1}{x^2} - \frac{GM_2}{(a - x)^2} - \omega^2\left(x - a\frac{M_2}{M}\right).$$

The Roche potential at the centre of mass, i.e. at $x = x_c = aM_2/M$, is (note that $a - x_c = aM_1/M$)

$$\Phi_R(x_c) = \frac{GM_1 M}{M_2 a} - \frac{GM_2 M}{M_1 a} - 0 = \frac{GM}{a}\left(\frac{M_2^2 - M_1^2}{M_1 M_2}\right).$$

The gradient at $x = x_c$ is

$$\frac{d\Phi_R(x)}{dx} = \frac{GM_1 M^2}{M_2^2 a^2} - \frac{GM_2 M^2}{M_1^2 a^2} - 0 = \frac{GM^2}{a^2}\left(\frac{M_1}{M_2^2} - \frac{M_2}{M_1^2}\right).$$

So the force $\boldsymbol{F} = -m\,\mathrm{d}\Phi_R(x)/\mathrm{d}x$ has magnitude

$$F = \frac{GmM^2}{a^2}\left|\frac{M_1}{M_2^2} - \frac{M_2}{M_1^2}\right|$$

and it is in the $-x$-direction for $M_1 > M_2$ ($+x$-direction for $M_1 < M_2$).

Exercise 1.7 We have $1\,\mathrm{pc} = 3.086 \times 10^{16}\,\mathrm{m}$ and $1\,\mathrm{AU} = 1.496 \times 10^{11}\,\mathrm{m}$, so $1\,\mathrm{pc} = 2.063 \times 10^5\,\mathrm{AU}$. Therefore $l = 2 \times 10^3\,\mathrm{AU} \approx 10^{-2}\,\mathrm{pc}$.

Exercise 1.8 The size of the emitting region is

$$r \approx 30\ \text{light-days} = 30 \times 86\,400 \times 3 \times 10^8\,\mathrm{m} \approx 10^{15}\,\mathrm{m}.$$

From Equation 1.19, the mass is

$$M \simeq \frac{\langle v^2\rangle r}{G} = \frac{(6\times10^6)^2\,\mathrm{m^2\,s^{-2}} \times 10^{15}\,\mathrm{m}}{6.673 \times 10^{-11}\,\mathrm{N\,m^2\,kg^{-2}}} = 5.4 \times 10^{38}\,\mathrm{kg} \simeq 3 \times 10^8\,\mathrm{M_\odot}.$$

Exercise 1.9 (a) We recall that $1\,\mathrm{M_\odot\,yr^{-1}} = 6.31 \times 10^{22}\,\mathrm{kg\,s^{-1}}$ (Equation 1.4). From Equation 1.21 we find

$$(2 \times T_{\text{peak}})^4 \simeq \frac{3 \times 6.673 \times 10^{-11}\,\mathrm{N\,m^2\,kg^{-2}}}{8\pi \times 5.671 \times 10^{-8}\,\mathrm{J\,m^{-2}\,K^{-4}\,s^{-1}}}\ \frac{0.6 \times 1.99 \times 10^{30}\,\mathrm{kg} \times 10^{-9} \times 6.31 \times 10^{22}\,\mathrm{kg\,s^{-1}}}{(8.7 \times 10^6\,\mathrm{m})^3},$$

which gives $T_{\text{peak}} \approx 3.2 \times 10^4\,\mathrm{K}$.

(b) For the neutron star we have instead

$$T_{\text{peak}} \simeq 0.5 \times \left(\frac{3 \times 6.673 \times 10^{-11}\,\mathrm{N\,m^2\,kg^{-2}} \times 1.4 \times 1.99 \times 10^{30}\,\mathrm{kg} \times 10^{-8} \times 6.31 \times 10^{22}\,\mathrm{kg\,s^{-1}}}{8\pi \times 5.671 \times 10^{-8}\,\mathrm{J\,m^{-2}\,K^{-4}\,s^{-1}} \times (10^4\,\mathrm{m})^3}\right)^{1/4}$$

or $T_{\text{peak}} \approx 1.1 \times 10^7\,\mathrm{K}$.

Exercise 1.10 (a) Using Equations 1.21 and 1.11, we find

$$(2 \times T_{\text{peak}})^4 \simeq \frac{3GM\dot{M}}{8\pi\sigma(3R_S)^3} = \frac{3GM\dot{M}}{8\pi\sigma \times 3^3 \times (2GM/c^2)^3} = \frac{c^6\dot{M}}{576\,\pi G^2\sigma M^2}$$

$$= \frac{c^6 \times \mathrm{M_\odot\,yr^{-1}}}{576\,\pi G^2\sigma\,\mathrm{M_\odot^2}}\left(\frac{M}{\mathrm{M_\odot}}\right)^{-2}\left(\frac{\dot{M}}{\mathrm{M_\odot\,yr^{-1}}}\right)$$

$$= \frac{(2.998 \times 10^8\,\mathrm{m})^6 \times 6.31 \times 10^{22}\,\mathrm{kg\,s^{-1}}}{576\,\pi \times (6.673 \times 10^{-11}\,\mathrm{N\,m^2\,kg^{-2}})^2 \times 5.671 \times 10^{-8}\,\mathrm{J\,m^{-2}\,K^{-4}\,s^{-1}} \times (1.99 \times 10^{30}\,\mathrm{kg})^2}$$

$$\times \left(\frac{M}{\mathrm{M_\odot}}\right)^{-2}\left(\frac{\dot{M}}{\mathrm{M_\odot\,yr^{-1}}}\right).$$

This gives

$$T_{\text{peak}} \simeq 1.1 \times 10^9\,\mathrm{K}\left(\frac{M}{\mathrm{M_\odot}}\right)^{-1/2}\left(\frac{\dot{M}}{\mathrm{M_\odot\,yr^{-1}}}\right)^{1/4}. \tag{1.17}$$

Note that the actual peak temperature is slightly different from this value because general relativistic corrections have to be applied.

(b) For $M_1 = 10\,\mathrm{M_\odot}$ and $\dot{M} = 10^{-7}\,\mathrm{M_\odot\,yr^{-1}}$, Equation 1.17 becomes

$$T_{\mathrm{peak}} \simeq 1.1 \times 10^9\,\mathrm{K} \times 10^{-7/4} \times 10^{-1/2} = 6 \times 10^6\,\mathrm{K}.$$

(c) For $M_1 = 10^7\,\mathrm{M_\odot}$ and $\dot{M} = 1\,\mathrm{M_\odot\,yr^{-1}}$, Equation 1.17 becomes

$$T_{\mathrm{peak}} \simeq 1.1 \times 10^9\,\mathrm{K} \times 10^{-7/2} = 3 \times 10^5\,\mathrm{K}.$$

Exercise 1.11 We have $1\,\mathrm{eV} = 1.602 \times 10^{-19}\,\mathrm{J}$ and $T \simeq E_{\mathrm{ph}}/k$. So for $E_{\mathrm{ph}} = 1\,\mathrm{eV}$ the temperature is

$$T \simeq \frac{1\,\mathrm{eV}}{1.381 \times 10^{-23}\,\mathrm{J\,K^{-1}}} = \frac{1.602 \times 10^{-19}\,\mathrm{J}}{1.381 \times 10^{-23}\,\mathrm{J\,K^{-1}}} = 1.160 \times 10^4\,\mathrm{K},$$

which is of order $10^4\,\mathrm{K}$.

Exercise 1.12 (a) We make use of Equation 1.25 (rather than the rule of thumb) to work out the typical photon energy:

$$\frac{E_{\mathrm{ph}}}{\mathrm{eV}} = 2.70 \times \frac{1.381 \times 10^{-23}\,\mathrm{J\,K^{-1}}}{1.602 \times 10^{-19}\,\mathrm{J}} \times T = 2.3 \times 10^{-4}\,\mathrm{K^{-1}} \times T.$$

For $T = 3.2 \times 10^4\,\mathrm{K}$ (white dwarf) this gives $E_{\mathrm{ph}} = 7.4\,\mathrm{eV}$, while for $T = 1.1 \times 10^7\,\mathrm{K}$ (neutron star) we obtain $E_{\mathrm{ph}} = 2.5\,\mathrm{keV}$ (these temperatures were found in Exercise 1.9).

Also, for $T = 6 \times 10^6\,\mathrm{K}$ (stellar mass black hole) this gives $E_{\mathrm{ph}} = 1.4\,\mathrm{keV}$, while for $T = 3 \times 10^5\,\mathrm{K}$ (AGN) we obtain $E_{\mathrm{ph}} = 69\,\mathrm{eV}$ (these temperatures were found in Exercise 1.10).

(b) Photon energy and wavelength λ are related as

$$E_{\mathrm{ph}} = h\frac{c}{\lambda}.$$

With Equation 1.25 this gives

$$\lambda = \frac{hc}{2.7kT} = \frac{6.626 \times 10^{-34}\,\mathrm{J\,s} \times 2.998 \times 10^8\,\mathrm{m\,s^{-1}}}{2.7 \times 1.381 \times 10^{-23}\,\mathrm{J\,K^{-1}}} \times \frac{1}{T} = 5.3275 \times 10^{-3}\,\mathrm{mK} \times \frac{1}{T}.$$

For $T = 3.2 \times 10^4\,\mathrm{K}$ (white dwarf) this gives $\lambda = 1.7 \times 10^{-7}\,\mathrm{m}$. For $T = 1.1 \times 10^7\,\mathrm{K}$ (neutron star) we obtain $\lambda = 4.8 \times 10^{-10}\,\mathrm{m} = 0.48\,\mathrm{nm}$. These wavelengths are much shorter than the wavelengths of visible light ($\approx 400\text{–}800\,\mathrm{nm}$). The first is in the classical X-ray range, the second in the ultraviolet range.

For the accreting stellar mass black hole we find $\lambda = 0.9\,\mathrm{nm}$ (soft X-rays), while for the AGN we obtain $\lambda = 18\,\mathrm{nm}$ (near the ultraviolet/X-ray boundary).

Exercise 1.13 For $h\nu \gg kT$ we have $\exp(h\nu/kT) \gg 1$ and hence $\exp(h\nu/kT) - 1 \simeq \exp(h\nu/kT)$, so that the Planck function becomes the Wien tail (Equation 1.27).

For the case $h\nu \ll kT$ we introduce the quantity $x = h\nu/kT$. As $x \ll 1$ we can use the first-order expansion $\exp(x) \simeq 1 + x$ to obtain for the denominator in Equation 1.23 $\exp(h\nu/kT) - 1 \simeq h\nu/kT$. Hence

$$B_\nu(T) \simeq \frac{2h\nu^3}{c^2} \times \frac{kT}{h\nu} = 2kT\nu^2/c^2,$$

confirming Equation 1.28 for the Rayleigh–Jeans tail.

Exercise 2.1 The results are given in the following table.

q	$f(q)$ Eggleton	$f(q)$ Paczyński	$\Delta f/f$ in %
0.5	0.3208	0.3203	0.14
1.0	0.3789	0.3667	3.2
2.0	0.4400	0.4036	8.3

The last column denotes the difference between the f values calculated according to Equation 2.7 (column 2) and Equation 2.8 (column 3), divided by the value in column 2, expressed in %.

Given that Eggleton's approximation is accurate to within 1%, it is clear that Paczyński's relation is at most 4% off for $q < 1$, and even for $q = 2$ it is good to within 9%.

Exercise 2.2 (a) We solve Kepler's law for the period,

$$P_{\text{orb}}^2 = a^3 \frac{4\pi^2}{GM},$$

and multiply both the numerator and denominator on the right-hand side by $(R_{\text{L},2}/a)^3$. Hence

$$P_{\text{orb}}^2 = \frac{a^3 4\pi^2}{GM} \frac{(R_{\text{L},2}/a)^3}{(R_{\text{L},2}/a)^3} = \frac{R_{\text{L},2}^3 \, 4\pi^2}{GM} \frac{1}{(R_{\text{L},2}/a)^3}.$$

Inserting Paczyński's approximation for $R_{\text{L},2}/a$ gives

$$P_{\text{orb}}^2 \simeq \frac{R_{\text{L},2}^3 \, 4\pi^2}{GM} \frac{M}{0.462^3 M_2} = \frac{4\pi^2}{0.462^3 \, G} \frac{R_{\text{L},2}^3}{M_2}.$$

Taking the square root and noting that

$$\overline{\rho} = \frac{M_2}{(4\pi/3) \times R_{\text{L},2}^3}$$

(the stellar radius R_2 equals the Roche-lobe radius $R_{\text{L},2}$), we have

$$P_{\text{orb}} \approx \left(\frac{4\pi^2}{0.462^3 \, G}\right)^{1/2} \left(\frac{4\pi}{3}\overline{\rho}\right)^{-1/2} = \left(\frac{3\pi}{0.462^3 \, G}\right)^{1/2} \overline{\rho}^{-1/2}.$$

So

$$\frac{P_{\text{orb}}}{\text{h}} = \frac{P_{\text{orb}}}{3600\,\text{s}} = \frac{1}{3600\,\text{s}} \left(\frac{3\pi}{0.462^3 \times 6.673 \times 10^{-11}\,\text{N}\,\text{m}^2\,\text{kg}^{-2}}\right)^{1/2} \times \left(\frac{\overline{\rho}}{10^3\,\text{kg}\,\text{m}^{-3}} \times 10^3\,\text{kg}\,\text{m}^{-3}\right)^{-1/2}$$

and hence

$$\frac{P_{\text{orb}}}{\text{h}} \cong 10.5 \left(\frac{\overline{\rho}}{10^3\,\text{kg}\,\text{m}^{-3}}\right)^{-1/2},$$

as required.

(b) We used Paczyński's approximation for $R_{\text{L},2}/a$, so Equation 2.9 is valid only in the range of mass ratios q where this approximation is good, i.e. for $q \lesssim 0.8$.

Exercise 2.3 In the previous exercise we obtained the expression

$$P_{\rm orb}^2 \simeq \frac{4\pi^2}{0.462^3\,G}\,\frac{R_{\rm L,2}^3}{M_2}$$

with R_2 instead of $R_{\rm L,2}$, so

$$P_{\rm orb} \simeq \frac{2\pi}{0.462^{3/2}\,G^{1/2}}\,\frac{R_2^{3/2}}{M_2^{1/2}}$$

or

$$\frac{P_{\rm orb}}{\rm h} \simeq \frac{2\pi {\rm R}_\odot^{3/2}}{3600\,{\rm s}\times 0.462^{3/2}\,(G\,{\rm M}_\odot)^{1/2}}\left(\frac{R_2}{{\rm R}_\odot}\right)^{3/2}\left(\frac{M_2}{{\rm M}_\odot}\right)^{-1/2}.$$

With the dimensionless constant

$$k_2 = \frac{2\pi {\rm R}_\odot^{3/2}}{3600\,{\rm s}\times 0.462^{3/2}\,(G\,{\rm M}_\odot)^{1/2}} \approx 8.856$$

this becomes

$$\frac{P_{\rm orb}}{\rm h} \simeq k_2\left(\frac{R_2}{{\rm R}_\odot}\right)^{3/2}\left(\frac{M_2}{{\rm M}_\odot}\right)^{-1/2}.$$

Taking logs in this equation reproduces Equation 2.10, as $\log_{10} k_2 = 0.9472$.

Exercise 2.4 Consider two point masses M_1 and M_2 on circular orbits around the common centre of mass, with separation a. The masses M_1 and M_2 execute circular orbits with radii a_1 and a_2, respectively, and angular speed ω about the common centre of mass. The total orbital angular momentum in the system is then $J = M_1 a_1^2 \omega + M_2 a_2^2 \omega$. From $a = a_1 + a_2$ and $a_1 M_1 = a_2 M_2$ we find $a_1 = (M_2/M)a$ and $a_2 = (M_1/M)a$. Using Kepler's law we have $\omega^2 = 4\pi^2/P^2 = GM/a^3$, so

$$J = M_1\left(\frac{M_2}{M}\right)^2 a^2\frac{(GM)^{1/2}}{a^{3/2}} + M_2\left(\frac{M_1}{M}\right)^2 a^2\frac{(GM)^{1/2}}{a^{3/2}} = \frac{M_1 M_2 (Ga)^{1/2}}{M^{3/2}}(M_2+M_1)$$

$$= M_1 M_2\left(\frac{Ga}{M}\right)^{1/2},$$

which reproduces Equation 2.17.

Exercise 2.5 Paczyński's approximation for the Roche-lobe radius is

$$R_{\rm L,2} \approx 0.462\left(\frac{M_2}{M}\right)^{1/3} a. \qquad\qquad (\text{Eqn }2.8)$$

Taking the logarithmic derivative gives

$$\frac{\dot{R}_{\rm L,2}}{R_{\rm L,2}} = \frac{1}{3}\frac{\dot{M}_2}{M_2} - \frac{1}{3}\frac{\dot{M}}{M} + \frac{\dot{a}}{a} = \frac{1}{3}\frac{\dot{M}_2}{M_2} + \frac{\dot{a}}{a}, \qquad\qquad (2.18)$$

as $\dot{M} = 0$ for conservative mass transfer. To find the logarithmic derivative of a, we solve the expression for the orbital angular momentum (Equation 2.17) for a,

$$a = \frac{J^2 M}{GM_1^2 M_2^2},$$

and take its logarithmic derivative:

$$\frac{\dot{a}}{a} = 2\frac{\dot{J}}{J} + \frac{\dot{M}}{M} - 2\frac{\dot{M_1}}{M_1} - 2\frac{\dot{M_2}}{M_2} = 2\frac{\dot{J}}{J} + 2\frac{\dot{M_2}}{M_1} - 2\frac{\dot{M_2}}{M_2} \qquad (2.19)$$

($\dot{M} = 0$, $\dot{M_1} = -\dot{M_2}$). Substituting from Equation 2.19 into Equation 2.18 gives

$$\frac{\dot{R}_{L,2}}{R_{L,2}} = \left(\tfrac{1}{3} - 2\right)\frac{\dot{M_2}}{M_2} + 2\frac{M_2}{M_1}\frac{\dot{M_2}}{M_2} + 2\frac{\dot{J}}{J}.$$

Collecting terms gives

$$\frac{\dot{R}_{L,2}}{R_{L,2}} = 2\frac{\dot{J}}{J} + \left(2q - \tfrac{5}{3}\right)\frac{\dot{M_2}}{M_2}.$$

Comparing this with Equation 2.18 shows that $\zeta_L = 2q - 5/3$.

Exercise 2.6   The table below gives a representative radius of the $5\,M_\odot$ star at the beginning of the corresponding mass transfer case. The orbital period P_{orb} was calculated using Equation 2.10. The mass ratio is just larger than 1, so it is still acceptable to use Paczyński's approximation (Equation 2.8), and Equation 2.10 does indeed use this approximation.

Case	$\log_{10} R/R_\odot$	$R/R_\odot$	P_{orb}
A	0.5	3	22.3 h
B	1.0	10	5.2 d
C	2.0	100	165 d

If both stars formed at the same time, and this is the first time the system experiences mass transfer, then a mass ratio $q < 1$ is unphysical because the more massive binary component evolves faster and fills its Roche lobe first. So at the start of a case A, B or C mass transfer, the mass ratio is > 1. (There are exceptions, however, such as systems where very strong wind losses have reduced the mass of the primary so much that, at the point of first contact with its Roche lobe, it is less massive than the less evolved secondary star.)

Exercise 2.7   For the Sun we have

$$t_{\mathrm{KH}} = \frac{6.673 \times 10^{-11}\,\mathrm{N\,m^2\,kg^{-2}} \times (1.99 \times 10^{30}\,\mathrm{kg})^2}{6.96 \times 10^8\,\mathrm{m} \times 3.83 \times 10^{26}\,\mathrm{J\,s^{-1}}} = 9.91 \times 10^{14}\,\mathrm{s} = 3.1 \times 10^7\,\mathrm{yr}.$$

With $R \propto M$ and $L \propto M^4$, we also have

$$t_{\mathrm{KH}} \propto \frac{M^2}{RL} \propto \frac{M^2}{MM^4} \propto M^{-3},$$

so

$$t_{\mathrm{KH}} \simeq 3.1 \times 10^7\,\mathrm{yr}\left(\frac{M_2}{M_\odot}\right)^{-3}.$$

Therefore the Kelvin–Helmholtz time for a $0.5\,M_\odot$ main-sequence star is about 2.5×10^8 yr, while for a $5\,M_\odot$ main-sequence star it is about 2.5×10^5 yr.

Exercise 2.8   The mass transfer rate is

$$\frac{\dot{M_2}}{M_2} = \frac{2\dot{J}_{\mathrm{sys}}/J - (\dot{R}_2/R_2)_{\mathrm{nuc}}}{\zeta - \zeta_L}. \qquad \text{(Eqn 2.22)}$$

By assumption we have $\zeta - \zeta_L = 0 - \zeta_L \simeq 1$, $\dot{J}_{sys}/J = 0$, $(\dot{R}_2/R_2) = 1/t_{th}$, and therefore $-\dot{M}_2 = M_2/t_{th}$. The thermal time t_{th} is just the Kelvin–Helmholtz time,

$$t_{KH} \simeq 3.1 \times 10^7 \text{ yr} \left(\frac{M_2}{M_\odot} \right)^{-3}$$

(see Exercise 2.7). So, Equation 2.22 becomes

$$-\dot{M}_2(\text{case B}) \simeq 3 \times 10^{-8} \text{ M}_\odot \text{ yr}^{-1} \times \left(\frac{M_2}{M_\odot} \right)^4. \tag{2.20}$$

(Hence the case B transfer rate is $2 \times 10^{10}/3.1 \times 10^7 \approx 6 \times 10^2$ times larger than the case A rate; see Worked Example 2.2.) For $M_2 = 0.5 \text{ M}_\odot, 1 \text{ M}_\odot, 5 \text{ M}_\odot$ this is $2 \times 10^{-9}, 3 \times 10^{-8}, 2 \times 10^{-5} \text{ M}_\odot \text{ yr}^{-1}$, respectively.

Exercise 2.9   For conservative mass transfer and $\zeta = 1$, Equation 2.22 becomes

$$\frac{-\dot{M}_2}{M_2} = \frac{-\dot{J}_{GR}/J}{4/3 - M_2/M_1}.$$

According to Equation 2.11 we also have $M_2/M_\odot \simeq P_{orb}/8.8 \text{ h} = 0.23$. Hence Equation 2.25 becomes

$$\frac{\dot{J}_{GR}}{J} = -1.27 \times 10^{-8} \text{ yr}^{-1} \times \frac{1 \times 0.23}{1.23^{1/3}} \times 2^{-8/3} = -4.29 \times 10^{-10} \text{ yr}^{-1},$$

so putting this value into Equation 2.22

$$\frac{-\dot{M}_2}{0.23 \text{ M}_\odot} = \frac{4.29 \times 10^{-10} \text{ yr}^{-1}}{4/3 - 0.23/1}.$$

This gives $-\dot{M}_2 = 8.9 \times 10^{-11} \text{ M}_\odot \text{ yr}^{-1}$.

Exercise 2.10   Assuming conservative mass transfer, the stability criterion requires $q \lesssim 1$, or $M_2 < M_1 = 1.4 \text{ M}_\odot$. The longest orbital period is realized for the most massive donor that still allows stable mass transfer, so $P_{orb} \simeq 8.8 \text{ h} \times 1.4 = 12 \text{ h}$ (Equation 2.11). A note of caution: the actual radius of a $1.4 \text{ M}_\odot$ main-sequence star can be up to a factor of 2 larger than what was assumed in Equation 2.11, so the period could be up to $2^{1.5} \approx 3$ times longer than the value that we have just calculated.

Exercise 2.11   The orbital speed v of the companion is given by Equation 2.1, but applied to the star with mass M_1 instead of M_2. As by assumption $M_2 \ll M_1$, we can set $M_1/(M_1 + M_2) \simeq 1$, so

$$v = \left(\frac{GM_1}{a} \right)^{1/2},$$

where a is the orbital separation of the circular pre-supernova orbit. The escape speed of the companion from the binary is

$$v_{esc} = \left(\frac{2GM_1}{a} \right)^{1/2}. \tag{Eqn 1.10}$$

Immediately after a prompt supernova explosion, the orbital speed of the companion is still v as calculated above, but the escape speed has changed. The binary will remain bound if v is smaller than the new escape speed, i.e.

$$\left(\frac{GM_1}{a} \right)^{1/2} < \left(\frac{2G(M_1 - \Delta M)}{a} \right)^{1/2}.$$

Here ΔM is the mass ejected in the supernova explosion, so the primary has a post-supernova mass $M_1 - \Delta M$. Therefore $M_1 < 2(M_1 - \Delta M)$ or $\Delta M < M_1/2$.

Exercise 3.1 A rather famous differentially rotating body is the Sun. This can be seen when groups of sunspots move across the disc of the Sun. Sunspots at higher latitudes lag behind sunspots that are closer to the equatorial region. Hence the angular velocity in equatorial regions is larger than in polar regions.

An indirect example of differential rotation can be seen when sprinters in separate lanes follow the curve of a stadium (e.g. in a 400 m heat). Even if the athletes in the inner and outer lanes run at the same speed, if they start at the same point, the one in the inner lane will be ahead of the one in the outer lane as the inner lane is closer to the centre of the circle that defines the bend. The angular velocity of the inner sprinter is larger than that of the outer sprinter, so the group of sprinters 'rotates' differentially. (Of course, to compensate for this, the lanes are staggered so that the sprinter in the inner lane starts further back than the one in the outer lane.)

Exercise 3.2 The product rule gives

$$\frac{\partial(r\omega)}{\partial r} = r\frac{\partial\omega}{\partial r} + \omega\frac{\partial r}{\partial r} = r\frac{\partial\omega}{\partial r} + \omega,$$

and this is non-zero (i.e. equal to ω) even in the *absence* of shear. But viscous stresses exist only in the *presence* of shearing motion. If $\omega = $ constant, there is no shear, hence no stress, so we *must* have $\sigma_s = 0$ in this case. Therefore only the first term, $r\,\partial\omega/\partial r$, can contribute to the shear stress σ_s.

Exercise 3.3 Equation 3.8 reads

$$G_{\rm vis} = 2\pi r\,\nu_{\rm vis}\,\Sigma r^2\frac{\partial\omega}{\partial r}.$$

The unit of the right-hand side is

$$\rm m \times (m \times m\,s^{-1}) \times (kg\,m^{-2}) \times m^2 \times (s^{-1}\,m^{-1}).$$

Collecting terms, this is $\rm m^{5-3}\,s^{-2}\,kg = m^2\,s^{-2}\,kg$.

With torque $=$ force $\times$ distance, the corresponding unit is

$$\rm N \times m = (kg\,m\,s^{-2}) \times m = kg\,m^2\,s^{-2},$$

as above.

Exercise 3.4 In the case of Keplerian motion the angular speed is (Equation 3.3)

$$\omega = \left(\frac{GM}{r^3}\right)^{1/2}.$$

So the radius derivative is

$$\frac{d\omega}{dr} = (GM)^{1/2}\left(-\tfrac{3}{2}r^{-5/2}\right).$$

Inserting this into Equation 3.11 gives

$$D(r) = \tfrac{1}{2}\nu_{\rm vis}\,\Sigma r^2 GM\left(-\tfrac{3}{2}\right)^2 r^{-5}.$$

Hence

$$D(r) = \tfrac{9}{8}\nu_{\text{vis}}\,\Sigma\,\frac{GM}{r^3},$$

as required.

Exercise 3.5 Equation 3.17 describes the conservation of angular momentum in the disc:

$$r\frac{\partial}{\partial t}(\Sigma r^2\omega) + \frac{\partial}{\partial r}(rv_r\Sigma r^2\omega) = \frac{1}{2\pi}\frac{\partial G_{\text{vis}}}{\partial r}.$$

The two terms on the left-hand side describe the angular momentum balance when $\partial G_{\text{vis}}/\partial r = 0$, i.e. in the absence of the so-called source term on the right-hand side. In this case the angular momentum J of a disc ring between radii r and $r + \Delta r$ changes only if there is an imbalance between the angular momentum that flows into the ring via the mass that flows into the ring, and the angular momentum leaving the ring via the mass flowing out of the ring. We find the flow rate of angular momentum at radius r by multiplying the mass flow rate $\mathrm{d}M/\mathrm{d}t$ with the specific angular momentum that this mass has. The specific angular momentum is just $r^2\omega$, and $\mathrm{d}M/\mathrm{d}t$ is given by Equation 3.15 as

$$\dot{M}(r,t) = -2\pi r v_r \Sigma.$$

So the local flow rate of angular momentum is just

$$\dot{J} = -2\pi r v_r \Sigma r^2\omega.$$

To work out the net change ΔJ in the angular momentum J of the disc ring due to this mass flow in a small time interval Δt, we take the difference between the local flow rates at $r + \Delta r$ and r, and multiply it by Δt. This can be written as

$$\Delta J = \left[\dot{J}(r + \Delta r, t) - \dot{J}(r, t)\right] \times \Delta t \approx \Delta r\,\frac{\partial \dot{J}}{\partial r} \times \Delta t.$$

Hence

$$\frac{\Delta J}{\Delta t} = \Delta r\,\frac{\partial \dot{J}}{\partial r} = -\Delta r\,\frac{\partial(2\pi r v_r \Sigma r^2\omega)}{\partial r}.$$

This becomes

$$\frac{\Delta J}{\Delta t} = -2\pi\,\Delta r\,\frac{\partial(v_r r\Sigma r^2\omega)}{\partial r}. \tag{3.21}$$

On the other hand, the total angular momentum J in the disc ring is

$$J = \text{mass in the ring} \times \text{specific angular momentum} = 2\pi r\,\Delta r \times \Sigma \times r^2\omega.$$

Hence the time derivative of J can be written as

$$\frac{\partial J}{\partial t} = -2\pi r\,\Delta r\,\frac{\partial}{\partial t}(\Sigma r^2\omega). \tag{3.22}$$

Note that r and Δr are not affected by the partial derivative with respect to t, as by definition this has to be taken for fixed r. For small time intervals Δt, the expression $\Delta J/\Delta t$ in Equation 3.21 becomes the derivative $\partial J/\partial t$. Equating the right-hand side of Equation 3.22 with the right-hand side of Equation 3.21, and dividing by $2\pi\,\Delta r$, finally reproduces the first two terms in Equation 3.17.

Exercise 3.6 Equation 3.27 describes the luminosity of a disc ring with inner radius r_1 and outer radius r_2:

$$L(r_1, r_2) = \frac{3GM\dot{M}}{2} \left\{ \frac{1}{r_1} \left[1 - \frac{2}{3} \left(\frac{R_1}{r_1} \right)^{1/2} \right] - \frac{1}{r_2} \left[1 - \frac{2}{3} \left(\frac{R_1}{r_2} \right)^{1/2} \right] \right\}.$$

We obtain the luminosity of the whole disc if we set r_1 equal to the radius of the accreting object (or inner rim of the accretion disc, if this is different), and r_2 to infinity. This is appropriate for an idealized, infinitely extended disc. A real disc in, for example, a binary system is limited by the size of the Roche lobe of the accreting star. But even in that case the choice $r_2 = \infty$ is usually a rather good approximation, as $r_2 \gg r_1$. So, with $r_1 = R_1$ and $r_2 = \infty$ we have

$$L_{\text{disc}} = L(R_1, \infty) = \frac{3GM\dot{M}}{2} \left\{ \frac{1}{R_1} \left[1 - \frac{2}{3} \right] - \frac{1}{\infty} \left[1 - \frac{2}{3} \left(\frac{R_1}{\infty} \right)^{1/2} \right] \right\}.$$

Clearly the second term in curly brackets is identical to 0 (division by ∞). So

$$L_{\text{disc}} = \frac{3GM\dot{M}}{2} \left\{ \frac{1}{3R_1} - 0 \right\} = \frac{GM\dot{M}}{2R_1}.$$

Exercise 3.7 (a) Introducing T_* into Equation 3.28 gives

$$T_{\text{eff}}^4(r) = T_*^4 \left(\frac{R_1}{r} \right)^3 \left[1 - \left(\frac{R_1}{r} \right)^{1/2} \right]$$

or

$$\frac{T_{\text{eff}}^4(r)}{T_*^4} = \left(\frac{R_1}{r} \right)^3 \left[1 - \left(\frac{R_1}{r} \right)^{1/2} \right].$$

(b) Hence with $y = (T_{\text{eff}}/T_*)^4$ and $x = r/R_1$ we have

$$y(x) = x^{-3}(1 - x^{-1/2}) = x^{-3} - x^{-3.5}.$$

(c) The maximum value of y is reached at a point x_0 where $dy/dx = 0$. As

$$\frac{dy}{dx} = -3x^{-4} - (-3.5)x^{-4.5},$$

we have at the maximum

$$0 = -3x_0^{-4} - (-3.5)x_0^{-4.5}.$$

Solving for x_0, this becomes

$$3x_0^{-4} = 3.5x_0^{-4.5} \quad \text{or} \quad x_0^{1/2} = 3.5/3,$$

hence $x_0 = (7/6)^2$.

(d) The function $(T_{\text{eff}}/T_*)^4$ attains a maximum value at the same radius as T_{eff} itself does. This radius is $r_0 = R_1 x_0$. Hence

$$r_0 = R_1 \times \left(\tfrac{7}{6} \right)^2 = \tfrac{49}{36} R_1,$$

as requested.

(e) Inserting the value for x_0 in the expression for y gives

$$y(x_0) = x_0^{-3} - x_0^{-3.5} = \left(\tfrac{6}{7}\right)^6 - \left(\tfrac{6}{7}\right)^7 = \left(\tfrac{6}{7}\right)^6 \left(1 - \tfrac{6}{7}\right) = \left(\tfrac{6}{7}\right)^6 \times \tfrac{1}{7}.$$

The maximum temperature $T_{\mathrm{eff}} = y(x_0)^{1/4} T_*$ is therefore

$$T_{\mathrm{eff}} = \frac{(6/7)^{3/2}}{7^{1/4}} T_* \simeq 0.488 T_*.$$

Exercise 3.8   The prime observational quantity is the energy flux through the surface area (Equation 3.11). In a steady-state disc this is *independent* of viscosity (Equation 3.24) because in a steady state, the viscosity *must* adjust itself to obey the equilibrium condition for the surface density and mass accretion rate expressed in Equation 3.23. So no matter what mechanism is causing the viscosity, the value of $\nu_{\mathrm{vis}} \Sigma$ is always the same.

As a further consequence the surface temperature of a steady-state disc, which is in principle accessible via the emitted spectrum, is also *independent* of the viscosity (Equation 3.28).

Exercise 3.9   We have

$$\frac{H}{r} \simeq \frac{c_s}{v_K} \tag{Eqn 3.35}$$

so we need to estimate the sound speed and the Keplerian speed at r_o. From Equation 3.32 with $T = 10^4$ K we have $c_s \simeq 10^4 \, \mathrm{m\,s^{-1}}$, while from Equation 1.5 we obtain

$$v_K = \left(\frac{GM}{r}\right)^{1/2}$$
$$= \left(\frac{6.673 \times 10^{-11} \, \mathrm{N\,m^2\,kg^{-2}} \times 1.99 \times 10^{30} \, \mathrm{kg}}{0.5 \times 6.96 \times 10^8 \, \mathrm{m}}\right)^{1/2}$$
$$= 6.18 \times 10^5 \, \mathrm{m\,s^{-1}}.$$

So we have

$$\tan \delta = \frac{H}{r} \simeq \frac{10^4}{6.18 \times 10^5} = 0.016, \tag{3.23}$$

so $\delta \simeq 0.92°$. The disc is indeed rather flat!

Exercise 4.1   From Equations 4.7 and 3.6 we find

$$t_{\mathrm{th}} \simeq \frac{c_s^2}{\nu_{\mathrm{vis}} GM/r^3} = \frac{c_s^2}{\alpha c_s H GM/r^3} = \frac{c_s r^3}{\alpha H GM}.$$

Noting Equation 3.35, we also have $H/r = c_s/v_K$, hence

$$t_{\mathrm{th}} \simeq \frac{c_s r^2}{\alpha GM} \frac{r}{H} = \frac{c_s r^2}{\alpha GM} \frac{v_K}{c_s} = \frac{r^2}{\alpha GM} \left(\frac{GM}{r}\right)^{1/2} = \frac{1}{\alpha} \left(\frac{r^3}{GM}\right)^{1/2} = \frac{1}{\alpha} \frac{1}{\omega_K} = \frac{1}{\alpha} t_{\mathrm{dyn}},$$

as required. We have used the identity for the Keplerian angular speed, $\omega_K = (GM/r^3)^{1/2}$.

Exercise 4.2   As $\nu_{\mathrm{vis}} = $ constant, we can move it to the front, and Equation 4.1 becomes

$$\frac{\partial \Sigma}{\partial t} = \frac{3\nu_{\mathrm{vis}}}{r} \frac{\partial}{\partial r} \left\{ r^{1/2} \frac{\partial}{\partial r} (\Sigma r^{1/2}) \right\}.$$

Using the product rule on the inner derivative gives

$$\frac{\partial \Sigma}{\partial t} = \frac{3\nu_{\mathrm{vis}}}{r} \frac{\partial}{\partial r} \left\{ r^{1/2} \left[\Sigma \frac{\partial}{\partial r} (r^{1/2}) + \frac{\partial \Sigma}{\partial r} r^{1/2} \right] \right\} = \frac{3\nu_{\mathrm{vis}}}{r} \frac{\partial}{\partial r} \left\{ r^{1/2} \left[\Sigma \frac{1}{2r^{1/2}} + \frac{\partial \Sigma}{\partial r} r^{1/2} \right] \right\}.$$

Factoring in $r^{1/2}$ gives

$$\frac{\partial \Sigma}{\partial t} = \frac{3\nu_{\mathrm{vis}}}{r} \frac{\partial}{\partial r} \left\{ \frac{\Sigma}{2} + \frac{\partial \Sigma}{\partial r} r \right\}.$$

Using the sum rule, this becomes

$$\frac{\partial \Sigma}{\partial t} = \frac{3\nu_{\mathrm{vis}}}{r} \left\{ \frac{\partial}{\partial r} \left(\frac{\Sigma}{2} \right) + \frac{\partial}{\partial r} \left(\frac{\partial \Sigma}{\partial r} r \right) \right\}.$$

Now using the product rule again gives

$$\frac{\partial \Sigma}{\partial t} = \frac{3\nu_{\mathrm{vis}}}{r} \left\{ \frac{1}{2} \frac{\partial \Sigma}{\partial r} + \frac{\partial^2 \Sigma}{\partial r^2} r + \frac{\partial \Sigma}{\partial r} \right\} = \frac{3\nu_{\mathrm{vis}}}{r} \left\{ \frac{3}{2} \frac{\partial \Sigma}{\partial r} + \frac{\partial^2 \Sigma}{\partial r^2} r \right\}.$$

Thus we finally obtain Equation 4.12:

$$\frac{\partial \Sigma}{\partial t} = \frac{9\nu_{\mathrm{vis}}}{2r} \frac{\partial \Sigma}{\partial r} + 3\nu_{\mathrm{vis}} \frac{\partial^2 \Sigma}{\partial r^2}.$$

Exercise 4.3 The viscous time $t_{\mathrm{vis}} = r_{\mathrm{c}}^2 / \nu_{\mathrm{vis}}$ is an appropriate estimate for the time it takes the torus at the circularization radius to spread into a disc-like structure. With $\nu_{\mathrm{vis}} = \alpha H c_{\mathrm{s}}$ and $H \simeq c_{\mathrm{s}}/\omega_{\mathrm{K}}$ (Equation 3.34) we obtain

$$\nu_{\mathrm{vis}} = \alpha \frac{c_{\mathrm{s}}^2}{(GM/r_{\mathrm{c}}^3)^{1/2}},$$

so

$$t_{\mathrm{vis}} \simeq \frac{r_{\mathrm{c}}^2}{\alpha c_{\mathrm{s}}^2} \times \left(GM/r_{\mathrm{c}}^3 \right)^{1/2} = \frac{(GMr_{\mathrm{c}})^{1/2}}{\alpha c_{\mathrm{s}}^2}.$$

Then using Equation 3.32 for the sound speed, we have

$$t_{\mathrm{vis}} \simeq \frac{\left(6.673 \times 10^{-11}\,\mathrm{N\,m^2\,kg^{-2}} \times 0.6 \times 1.99 \times 10^{30}\,\mathrm{kg} \times 0.2 \times 6.96 \times 10^8\,\mathrm{m} \right)^{1/2}}{0.3 \times (10^4\,\mathrm{m\,s^{-1}})^2}$$

$$= 3.5 \times 10^6\,\mathrm{s}.$$

This is about 40 days, i.e. a little over a month.

Exercise 4.4 A stability analysis studies the reaction of a physical system (e.g. an accretion disc) to perturbations. Initially the system is assumed to be in equilibrium. Then a perturbation is applied to the system, and the reaction of the system is calculated. The stability analysis is said to be linear if the initial perturbations are sufficiently small, so that the resulting change of other quantities can be described by the first (linear) term in the corresponding Taylor expansion with respect to the perturbing quantity. The stability analysis is local if the reaction of the system is studied only in the immediate vicinity of a given point in the system. Therefore the reaction at this point is assumed to be determined by its immediate vicinity only, not by events far away from this point.

Exercise 4.5 With Kramers' opacity $\kappa_R \propto \rho T^{-3.5}$, the denominator of Equation 3.41 scales as

$$\kappa_R \rho H \propto \rho^2 T^{-3.5} H,$$

so that with $\rho = \Sigma/H$,

$$\kappa_R \rho H \propto \Sigma^2 \frac{1}{H^2} T^{-3.5} H \propto \Sigma^2 H^{-1} T^{-3.5} \propto \Sigma^2 T^{-4},$$

where for the last step we have used $H \propto T^{1/2}$ (Equation 3.36). With this,

$$F(H) \propto \frac{T^4}{\Sigma^2 T^{-4}} \propto T^8 \Sigma^{-2},$$

as required.

Exercise 4.6 (a) For $r \gg R_1$ the temperature profile of a steady-state disc is

$$T_{\text{eff}}^4(r) = \frac{3GM_1\dot{M}}{8\pi\sigma r^3} \qquad\qquad \text{(Eqn 3.28)}$$

(with M_1 as the mass of the central accretor, the white dwarf). Setting $r = r_D$, $T_{\text{eff}}(r_D) = T_H$ and solving for $\dot{M}$ gives

$$\dot{M} = \frac{8\pi\sigma T_H^4 r_D^3}{3GM_1}.$$

To determine r_D we note that

$$R_{L,1} \simeq 0.462 \left(\frac{M_1}{M_1 + M_2}\right)^{1/3} a \qquad\qquad \text{(Eqn 2.8)}$$

can be used here as $1/q = M_1/M_2 \simeq 1$ (but note that in general, short-period CVs would have $1/q \gg 1$ in which case Equation 2.8 is *not* a good approximation). Therefore

$$r_D^3 = (0.5R_{L,1})^3 \simeq 0.231^3 \frac{M_1}{M} a^3.$$

With Kepler's law $a^3 = G(M_1 + M_2)P_{\text{orb}}^2/4\pi^2$, this becomes

$$r_D^3 \simeq 0.231^3 \frac{GM_1}{4\pi^2} P_{\text{orb}}^2.$$

Inserting into the above expression for $\dot{M}$, we have

$$\dot{M} = \frac{0.231^3 \times 8\pi\sigma \times T_H^4 GM_1}{3 \times 4\pi^2 GM_1} P_{\text{orb}}^2$$

or

$$\dot{M} = \frac{0.231^3 \times 2\sigma \times T_H^4}{3\pi} P_{\text{orb}}^2.$$

This is a lower limit for the mass transfer rate if the disc is meant to be stable. Note that it scales as $\dot{M} \propto P_{\text{orb}}^2$.

(b) We have

$$\frac{\dot{M}}{M_\odot\,\text{yr}^{-1}} = \frac{0.231^3 \times 2 \times 5.671 \times 10^{-8}\,\text{J}\,\text{m}^{-2}\,\text{K}^{-4}\,\text{s}^{-1} \times (6 \times 10^3\,\text{K})^4 \times (3600\,\text{s})^2}{3\pi \times 6.31 \times 10^{22}\,\text{kg}\,\text{s}^{-1}} \left(\frac{P_{\text{orb}}}{\text{h}}\right)^2$$

$$= 4 \times 10^{-11} \left(\frac{P_{\text{orb}}}{\text{h}}\right)^2,$$

so $\dot{M} \simeq 4 \times 10^{-10} \, M_\odot \, \mathrm{yr}^{-1}$ at $P_{\mathrm{orb}} = 3 \, \mathrm{h}$. In nova-like systems with periods longer than 3 h, the observationally estimated mass accretion rate is a few times $10^{-9} \, M_\odot \, \mathrm{yr}^{-1}$, while for shorter periods the rate is thought to be as low as a few times $10^{-11} \, M_\odot \, \mathrm{yr}^{-1}$ — and the vast majority of short-period cataclysmic variables ($P_{\mathrm{orb}} \lesssim 2 \, \mathrm{h}$) are indeed dwarf novae.

Exercise 5.1   Cataclysmic variables are ideal laboratories for the study of accretion phenomena for the following reasons:

- The mass donor is faint and does not swamp the optical and ultraviolet radiation emitted by the accretion flow itself.

- The irradiation of the accretion disc by the hot accreting white dwarf is negligible.

- The size of the orbit is compact enough so that orbital changes can be observed within hours — a convenient timescale for human observers.

- Eclipses and radial velocity studies allow one to map the accretion flow.

- Major brightness variations of the disc due to thermal and viscous evolution occur on a convenient timescale of weeks to months.

Exercise 5.2   (a) The accretion luminosity is given by Equation 1.3:

$$L_{\mathrm{acc}} = \frac{GM\dot{M}}{R}.$$

We set $\dot{M} = 10^{-9} \, M_\odot \, \mathrm{yr}^{-1}$, $M = 1 \, M_\odot$ and $R = 8.7 \times 10^6 \, \mathrm{m}$. Then

$$L_{\mathrm{acc}} = (6.673 \times 10^{-11} \, \mathrm{N \, m^2 \, kg^{-2}}) \times (1.99 \times 10^{30} \, \mathrm{kg}) \times 10^{-9} \times 1.99 \times 10^{30} \, \mathrm{kg}$$
$$\times (365.25 \times 24 \times 3600 \, \mathrm{s})^{-1}/(8.7 \times 10^6 \, \mathrm{m})$$
$$= 9.6 \times 10^{26} \, \mathrm{N \, m \, s^{-1}}$$
$$\approx 10^{27} \, \mathrm{J \, s^{-1}}.$$

The solar luminosity is $L_\odot \approx 4 \times 10^{26} \, \mathrm{J \, s^{-1}}$. Hence, using the definition for astronomical magnitudes, for the difference between the absolute magnitude M_{CV} of the CV and the absolute magnitude M_{Sun} (not to be confused with the solar mass!) we have

$$M_{\mathrm{CV}} - M_{\mathrm{Sun}} = -2.5 \log_{10} \left(\frac{L_{\mathrm{acc}}}{L_\odot} \right).$$

Hence

$$M_{\mathrm{CV}} - 4.83 = -2.5 \log_{10} \left(\frac{10^{27}}{4 \times 10^{26}} \right),$$

which gives

$$M_{\mathrm{CV}} = 4.83 - 2.5 \log_{10}(2.5) = 3.84.$$

Using the distance modulus (with zero extinction)

$$m = M_{\mathrm{CV}} - 5 + 5 \log_{10}(d/\mathrm{pc}),$$

we find the apparent magnitude of the CV when it is located at a distance $d = 1000 \, \mathrm{pc}$:

$$m = 3.84 - 5 + 5 \log_{10}(1000) = 3.84 - 5 + 15.$$

Hence $m = 13.84$.

(b) With a transfer rate $\dot{M} = 10^{-9} \, \text{M}_\odot \, \text{yr}^{-1}$, this CV is one of the brighter ones anyway, and still its apparent magnitude, at a distance of $1000 \, \text{pc}$, is only about 14. (Note: In reality a bolometric correction should also be applied as the calculation here leads to the bolometric magnitude not the visual magnitude. This correction of about 2 magnitudes would make the CV even fainter than calculated in the V band.) This CV would only just be observable with a 12-inch telescope. Larger telescopes and deeper surveys of the sky would of course easily detect such a system, and also much fainter CVs, but the problem then is to distinguish the CVs in the survey field from the much more numerous ordinary stars.

Exercise 5.3 For the compact binary disc, the inner disc radius is close to the compact star's radius, i.e. $r_{\text{in}} \simeq 10^{-2} \text{R}_\odot$, which is $10^9 \, \text{m}$ for a white dwarf and $10^4 \, \text{m}$ for a neutron star. For the supermassive black hole accretor, we assume that the inner disc radius is close to the last stable circular orbit, i.e.

$$r_{\text{in}} \simeq 3\frac{2GM}{c^2} \approx \frac{6 \times 7 \times 10^{-11} \, \text{N kg}^{-2} \, \text{m}^2 \times 10^8 \times 2 \times 10^{30} \, \text{kg}}{(3 \times 10^8 \, \text{m s}^{-1})^2} \approx 10^{12} \, \text{m}.$$

The outer radius for the discs in binaries is a fraction of the Roche-lobe radius of the accretor, i.e. of order the orbital separation. For short-period systems this is $\lesssim 1 \, \text{R}_\odot \simeq 10^9 \, \text{m}$. For AGN, this is perhaps $\gtrsim 10^{-2} \, \text{pc} \simeq 10^{14} \, \text{m}$.

Hence we have $r_{\text{out}}/r_{\text{in}} \simeq 10^2$ for cataclysmic variables and AGN, and $r_{\text{out}}/r_{\text{in}} \simeq 10^5$ for LMXBs.

Exercise 5.4 The surface pattern arises from lines that connect points in the disc with constant magnitude of the radial velocity, i.e. constant magnitude of the y-component of the orbital velocity v.

Consider now two points, A and B, in the accretion disc that are mirror-symmetric with respect to the y-axis. If point A has coordinates (x_0, y_0), then point B must have coordinates $(-x_0, y_0)$. The symmetry with respect to the y-axis arises because v_y at A has the same magnitude but opposite sign to v_y at B. Therefore the only difference between a point 'to the left' (B) and 'to the right' (A) of the y-axis is that the plasma to the left is approaching, while the plasma to the right is receding from the observer (if the orbital motion is anticlockwise).

The situation is similar if we consider two points, A and C, in the accretion disc that are mirror-symmetric with respect to the x-axis. As point A has coordinates (x_0, y_0), point C must have coordinates $(x_0, -y_0)$. The symmetry with respect to the x-axis arises because v_y at A has the same magnitude *and* sign as v_y at C! The velocities at A and C differ only in the sign of the x-component of v.

Exercise 5.5 (a) (i) Disc plasma at a distance r from the accretor with mass M has the Keplerian speed $v_K = (GM/r)^{1/2}$ (Equation 1.5). For an edge-on system ($i = 90°$), the line-of sight velocity $v_\parallel$ varies with azimuth ϕ as $v_\parallel = v_K \cos \phi$. So the lines of constant line-of-sight velocity in Figure 5.3 are defined by the relation

$$\frac{\cos \phi}{r^{1/2}} = \frac{v_\parallel}{(GM)^{1/2}}.$$

(ii) Converting the polar coordinates (r, ϕ) into Cartesian coordinates (x, y) (see Figure S5.1), we have $\cos \phi = x/r$ and $r^2 = x^2 + y^2$, so

$$\frac{v_\parallel}{(GM)^{1/2}} = \frac{x}{r^{3/2}} = \frac{x}{(x^2 + y^2)^{3/4}}.$$

Solving for y,

$$x^2 + y^2 = \left(x \times \frac{(GM)^{1/2}}{v_\parallel} \right)^{4/3} = x^{4/3} \times \left(\frac{GM}{v_\parallel^2} \right)^{2/3}$$

or

$$y = \left[x^{4/3} \times \left(\frac{GM}{v_\parallel^2} \right)^{2/3} - x^2 \right]^{1/2} .$$

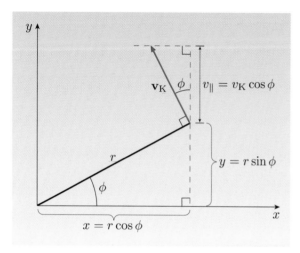

Figure S5.1 Cartesian and polar coordinates for Exercise 5.5.

(b) From the last equation we see that $y(x_0) = 0$ if $x_0^{4/3} \times (GM/v_\parallel^2)^{2/3} = x_0^2$, or $x_0^{2/3} = (GM/v_\parallel^2)^{2/3}$, hence $x_0 = GM/v_\parallel^2$.

(c) Alternatively, for $y = 0$ we always have $x = r$. Therefore $v_\parallel = v_K(r)$, i.e. the line-of-sight velocity is just the Kepler speed as the disc plasma moves straight towards (or directly away from) the observer. This gives $r = GM/v_K(r)^2$, which is equivalent to the expression for x_0 that we have just found.

Exercise 5.6 (a) The Keplerian speed at distance r from the white dwarf with mass M is (Equation 1.5)

$$v_K = \left(\frac{GM}{r} \right)^{1/2} .$$

For $M = 0.8 \, M_\odot$ ($1 \, M_\odot = 1.99 \times 10^{30}$ kg) and $R = R_{out} = 3.0 \times 10^8$ m, we therefore have the velocity

$$v_{K, \, out} = \left(\frac{6.67 \times 10^{-11} \, N \, m^2 \, kg^{-2} \times 0.8 \times 1.99 \times 10^{30} \, kg}{3.0 \times 10^8 \, m} \right)^{1/2}$$

$$= 5.95 \times 10^5 \, m \, s^{-1} = 595 \, km \, s^{-1}.$$

For the inner edge of the accretion disc we set $R = R_{in} = 7 \times 10^6$ m (comparable to the white dwarf radius). As

$$v_{K, \, in} = v_{K, \, out} \left(\frac{R_{out}}{R_{in}} \right)^{1/2} ,$$

we have

$$v_{K, \text{in}} = 5.95 \times 10^5 \, \text{m s}^{-1} \times \left(\frac{3.0 \times 10^8}{7.0 \times 10^6}\right)^{1/2}$$

$$= 3.90 \times 10^6 \, \text{m s}^{-1} = 3900 \, \text{km s}^{-1}.$$

(b) The Doppler shift is given by Equation 5.1

$$\frac{\Delta\lambda}{\lambda_{\text{em}}} = \frac{v_{\parallel}}{c}.$$

Hence at the outer edge of the disc the Doppler shift is

$$\Delta\lambda_{\text{out}} = \frac{5.95 \times 10^5}{3.0 \times 10^8} \times 656 \, \text{nm} = 1.3 \, \text{nm},$$

while at the inner edge of the disc the Doppler shift is

$$\Delta\lambda_{\text{in}} = \frac{3.90 \times 10^6}{3.0 \times 10^8} \times 656 \, \text{nm} = 8.5 \, \text{nm}.$$

Exercise 5.7 The hot spot appears brightest when it faces the observer, i.e. immediately before phase 0 (in Figure 5.4 this is at phase 0.875). The binary is at phase 0 when the secondary star is closest to the observer. At the opposite phase, close to phase 0.5, the hot spot is facing away from the observer. Hardly any light from the spot reaches the observer as the disc is in the way. This variable contribution from the hot spot gives rise to an orbital 'hump' in the optical light curve. The hump is most pronounced when the system is seen nearly edge-on. If we see the system face-on, there is no such hump. In this case the hot spot always contributes roughly the same (small) amount to the total light.

Exercise 5.8 The 'shadow' in the figure indicates those regions on the accretion disc from where the Earth (i.e. the telescope that collects photons emitted from the disc) cannot be seen because it is obscured by the donor star. The shadow is long if the inclination is high. In a system seen edge-on (inclination 90°), the shadow formally has an infinite length, while in a system seen face-on (inclination 0°), there is no shadow.

Exercise 5.9 The Keplerian speed for accretion disc material is given by Equation 1.5: $v_K = (GM/r)^{1/2}$, i.e. the speed increases with decreasing r. In a velocity map, the surface brightness of the accretion disc is plotted as a function of the x- and y-components of the velocity v of the emitting material. Hence the rapidly moving material from the inner regions of the accretion disc will appear at large values of v_x and v_y, while the slowly moving material from the outer regions of the accretion disc will appear at small values of v_x and v_y.

Exercise 5.10 (a) The shift between the continuum and line flux is about 40 days.

(b) The emission line light curve needs to be shifted about 40 days earlier in order to correlate with the continuum light curve. The emission line light curve therefore lags about 40 days after the continuum light curve.

(c) This indicates that in this case, the emission line flux could be caused by reprocessing or reflection of the continuum flux after an interval of about 40 days.

This may represent the light travel time from the site of emission of the continuum to the site of emission of the emission line light.

Exercise 6.1   The energy of the photon is given by

$$E = h\nu = \frac{hc}{\lambda}.$$

Hence the frequency ν for a 1 keV photon is given by

$$\nu = \frac{E}{h} = \frac{10^3 \, \text{eV} \times 1.602 \times 10^{-19} \, \text{J eV}^{-1}}{6.626 \times 10^{-34} \, \text{J s}} = 2.4 \times 10^{17} \, \text{Hz}.$$

Also,

$$\lambda = \frac{hc}{E} = \frac{(6.626 \times 10^{-34} \, \text{J s}) \times (2.998 \times 10^8 \, \text{m s}^{-1})}{10^3 \, \text{eV} \times 1.602 \times 10^{-19} \, \text{J eV}^{-1}}$$
$$= 1.2 \times 10^{-9} \, \text{m} = 1.2 \, \text{nm}.$$

Exercise 6.2   (a) Using Equation 6.9, the Eddington limit becomes

$$L_{\text{Edd}} = \frac{4\pi \times 6.673 \times 10^{-11} \, \text{N m}^2 \, \text{kg}^{-2} \times 1.673 \times 10^{-27} \, \text{kg} \times 2.998 \times 10^8 \, \text{m s}^{-1}}{6.652 \times 10^{-29} \, \text{m}^2} \times M$$
$$= 6.322 \times (1.99 \times 10^{30}) \times \left(\frac{M}{M_\odot}\right) \text{W}$$
$$= 1.26 \times 10^{31} \left(\frac{M}{M_\odot}\right) \text{W}.$$

(b) In this case, $M = 1.4 \, M_\odot$, and $L_{\text{Edd}} = 1.8 \times 10^{31} \, \text{W}$.

(c) For a $10 \, M_\odot$ black hole, $L_{\text{Edd}} = 1.3 \times 10^{32} \, \text{W}$.

Exercise 6.3   (a) First we need to calculate the cross-section at 1 keV:

$$\sigma(E) = (c_0 + c_1 \times E + c_2 \times E^2)E^{-3} \times 10^{-24} \, \text{cm}^2,$$

where E is measured in keV. Hence

$$\sigma(1 \, \text{keV}) = (120.6 + (169.3 \times 1) + (-47.7 \times 1^2)) \times 1^{-3} \times 10^{-24} \, \text{cm}^{-2},$$

giving $\sigma(1 \, \text{keV}) = 2.42 \times 10^{-22} \, \text{cm}^{-2}$.

The fraction of radiation transmitted through the absorber, $f_{\text{trans}}(E)$, is given by Equation 6.15. For $N_{\text{H}} = 1.5 \times 10^{22}$ atom cm^{-2},

$$f_{\text{trans}}(1 \, \text{keV}) = \exp\left[-(1.5 \times 10^{22}) \times (2.42 \times 10^{-22})\right] = 0.0265.$$

The fraction of energy absorbed (f_{abs}) is $1 - f_{\text{trans}}$, i.e. $f_{\text{abs}} \simeq 97\%$.

(b) For 5 keV photons,

$$\sigma(5 \, \text{keV}) = (433 - (2.4 \times 5) + (0.75 \times 5^2)) \times 5^{-3} \times 10^{-24} \, \text{cm}^{-2},$$

hence $\sigma(5 \, \text{keV}) = 3.5 \times 10^{-24} \, \text{cm}^2$. Therefore

$$f_{\text{abs}} = 1 - \exp\left[-(1.5 \times 10^{22}) \times (3.5 \times 10^{-24})\right] = 5\%.$$

Exercise 6.4   (a) The lowest frequency is given by

$$f_{\text{min}} = \frac{1}{N \, \Delta T} = \tfrac{1}{4} \, \text{Hz}.$$

The highest frequency is given by

$$f_{\text{max}} = \frac{1}{2\,\Delta T} = \tfrac{1}{2}\,\text{Hz}.$$

There are $N/2$ frequencies in the PDS, i.e. 2.

(b) Equation 6.19 tells us that the power for frequency k, $P(f_k)$, is given by

$$P(f_k) = C\left(\left[\sum_{j=1}^{N} I_j \cos(2\pi f_k t_j)\right]^2 + \left[\sum_{j=1}^{N} I_j \sin(2\pi f_k t_j)\right]^2\right),$$

where C is a constant, $k = 1, 2$, $j = 1, 2, 3, 4$.

First, we shall tackle $k = 1$, where $f_k = \tfrac{1}{4}$:

$$P(\tfrac{1}{4}) = C\left(\left[\sum_{j=1}^{4} I_j \cos(2\pi \times \tfrac{1}{4} \times t_j)\right]^2 + \left[\sum_{j=1}^{4} I_j \sin(2\pi \times \tfrac{1}{4} \times t_j)\right]^2\right).$$

Summing the cosine terms:

$$\sum_{j=1}^{4} I_j \cos(2\pi \times \tfrac{1}{4} \times t_j) = (3 \times 0) + (1 \times -1) + (3 \times 0) + (1 \times 1) = 0.$$

Now summing the sine terms:

$$\sum_{j=1}^{4} I_j \sin(2\pi \times \tfrac{1}{4} \times t_j) = (3 \times 1) + (1 \times 0) + (3 \times -1) + (1 \times 0) = 0.$$

Since both the sine and cosine terms are zero, $P(\tfrac{1}{4}) = 0$.

Now we shall tackle $k = 2$, where $f_k = \tfrac{1}{2}$:

$$P(\tfrac{1}{2}) = C\left(\left[\sum_{j=1}^{4} I_j \cos(2\pi \times \tfrac{1}{2} \times t_j)\right]^2 + \left[\sum_{j=1}^{4} I_j \sin(2\pi \times \tfrac{1}{2} \times t_j)\right]^2\right).$$

Summing the cosine terms:

$$\sum_{j=1}^{4} I_j \cos(2\pi \times \tfrac{1}{2} \times t_j) = (3 \times -1) + (1 \times 1) + (3 \times -1) + (1 \times 1) = -4.$$

Summing the sine terms:

$$\sum_{j=1}^{4} I_j \sin(2\pi \times \tfrac{1}{2} \times t_j) = (3 \times 0) + (1 \times 0) + (3 \times 0) + (1 \times 0) = 0.$$

Hence

$$P(\tfrac{1}{2}) = C\left[(-4)^2 + (0)\right]^2 = C \times 16.$$

This tells us that the light curve is produced by a cosine with frequency $\tfrac{1}{2}$.

Exercise 6.5   (a) The detected flux F is given by

$$F = \frac{L}{4\pi d^2}, \qquad\qquad\qquad\qquad \text{(Eqn 6.20)}$$

so

$$L = F \times 4\pi d^2.$$

The distance d is estimated to be 8 kpc, i.e. $8000 \times 3.086 \times 10^{16}$ m. Hence

$$L = (4.5 \times 10^{-11}\,\mathrm{W\,m^{-2}}) \times 4\pi \times (8000 \times 3.086 \times 10^{16}\,\mathrm{m})^2$$
$$= 3.4 \times 10^{31}\,\mathrm{W}.$$

This luminosity is higher than expected for a $1.4\,\mathrm{M_\odot}$ neutron star accreting hydrogen, by about a factor of 2 (see Exercise 6.2), but is in line with observations of LMXBs in globular clusters.

(b) The X-ray peak of a radius-expanding burst corresponds to the photosphere shrinking down to normal size, so the black body radius gives us the radius of the neutron star, R_{NS}. For black body radiation,

$$L = 4\pi R_{\mathrm{NS}}^2 \sigma T^4.$$

First, we need to work out the temperature corresponding to $kT = 2.1$ keV:

$$T = \frac{2.1 \times 1000 \times (1.602 \times 10^{-19})\,\mathrm{J}}{1.381 \times 10^{23}\,\mathrm{J\,K^{-1}}} = 2.4 \times 10^7\,\mathrm{K}.$$

We then obtain R_{NS} using the Stefan–Boltzmann law:

$$R_{\mathrm{NS}} = \left(\frac{3.4 \times 10^{31}\,\mathrm{W}}{4 \times \pi \times 5.671 \times 10^{-8}\,\mathrm{W\,m^{-2}\,K^{-4}} \times (2.4 \times 10^7\,\mathrm{K})^4} \right)^{1/2} \approx 1.2 \times 10^4\,\mathrm{m} = 12\,\mathrm{km}.$$

Exercise 6.6 We estimate the donor mass from the approximate relation

$$P_{\mathrm{orb}} \simeq 8.8\,\mathrm{h}\,\frac{M_2}{\mathrm{M_\odot}}. \qquad \text{(Eqn 2.11)}$$

This gives

$$\frac{M_2}{\mathrm{M_\odot}} \simeq \frac{P_{\mathrm{orb}}}{8.8\,\mathrm{h}} = \frac{4.4}{8.8} = 0.5.$$

So $M_2 = 0.5\,\mathrm{M_\odot}$. With $R_2/\mathrm{R_\odot} \simeq M_2/\mathrm{M_\odot}$, the radius of the donor star is $0.5\,\mathrm{R_\odot}$, and as the donor is Roche-lobe filling, this is also equal to the Roche-lobe radius, $R_{\mathrm{L},2} \approx 3.5 \times 10^8$ m.

Since the LMXB exhibits X-ray bursts, it must contain a neutron star primary, so we may assume a mass of $M_1 = 1.4\,\mathrm{M_\odot}$. Therefore the mass ratio of the system is $q = M_2/M_1 \approx 0.5/1.4 = 0.36$.

We find the neutron star's Roche-lobe radius by noting that

$$\frac{R_{\mathrm{L},1}}{R_{\mathrm{L},2}} = \frac{f_1(1/q)}{f_2(q)},$$

where $f(q)$ is given by Eggleton's approximation

$$f(q) \simeq \frac{0.49q^{2/3}}{0.6q^{2/3} + \log_e(1 + q^{1/3})}. \qquad \text{(Eqn 2.7)}$$

For $q = 0.36$ we have $1/q = 2.8$ and so $f(2.8)/f(0.36) \approx 1.6$. The Roche-lobe radius of the neutron star is therefore $R_{\mathrm{L},1} \simeq 1.6 \times R_{\mathrm{L},2} = 5.6 \times 10^8$ m.

To estimate the size of the corona, we rearrange Equation 6.21 to get

$$D_{ADC} = \frac{2\pi R_{disc} \Delta T_{ing}}{P_{orb}}.$$

R_{disc} is estimated to be 30–50% of the Roche-lobe radius. So the smallest corona size is obtained by taking $R_{disc} = 0.3 \times R_{L,1}$. Then

$$D_{ADC} = \frac{2\pi \times 0.3 \times (5.6 \times 10^8)\,\text{m} \times 2000\,\text{s}}{4.4 \times 3600\,\text{s}}$$

$$= 1.3 \times 10^8\,\text{m}.$$

For $R_{disc} = 0.5 \times R_{L,1}$, the corona is larger by a factor $0.5/0.3 = 1.7$. Hence the corona has a diameter of $\sim$130 000–230 000 km.

Exercise 6.7   We have the relation $H/r = \tan\delta$ (see Figures 6.22 and 6.23). Hence $\delta = \tan^{-1}(0.2) = 11.3°$.

Exercise 6.8   We have to calculate the radius in the disc where the orbital frequency equals the QPO frequency f. The angular speed ω at distance r from the accretor with mass M is

$$\omega = \left(\frac{GM}{r^3}\right)^{1/2}. \tag{Eqn 3.3}$$

Rearranging this to solve for r yields

$$r = \left(\frac{GM}{\omega^2}\right)^{1/3}.$$

Since $\omega = 2\pi f$,

$$r = \left[\frac{(6.673 \times 10^{-11}\,\text{N}\,\text{m}^2\,\text{kg}^{-2}) \times (1.4 \times 1.99 \times 10^{30}\,\text{kg})}{(2 \times \pi \times 6.0\,\text{s}^{-1})^2}\right]^{1/3}.$$

Hence $r = 5.2 \times 10^5$ m, or $\sim$520 km. This is only about 30 times the neutron star radius.

Exercise 6.9   Observations of Galactic LMXBs with known neutron star primaries suggest that transitions from the low state to the high state occur at $0.1L_{Edd}$ or less. For a neutron star with a mass of $2.1\,\text{M}_\odot$,

$$L_{Edd} \simeq 1.26 \times 10^{31} \times \frac{2.1\,\text{M}_\odot}{1\,\text{M}_\odot} = 2.7 \times 10^{31}\,\text{W}.$$

Since the transition occurs at $\sim 0.1L_{Edd}$ or less, we do not expect to see neutron star LMXBs exhibit the low state at 0.01–1000 keV luminosities higher than $\sim 2.7 \times 10^{30}$ W.

Exercise 6.10   (a) The Eddington luminosity is given by

$$L_{Edd} = 1.26 \times 10^{31} \left(\frac{M}{\text{M}_\odot}\right)\,\text{W}, \tag{Eqn 6.10}$$

while the accretion luminosity for a Schwarschild black hole can be written as $L_{acc} = \eta_{acc} c^2 \dot{M}$ (Equation 1.9), with $\eta_{acc} = 0.057$ (see Subsection 1.2.1). Equating these luminosities and solving for $\dot{M}$ gives

$$\dot{M}_{Edd} = \frac{1.26 \times 10^{31}\,\text{W}}{0.057 \times c^2} \left(\frac{M}{\text{M}_\odot}\right)$$

or

$$\dot{M}_{Edd} = 3.9 \times 10^{-8} \, M_\odot \, \text{yr}^{-1} \left(\frac{M}{M_\odot} \right). \tag{6.24}$$

(b) Inserting Equation 6.24 into Equation 1.17 (in the solution to Exercise 1.10) gives

$$T_{peak} = 1.5 \times 10^7 \, \text{K} \left(\frac{M}{M_\odot} \right)^{-1/4}. \tag{6.25}$$

(c) For a $100 \, M_\odot$ black hole accreting at the Eddington limit, this is $T_{peak} \simeq 5 \times 10^6 \, \text{K}$, a factor of 2–4 lower than the observed value.

Exercise 6.11   (a) If the source is unbeamed, its luminosity would be

$$L = F \times 4\pi d^2$$
$$= (3.5 \times 10^{-15} \, \text{W}) \times 4\pi \times (3.26 \times 10^6 \, \text{pc} \times 3.059 \times 10^{16} \, \text{m} \, \text{pc}^{-1})^2.$$

So $L = 4.5 \times 10^{32} \, \text{W}$.

(b) From Equation 6.10,

$$L_{Edd} = 1.26 \times 10^{31} \left(\frac{10 \, M_\odot}{M_\odot} \right) = 1.26 \times 10^{32} \, \text{W}.$$

The source therefore exceeds its Eddington luminosity unless it is beamed by at least the minimum beaming factor

$$b_{min} = \frac{4.5 \times 10^{32} \, \text{W}}{1.26 \times 10^{32} \, \text{W}} = 3.6.$$

(c) From Equation 6.25, b_{min} is also defined as $b_{min} = 4\pi/\Delta\Omega$. Hence

$$\Omega = \frac{4\pi}{b_{min}} = \frac{4 \times 3.141}{3.6} = 3.5 \, \text{sr}.$$

This is a fraction $3.5/4\pi = 28\%$ of the whole sphere. If the beaming factor is larger than b_{min}, then $\Delta\Omega$ decreases, hence this is a maximum solid angle.

Exercise 7.1   (a) At a mass loss rate of $10^{-5} \, M_\odot \, \text{yr}^{-1}$, the Wolf–Rayet star will lose $\Delta M = \dot{M} \, \Delta t_{wind} \simeq 10^{-5} \times 0.1 \times 5 \times 10^6 \, M_\odot = 5 \, M_\odot$, which is a large fraction of its initial mass!

(b) At a mass loss rate of $10^{-14} \, M_\odot \, \text{yr}^{-1}$, the Sun will lose just $10^{-4} \, M_\odot$ over its lifetime, which is a tiny fraction of its mass.

Exercise 7.2   Taking 6000 Å as a typical optical wavelength, the application of Equation 5.1 suggests a speed of $c/150$ or $2000 \, \text{km} \, \text{s}^{-1}$. The component along the line of sight coincides with the actual velocity as the outflow is spherically symmetric (a wind).

Exercise 7.3   In 27 days from 3rd April to 30th April, the two jets appear in the sky to have moved apart by about 700 milliarcsec, which corresponds to $0.2 \times 0.7 \, \text{ly} = 0.14 \, \text{ly}$. Thus the speed at which the two bright spots appear to move apart is $v = (0.14 \, \text{ly})/(27 \, \text{days})$. We may express this directly in terms of the speed of light by noting that a light-year is the distance that light travels in

1 year. Thus $v = (0.14 \times 365c)/27 \simeq 1.9c$. The left bright spot has moved further from the centre than the right spot. It is roughly at twice the distance from the cross than the right spot. Thus the left spot has moved from the centre at about $1.3c$, and the right at about $0.6c$. The left spot appears to be moving at a speed faster than light!

Exercise 7.4   Solving Equation 7.1 for V/c gives

$$\frac{V}{c} = \left(1 - \frac{1}{\gamma^2}\right)^{1/2}.$$

Using the first-order expansion $(1 + x)^{1/2} \simeq 1 + x/2$ $(x \ll 1)$ with $x = -1/\gamma^2$, this becomes

$$\frac{V}{c} \simeq 1 - \frac{1}{2\gamma^2},$$

as required.

Exercise 7.5   (a) Equation 7.13 gives the maximum value of α for observing a star that is at right angles to the ecliptic ($\theta = \pi/2$). Thus $\alpha_{max} = v/c$. Converting the angle to radians, we get $\alpha = 21\pi/(180 \times 60 \times 60)$ rad $= 1.0 \times 10^{-4}$ rad. So $v/c \approx 10^{-4}$ and $v \approx 30\,\text{km s}^{-1}$.

(b) The Earth's rotation period is $P_\oplus = 24\,\text{h}$. The rotational velocity of a point at the surface of the Earth's equator (of radius $R_\oplus$) is

$$v = 2\pi R_\oplus/P_\oplus = 2\pi \times 6.38 \times 10^6/(24 \times 60 \times 60)\,\text{m s}^{-1} \approx 464\,\text{m s}^{-1}.$$

We may use this in Equation 7.13 to calculate α_{max}. Alternatively, we may exploit the result of part (a): the ratio of the two speeds is equal to the ratio of the aberration angles in the two situations. Thus $\alpha_{\oplus,max}/21'' = 464/30\,000$, giving $\alpha_{\oplus,max} \approx 0.32''$.

Exercise 7.6   The transformation equation for x is Equation 7.7

$$x = \gamma(x' + Vt'),$$

while for t it is Equation 7.10: $t = \gamma(t' + Vx'/c^2)$. Hence we have

$$\mathrm{d}x = \gamma(\mathrm{d}x' + V\,\mathrm{d}t') \qquad \text{and} \qquad \mathrm{d}t = \gamma\left(\mathrm{d}t' + \frac{V\,\mathrm{d}x'}{c^2}\right).$$

So we obtain for the velocity

$$v_x = \frac{\mathrm{d}x}{\mathrm{d}t} = \frac{\mathrm{d}x' + V\mathrm{d}t'}{\mathrm{d}t' + (V/c^2)\,\mathrm{d}x'} = \frac{\mathrm{d}x'/\mathrm{d}t' + V}{1 + (V/c^2)(\mathrm{d}x'/\mathrm{d}t')} = \frac{v_x' + V}{1 + Vv_x'/c^2}.$$

(The corresponding transformations for v_y and v_z can be obtained in a similar way.)

Exercise 7.7   (a) v_{ap} attains a maximum for $\mathrm{d}v_{app}/\mathrm{d}\theta = 0$. From Equation 7.24, we get

$$\frac{\mathrm{d}v_{ap}}{\mathrm{d}\theta} = \frac{V\cos\theta}{1 - (V/c)\cos\theta} - \frac{V\sin\theta}{(1 - (V/c)\cos\theta)^2}\frac{V}{c}\sin\theta,$$

and applying the condition for a maximum

$$0 = \left(1 - \frac{V}{c}\cos\theta_{max}\right)V\cos\theta_{max} - \frac{V^2}{c}\sin^2\theta_{max}.$$

Manipulating this further gives

$$\cos\theta_{max} - \frac{V}{c}\cos^2\theta_{max} - \frac{V}{c}\sin^2\theta_{max} = 0 \quad\text{or}\quad \cos\theta_{max} - \frac{V}{c}(\cos^2\theta_{max} + \sin^2\theta_{max}) = 0,$$

thus

$$\cos\theta_{max} = \frac{V}{c} \quad\text{and so}\quad \theta_{max} = \cos^{-1}(V/c).$$

(b) From $\cos\theta_{max} = (V/c)$, we get

$$\sin\theta_{max} = \sqrt{1 - (V/c)^2} = \frac{1}{\gamma}.$$

Equation 7.24 thus gives

$$v_{ap} = \frac{V\sin\theta_{max}}{1 - (V/c)\cos\theta_{max}} = \frac{V}{\gamma[1 - (V^2/c^2)]} = \frac{V\gamma^2}{\gamma} = \gamma V.$$

Exercise 7.8 (a) For $\theta = \pi/2$, Equation 7.28 gives $\mathcal{D} = 1/\gamma$.

(b) For $\theta = 0$, Equation 7.28 becomes $\mathcal{D} = [\gamma(1 - V/c)]^{-1}$.

From the definition of γ (Equation 7.1), we have

$$\gamma^2 = [(1 + V/c)(1 - V/c)]^{-1}.$$

So $\gamma(1 - V/c) = [\gamma(1 + V/c)]^{-1}$.

Thus $\mathcal{D} = \gamma(1 + V/c)$, which for highly relativistic speeds becomes $\mathcal{D} \simeq 2\gamma$.

(c) For a source moving at right angles to the observer, the received frequency is redshifted: $\nu_{rec} = \nu'_{em}/\gamma$.

For a source moving towards the observer, the received frequency is blueshifted: $\nu_{rec} \simeq 2\gamma\nu'_{em}$.

Exercise 7.9 For the given values of the parameters, $\gamma_1 = 1$, $m_1 \simeq m_2/\gamma_2$, $\gamma_b = \gamma_2/2$ and $E_{th} \simeq \gamma_2 m_1 c^2/4$, Equation 7.39 gives

$$\varepsilon = \frac{\gamma_2 m_1 c^2/4}{(m_1 + \gamma_2 m_2)c^2} = \frac{\gamma_2/4}{1/\gamma_2 + \gamma_2} = \frac{1}{4}\left(\frac{\gamma_2}{1/\gamma_2 + \gamma_2}\right).$$

As $\gamma_2 \gg 1$, we therefore have $\varepsilon \approx \frac{1}{4} = 25\%$.

Exercise 7.10 If the electrons execute a circular motion with speed v and radius r, we have $|\boldsymbol{v} \times \boldsymbol{B}| = vB_\perp$, so

$$evB_\perp = \frac{m_e v^2}{r}. \tag{7.26}$$

Hence the circular speed is $v = eB_\perp r/m_e$. The frequency is given by $\nu_{cy} = v/(2\pi r)$, so

$$\nu_{cy} = \frac{eB_\perp r}{m_e 2\pi r} = \frac{eB_\perp}{2\pi m_e}. \tag{7.27}$$

Exercise 7.11 A comparison between the power emitted by a single electron by each mechanism (Equations 7.53 and 7.59) shows that $P_{sy}/P_{ic} = U_B/U_{rad}$. So it is the relative intensity of the respective underlying fields that determines which component dominates.

Exercise 8.1 Solving Equation 7.49 for E, we get $E = 4\pi d^2 S$. The distance has to be converted to m, using $1\,\text{pc} = 3 \times 10^{16}\,\text{m}$.

(a) For a Galactic halo source,

$$d \approx 50\,\text{kpc} = 50 \times 10^3\,\text{pc} \times 3 \times 10^{16}\,\text{m}\,\text{pc}^{-1} = 1.5 \times 10^{21}\,\text{m}$$

and

$$E \approx 4\pi \times (1.5 \times 10^{21}\,\text{m})^2 \times 10^{-9}\,\text{J}\,\text{m}^{-2} \approx 3 \times 10^{34}\,\text{J}.$$

(b) Similarly, for the cosmological source,

$$d \approx 9 \times 10^{25}\,\text{m} \approx 10^{26}\,\text{m}$$

and

$$E \approx 4\pi \times (10^{26}\,\text{m})^2 \times 10^{-9}\,\text{J}\,\text{m}^{-2} \approx 10^{44}\,\text{J}.$$

We have

$$M_\odot c^2 \approx 2 \times 10^{30}\,\text{kg} \times (2.998 \times 10^8\,\text{m}\,\text{s}^{-1})^2 \approx 2 \times 10^{47}\,\text{J}.$$

The energy implied for a Galactic halo source is about 10 times higher than for an X-ray burst, and much less compared to a supernova or the mass energy of a solar mass. The energy in a source at a cosmological distance is comparable to the energy in a supernova, and $1/1000$ of the mass energy available in 1 solar mass.

Exercise 8.2 The shorter timescale, $\Delta t_{\text{var}} \simeq 1\,\text{ms}$, provides the more stringent constraint, $\Delta r \lesssim c\,\Delta t_{\text{var}} \approx 300\,\text{km}$. This suggests a compact stellar object (neutron star or black hole).

Exercise 8.3 We need to estimate n_γ for use in Equation 8.3. From Equation 7.50, $n_\gamma \simeq L_\gamma/(4\pi(\Delta r)^2 ch\nu)$, and using $\Delta r = c\,\Delta t_{\text{var}}$, we get $\tau_{\gamma\gamma} \simeq \sigma_T L_\gamma/(4\pi c^2 \Delta t_{\text{var}}\,h\nu)$. Substituting values in and converting MeV to SI units by $1\,\text{eV} = 1.602 \times 10^{-19}\,\text{J}$, we obtain

$$\tau_{\gamma\gamma} \approx \frac{0.665 \times 10^{-28}\,\text{m}^{-2} \times 10^{44}\,\text{J}\,\text{s}^{-1}}{4\pi \times (2.998 \times 10^8\,\text{m}\,\text{s}^{-1})^2 \times 10 \times 10^{-3}\,\text{s} \times 0.5 \times 10^6\,\text{eV} \times 1.602 \times 10^{-19}\,\text{J}\,\text{eV}^{-1}}$$
$$\approx 7 \times 10^{12}.$$

Exercise 8.4 The source reached the apparent size R_{def} in 4 weeks' time. 4 weeks corresponds to $\Delta t = 4 \times 7 \times 24 \times 60^2\,\text{s} = 2.4 \times 10^6\,\text{s}$. An estimate of the speed V of expansion is obtained from $R_{\text{def}} \simeq V\,\Delta t$, thus

$$V \approx \frac{10^{15}\,\text{m}}{2.4 \times 10^6\,\text{s}} \approx 4.13 \times 10^8\,\text{m}\,\text{s}^{-1} = \frac{4.13 \times 10^8}{3 \times 10^8}\,c \approx 1.4c,$$

an apparent superluminal expansion as in AGN jets and microquasars.

Exercise 8.5 Using Equation 7.36 we obtain

$$r_{\text{dis}} \simeq 2\gamma^2 c\,\Delta t_{\text{var}} = 2 \times 300^2 \times (2.998 \times 10^8\,\text{m}\,\text{s}^{-1}) \times (10 \times 10^{-3}\,\text{s}) = 5.4 \times 10^{11}\,\text{m} \approx 5 \times 10^{11}\,\text{m}.$$

Exercise 8.6 (a) $1\,\text{cm} = 10^{-2}\,\text{m}$. Therefore $1\,\text{cm}^3 = 10^{-6}\,\text{m}^3$, which gives $1\,\text{cm}^{-3} = 10^6\,\text{m}^{-3}$.

(b) Substituting values in Equation 8.12, we obtain

$$r_{dec} \approx \left(\frac{3}{4\pi \times 1.67 \times 10^{-27}\,\text{kg} \times (2.998 \times 10^8\,\text{m s}^{-1})} \right)^{1/3} \times \left(\frac{10^{44}\,\text{J}}{10^6\,\text{m}^{-3} \times 300^2} \right)^{1/3}$$

$$\approx 1.2 \times 10^{14}\,\text{m},$$

which is about 0.004 pc or 800 AU.

Exercise 8.7 We may use the result of Exercise 8.6, $r_{dec} \approx 1.2 \times 10^{14}$ m, in Equation 8.14:

$$t_{dec} \approx \frac{1.2 \times 10^{14}\,\text{m}}{(2 \times 300^2) \times (2.998 \times 10^8\,\text{m s}^{-1})} \approx 2\,\text{s}.$$

Exercise 8.8 For $\Theta_J = 10° \approx 0.17$ rad, the subtended solid angle of one jet is $\Delta\Omega = 2\pi(1 - \cos(\Theta_J/2))$ (Equation 7.20). For two jets the solid angle is

$$\Delta\Omega = 4\pi(1 - \cos 5°) = 4.8 \times 10^{-2}.$$

(a) The required energy is $4\pi/\Delta\Omega \approx 260$ times less than estimated in Exercise 8.1.

(b) As the emission is confined to a narrow jet, on average only 1 out of a few hundred GRBs will be detected by an observer on Earth. Hence the required source rate is a few hundred times higher than estimated in Worked Example 8.2, where it was assumed that GRBs emit isotropically.

Acknowledgements

Grateful acknowledgement is made to the following sources:

Figures

Cover image: A multi-wavelength composite NASA image of the spiral galaxy M81, with X-ray data from the Chandra X-ray Observatory (blue), optical data from the Hubble Space Telescope (green), infrared data from the Spitzer Space Telescope (pink) and ultraviolet data from GALEX (purple). (Credit: X-ray: NASA/CXC/Wisconsin/D.Pooley & CfA/A.Zezas; Optical: NASA/ESA/CfA/A.Zezas; UV: NASA/JPL-Caltech/CfA/J.Huchra et al.; IR: NASA/JPL-Caltech/CfA). Superimposed is an artist's representation of M33 X-7, a black hole binary with a high-mass companion star. Credit: NASA/CXC/M.Weiss;

Figure 1.1: National Optical Astronomy Observatory; Figure 1.5: Source Unknown; Figure 1.9: adapted from Pringle, J. E. and Wade, R. A. (1985) 'Introduction', *Interacting Binary Stars*, Cambridge Astrophysics Series, CUP; Figure 1.11: Courtesy of Dan Rolfe, University of Leicester; Figure 1.12: Adapted from a figure created by Jerry Orosz, University of Utrecht; Figure 1.13: adapted from Ferrarese, L. Ford, H. (1996), *Space Science Reviews*, **116**, 523–624; Figure 1.14: Jun-Hui Zhao and W. M. Goss/AOC/NRAO; Figure 1.15: European Southern Observatory; Figures 1.17 and 1.18: Rob Hynes, University of Southampton; Figure 2.5: LISA International Science Team; Figure 2.8: Bulik, T. (2007) 'Black holes go extragalactic', *Nature*, **449**, 799, *Nature* Publishing Group; Figure 2.9: Bruce Balick (University of Washington), Vincent Icke (Leiden University, The Netherlands), Garrelt Mellema (Stockholm University), and NASA/ESA; Figure 2.10: courtesy of Bart Willems; Figure 2.11: adapted from Hobbs, G. et al. (2005) 'A statistical study of 233 pulsar proper motions', *Monthly Notices of The Royal Astronomical Society*, **360**, 974, Blackwell Science Limited; Figure 2.12: adapted from Danzmann K, Rüdiger A. (2003), *Class. Quantum Grav.* **20**, S1–S9; Figure 4.5: Constructed from observations made by the AAVSO, courtesy of John Cannizzo; Figures 5.1 and 5.4: Courtesy of Dan Rolfe, University of Leicester; Figure 5.2: adapted from Gilliland, R. L. et al. (1986) 'WZ Sagittae: time-resolved spectroscopy during quiescence', *Astrophysical Journal*, **301**, 252, The American Astrophysical Society; Figure 5.3: adapted from Horne, K. and Marsh, T. R. (1986) 'Emission line formation in accretion discs', *Monthly Notices of the Royal Astronomical Society*, **218**, 761, Blackwell Science Limited; Figure 5.5: adapted from Frank, J., King, A. and Raine, D. (2002) *Accretion Power in Astrophysics*, 3rd ed. CUP; Figure 5.6: adapted from Schoembs, R. et al. (1987) 'Simultaneous multicolour photometry of OY Carinae during quiescence', *Astronomy and Astrophysics*, **181**, 50, Springer–Verlag GmBH & Co; Figure 5.7: Courtesy of Raymundo Baptista, UFSC, Brazil; Figure 5.8: adapted from Tom Marsh, University of Southampton; Figure 5.9: adapted from Steeghs, D. (1999) Ph.D. thesis from University of St Andrews: 'Spiral waves in accretion discs', Danny Steeghs, University of Warwick; Figures 5.11 and 5.12: Danny Steeghs, University of Warwick; Figures 5.17: The Anglo–Australian Observatory/David Malin; Figure 6.1: Image courtesy of ESA; Figure 6.2: Josef Pöpsel; Figure 6.3: Barnard, R. Shaw-Greening, L. and Kolb, U. (2008) 'A multicoloured survey of NGC253 with XMM-Newton: testing the methods used for creating luminosity functions from low-count data', *Monthly*

Notices of the Royal Astronomical Society, **388**, 849, Blackwell Science Limited; Figure 6.5: D. de Chambure, XMM-Newton Project, ESA/ESTEC; Figure 6.8: adapted from Titarchuk, L, (1994) *Astrophysical Journal*, **434**, 570; Figure 6.9: adapted from Morrison, R. and McCammon, D. (1983) 'Interstellar photoelectric absorption cross sections, 0.03-10 keV', *Astrophysical Journal*, **270**, 119, The American Astronomical Society; Figure 6.15: adapted from Revnivtsev, M. et al. (2000) 'High frequencies in the power spectrum of Cyg X-1 in the hard and soft spectral states', *Astronomy and Astrophysics*, **363**, 1013, EDP Sciences; Figure 6.16: adapted from Strohmayer, T. and Bildsten, L. in Lewin, W. H. G. and van der Klis, M. (2006) Compact Stellar X-ray Sources, CUP; Figure 6.18: adapted from Boirin, L. et al. (2004) 'Discovery of X-ray absorption lines from the low-mass X-ray binaries 4U 1916-053 and X 1254-690 with XMM Newton', *Nuclear Physics*, Section B, **132**, 506, Elsevier Science; Figures 6.19 and 6.20: adapted from Church, M. J. and Balucinska-Church, M. (1995) 'A complex continuum model for the low-mass X-ray binary dipping sources: application to X 1624-49', *Astronomy and Astrophysics*, **300**, 441, EDP Sciences; Figure 6.23: Wijers, R. A. M. J. and Pringle, J. E. (1999) 'Warped accretion discs and the long periods in X-ray binaries', *Monthly Notices of the Royal Astronomical Society*, **308**, 207, Blackwell Publishers; Figure 6.25: adapted from van der Klis, M. et al. (1996) 'Discovery of submillisecond quasi-periodic oscillations in the X-ray flux of Scorpius X-1', *Astrophysical Journal*, **469**, 1, The American Astronomical Society; Figure 6.26: adapted from Done, C. and Gierlinski, M. (2003) 'Observing the effects of the event horizon in black holes', *Monthly Notices of the Royal Astronomical Society*, **342**, 1041, Blackwell Publishers; Figure 6.27: NASA Marshall Space Flight Center (NASA-MSFC); Figure 6.28: adapted from McHardy, I. (2006) 'Long timescale X-ray variability of 3C273: similarity to Seyfert galaxies and Galactic binary systems', Blazar Variability Workshop II: Entering the GLAST Era, ASP Conference Series, **350**, 94, Astronomical Society of the Pacific; Figure 6.29: adapted from Brenneman, L. W. and Reynolds, C. S. (2006) 'Constraining black hole spin via X-ray spectroscopy', *Astrophysical Journal*, **652**, 1028, The American Astronomical Society; Figures 7.1 and 7.2b: Images courtesy of NRAO/AUI; Figure 7.2a: Ann Wehrle and Steve Unwin/NASA; Figure 7.8: Marscher et al., Wolfgang Steffen, Cosmovision, NRAO/AUI/NSF; Figures 8.1 and 8.4: Courtesy of the Gamma-Ray Astronomy Team at the National Space Science and Technology Center (NSSTC), http://gammaray.nsstc.nasa.gov/batse/grb/; Figure 8.2 adapted from Schaeffer, B. et al. (1994), *The Astrophysical Journal Supplement Series*, **92**, 285, The American Astronomical Society; Figures 8.3a and 8.5: adapted from from Ph.D. thesis 'Spectral studies of gamma-ray burst prompt emission', by Y. Kaneko (2005) University of Alabama, Huntsville; Figure 8.3b: Fishman, G. J. (1999) 'Observed properties of gamma-ray bursts', *Astronomy and Astrophysics Supplement Series*, **138**, 395–398, The European Southern Observatory; Figure 8.6: adapted from Liang, E. W. et al. (2006) 'Testing the curvature effect and internal origin of gamma-ray burst prompt emissions and X-ray flare with swift data', *Astrophysical Journal*, **646**, 351, The American Astronomical Society; Figure 8.7: adapted from Stanek, K. Z. et al. (1999) 'BVRI Observations of the optical afterglow of GRB 990510', *Astrophysical Journal*, **522**, 39 The American Astronomical Society; Figure 8.8a: Taken from www.tls-tautenburg.de/research/klose/GRB030329.html; Figure 8.8b: Taken from http://www.mpe.mpg.de/~jcg/grb030329.html; Figure 8.10: adapted from

Panaitescu, A. (2009) 'Gamma-Ray Burst afterglows: theory and observations', *American Institute of Physics Conference Proceedings*, **1133**, 127–138, American Institute of Physics; Figure 8.12: adapted from an original sketch by M. Ruffert and T. H. Janka; Figure 8.13: adapted from Zhang, W., Woosley, S. E. and MacFadyen, A. I. (2003) 'Relativistic jets in collapsars', *Astrophysical Journal*, **586**, 356, American Astronomical Society; Figure 8.14a: adapted from the original produced by Edo Berger (Harvard-Smithsonian CfA).

Every effort has been made to contact copyright holders. If any have been inadvertently overlooked the publishers will be pleased to make the necessary arrangements at the first opportunity.

Index

Items that appear in the Glossary have page numbers in **bold type**. Ordinary
index items have page numbers in Roman type.